MICROBIOLOGY FOR FOOD AND HEALTH

Technological Developments and Advances

MICROBIOLOGY FOR FOOD AND HEALTH

Technological Developments and Advances

Edited by

Deepak Kumar Verma, PhD

Ami R. Patel, PhD

Prem Prakash Srivastav, PhD

Balaram Mohapatra, PhD

Alaa Kareem Niamah, PhD

AAP | APPLE ACADEMIC PRESS

Apple Academic Press Inc.
4164 Lakeshore Road
Burlington ON L7L 1A4
Canada

Apple Academic Press Inc.
1265 Goldenrod Circle NE
Palm Bay, Florida 32905
USA

International Standard Book Number-13: 978-1-77188-813-4 (Hardcover)
International Standard Book Number-13: 978-0-42927-617-0 (eBook)

Library and Archives Canada Cataloguing in Publication

Title: Microbiology for food and health : technological developments and advances / edited by Deepak Kumar Verma, Ami R. Patel, Prem Prakash Srivastav, Balaram Mohapatra, Alaa Kareem Niamah.

Names: Verma, Deepak Kumar, 1986- editor. | Patel, Ami R., editor. | Srivastav, Prem Prakash, editor. | Mohapatra, Balaram, editor. | Niamah, Alaa Kareem, editor.

Description: Includes bibliographical references and index.

Identifiers: Canadiana (print) 20190215712 | Canadiana (ebook) 20190215828 | ISBN 9781771888134 (hardcover) | ISBN 9780429276170 (ebook)

Subjects: LCSH: Dairy microbiology. | LCSH: Food—Microbiology.

Classification: LCC QR121 .M53 2020 | DDC 641.3/7—dc23

CIP data on file with US Library of Congress

Apple Academic Press also publishes its books in a variety of electronic formats. Some content that appears in print may not be available in electronic format. For information about Apple Academic Press products, visit our website at **www.appleacademicpress.com** and the CRC Press website at **www.crcpress.com**

About the Editors

Deepak Kumar Verma, PhD
Research Scholar, Department of Agricultural and Food Engineering, Indian Institute of Technology Kharagpur, West Bengal, India

Deepak Kumar Verma is an agricultural science professional and is currently a PhD Research Scholar in the specialization of food processing engineering in the Agricultural and Food Engineering Department, Indian Institute of Technology, Kharagpur (WB), India. In 2012, he received a DST-INSPIRE Fellowship for PhD study by the Department of Science & Technology (DST), Ministry of Science and Technology, Government of India. Mr. Verma is currently working on the research project "Isolation and Characterization of Aroma Volatile and Flavoring Compounds from Aromatic and Non-Aromatic Rice Cultivars of India." His previous research work included "Physico-Chemical and Cooking Characteristics of Azad Basmati (CSAR 839-3): A Newly Evolved Variety of Basmati Rice (*Oryza sativa* L.)". He earned his BSc degree in agricultural science from the Faculty of Agriculture at Gorakhpur University, Gorakhpur, and his MSc (Agriculture) in Agricultural Biochemistry in 2011. He also received an award from the Department of Agricultural Biochemistry, Chandra Shekhar Azad University of Agricultural and Technology, Kanpur, India. Apart from his area of specialization in plant biochemistry, he has also built a sound background in plant physiology, microbiology, plant pathology, genetics and plant breeding, plant biotechnology and genetic engineering, seed science and technology, food science and technology etc. In addition, he is member of different professional bodies, and his activities and accomplishments include conferences, seminar, workshop, training, and also the publication of research articles, books, and book chapters.

Ami R. Patel, PhD

Assistant Professor, Division of Dairy and Food Microbiology, Mansinhbhai Institute of Dairy and Food Technology, Gujarat, India

Ami R. Patel, PhD, is an Assistant Professor in the Division of Dairy and Food Microbiology at the Mansinhbhai Institute of Dairy and Food Technology-MIDFT, Dudhsagar Dairy Campus, Mehsana, Gujarat, India. In 2002, she earned her BSc (Microbiology) and MSc (Microbiology) degrees from Sardar Patel University, Vallabh Vidyanagar, Gujarat, India, and received her doctorate degree in Dairy Microbiology from the Dairy Department of SMC College of Dairy Science, Anand Agricultural University, Anand, Gujarat, India. Professor Patel has expertise in specialized areas that involve the isolation, screening, and characterization of exopolysaccharides from potential probiotic cultures and employing them for food and health applications. In addition, she is engaged with teaching undergraduate and postgraduate students and research. She has a sound background in food microbiology, microbial biotechnology, food biotechnology, food science and technology, clinical microbiology and immunology, etc. She has authored a number of peer-reviewed papers and technical articles in international and national journals, as well as book chapters, books, proceedings, and technical bulletins. She has received a number of awards and honors. She also serves as an expert reviewer for several scientific journals.

Prem Prakash Srivastav, PhD

Associate Professor, Food Science and Technology, Agricultural and Food Engineering Department, Indian Institute of Technology Kharagpur, West Bengal, India

Prem Prakash Srivastav, PhD, is Associate Professor of Food Science and Technology in the Agricultural and Food Engineering Department at the Indian Institute of Technology Kharagpur, West Bengal, India. He teaches various undergraduate-, postgraduate-, and PhD-level courses and has guided many research projects at the PhD, master's, and undergraduate levels. His research interests include the development of specially designed convenience, functional and therapeutic foods; extraction of nutraceuticals; and the development of various low-cost food processing machineries. He has organized many sponsored short-term courses and has completed sponsored research projects and consultancies. He has published various research papers in peer-reviewed international and national journals, proceedings, technical bulletins, and monographs. Other publications include books and book chapters along with many patents. He has attended and chaired international and national conferences and has delivered many invited lectures at various summer/winter schools. He graduated from Gorakhpur University, India, and received his MSc degree with major in Food Technology and a minor in Process Engineering from G. B. Pant University of Agriculture and Technology, Pantnagar, India. He was awarded his PhD by the Indian Institute of Technology, Kharagpur, India.

Balaram Mohapatra, PhD
PhD Research Scholar, Department of Biotechnology, Indian Institute of Technology Kharagpur, West Bengal, India

Balaram Mohapatra is a biological science professional and is currently a PhD Research Scholar in the Department of Biotechnology at the Indian Institute of Technology Kharagpur, West Bengal, India. In 2012, he received a DST-INSPIRE Fellowship for PhD study by the Department of Science and Technology (DST), Ministry of Science and Technology, Government of India. Mr. Mohapatra earned his BSc degree in Botany from Utkal University and his MSc degree in Microbiology with first rank and also received a department topper award from the Department of Microbiology, College of Basic Science and Humanities, Odisha University of Agriculture and Technology, Bhubaneswar (Odisha), India. In addition to his area of specialization in microbiology, he has also built up a sound background in microbial biotechnology, molecular biology, biochemistry, genomics, etc. He is member of several professional bodiesm and his activities and accomplishments include conferences, seminars, workshops, and the publication of several research articles, books and book chapters.

Alaa Kareem Niamah, PhD
Assistant Professor, Department of Food Science, College of Agriculture, University of Basrah, Basra, Iraq

Alaa Kareem Niamah, PhD, is working as Assistant Professor in the Department of Food Science, College of Agriculture, University of Basrah, Basra, Iraq. He earned his BSc (food and dairy technology), his MSc (dairy microbiology with specialization in probiotic bacteria), and his PhD (dairy microbiology) degrees from the College of Agriculture, University of Basrah, Iraq. Dr. Niamah has expertise in the specialized area that involves purification, isolation, screening, and characterization of bacteriocins or peptides from lactic acid bacteria and probiotic yeasts and employing them for food and health applications. In addition, he is engaged with teaching undergraduate and postgraduate (MSc and PhD) students. His scientific background in dairy microbiology, fermented milk, dairy technology, probiotic bacteria, food biotechnology, natural antimicrobial, essential oils, etc. He has authored a number of peer-reviewed papers and technical articles in international and national journals, book chapters and books, and proceedings and technical bulletins. He has participated in many workshops and training courses around the world.

Contents

Contributors

Alaa Jabbar Abd Al-Manhel
Department of Food Science, College of Agriculture, University of Basrah, Basra City, Iraq, Tel.: +964-7808785772; E-mail: alaafood_13@yahoo.com

Yaovi Ameyapoh
Laboratory of Microbiology and Quality Control of Foodstuffs (LAMICODA), School of Biological and Food Techniques (ESTBA), University of Lomé, BP 1515, Lomé, Togo, Tel.: (+228) 90-10-81-03; E-mail: ameyapoh.blaise@gmail.com

Lamine Baba-Moussa
Laboratory of Molecular Biology and Typing in Microbiology, Department of Biochemistry and Cellular Biology, Faculty of Science and Technology (FAST), University of Abomey-Calavi; 05BP 1604 Cotonou (Benin), Tel.: (+229) 97-12-34-68; E-mail: laminesaid@yahoo.fr

Neha Bajwa
Department of Pharmaceutical Sciences & Technology, Maharaja Ranjit Singh Punjab Technical University, Bathinda–151001, Punjab, India, Tel.: +917988439553; E-mail: nehabajwa2765@gmail.com

Ashish Baldi
Department of Pharmaceutical Sciences & Technology, Maharaja Ranjit Singh Punjab Technical University, Bathinda–151001, Punjab, India, Tel.: +91-01642970971; +91-8968423848; Fax: +91-16362284197; E-mail: baldiashish@gmail.com

Roostita Balia
Professor, Faculty of Animal Husbandry, Universitas Padjadjaran, Sumedang–45363, West Java, Indonesia, Tel.: +62-811225814; E-mail: roostita@gmail.com

Vasilica Barbu
Assistant Professor, Faculty of Food Science and Engineering, Dunarea de Jos University of Galati, 47, Domneasca Street, 800008 Galati, Romania, Tel.: +40-336-130177; +40-742211955; Fax: +40-236461353; E-mail: vasilica.barbu@ugal.ro

Daniela Borda
Professor, Faculty of Food Science and Engineering, Dunarea de Jos University of Galati, 47, Domneasca Street, 800008 Galati, Romania, Tel.: +40-336-130177; +40-726224330; Fax: +40-236461353; E-mail: daniela.borda@ugal.ro

Marli Busanello
Department of Food Science and Technology, Londrina State University, Londrina-PR, 86057-970, Brazil, Tel.: + 55-43-99246724; E-mail: marlibusanello@gmail.com

Bouraïma Djery
Laboratory of Microbiology and Quality Control of Foodstuffs (LAMICODA), School of Biological and Food Techniques (ESTBA), University of Lomé, BP 1515, Lomé, Togo, Tel.: (+228) 90-92-43-67; E-mail: djerifr2002@gmail.com

Fernanda Silva Farinazzo
Department of Food Science and Technology, Londrina State University, Londrina-PR, 86057-970, Brazil, Tel.: +55-43-96298199; E-mail: fsfarinazzo@gmail.com

Maria Thereza Carlos Fernandes
Department of Food Science and Technology, Londrina State University, Londrina-PR, 86057-970, Brazil, Tel.: +55-43-99275533; E-mail: thereza.fernandes@hotmail.com

Marsilvio Lima de Moraes Filho
Department of Food Science and Technology, Londrina State University, Londrina-PR, 86057-970, Brazil, Tel.: + 55-43-96780592; E-mail: marsilviolimafilho@gmail.com

Sandra Garcia
Department of Food Science and Technology, Londrina State University, Londrina-PR, 86057-970, Brazil, Tel.: +55-43-33715966; +55-43-999771526; Fax: +55-43-33714080; E-mail: sgarcia@uel.br; gassandra15@gmail.com

Karla Bigetti Guergoletto
Department of Food Science and Technology, Londrina State University, Londrina-PR, 86057-970, Brazil, Tel.: + 55-43-99777505; E-mail: karla2901@gmail.com

Herlina
Faculty of Agro-Industrial Technology, Universitas Padjadjaran, Sumedang–45363, West Java, Indonesia, Tel.: +62-8597-7822-306; E-mail: herlinalim96@gmail.com

Cătălin Iancu
Researcher, Micreos Food Safety BV, Nieuwe Kanaal 7P, 6709 PA Wageningen, The Netherlands, Tel.: +31 (0) 888007151; +31-617938533; Fax: +31 (0) 888007151; E-mail c.iancu@micreos.com

Simplice D. Karou
Laboratory of Microbiology and Quality Control of Foodstuffs (LAMICODA), School of Biological and Food Techniques (ESTBA), University of Lomé, BP 1515, Lomé, Togo, Tel.: (+228) 90-70-19-25; E-mail: simplicekarou@hotmail.com

G. Kimmy
Department of Food Engineering & Technology, Sant Longowal Institute of Engineering and Technology Longowal, Sangrur–148106, Punjab, India, Tel.: +00-91-8699271602; E-mail: kishorigoyal09@gmail.com

Kumari Shanti Kiran
Department of Biotechnology, Indian Institute of Technology, Kharagpur–721302, West Bengal, India

Essozimna Kogno
Laboratory of Microbiology and Quality Control of Foodstuffs (LAMICODA), School of Biological and Food Techniques (ESTBA), University of Lomé, BP 1515, Lomé, Togo, Tel.: (+228) 91-33-98-27; E-mail: kognoserge@gmail.com

Chanchal Kumar
Department of Microbiology, Vallabhbhai Patel Chest Institute, University of Delhi, New Delhi–110007, India, Tel.: +91-9717448305; E-mail: ckumarbiotech@gmail.com

Widia Dwi Lestari
Faculty of Agro-Industrial Technology, Universitas Padjadjaran, Sumedang–45363, West Java, Indonesia, Tel.: +62-8180-9632-771; E-mail: widiadwibunzani@gmail.com

Carolina Saori Ishii Mauro
Department of Food Science and Technology, Londrina State University, Londrina-PR, 86057-970, Brazil, Tel.: +55-43-91549927; E-mail: carol.saori@gmail.com

Balaram Mohapatra
Department of Biotechnology, Indian Institute of Technology, Kharagpur – 721302, West Bengal, India, Tel.: +91-7679282452; E-mail: balarammohapatra09@gmail.com

Alaa Kareem Niamah
Department of Food Science, College of Agriculture, University of Basrah, Basra City, Iraq, Tel.: +00-96-47709042069; E-mail: alaakareem2002@hotmail.com

Anca Ioana Nicolau
Professor, Faculty of Food Science and Engineering, Dunarea de Jos University of Galati, 47, Domneasca Street, 800008 Galati, Romania, Tel.: +40-336-130177; +40-755746227; Fax: +40-236461353; E-mail: anca.nicolau@ugal.ro

Ami R. Patel
Division of Dairy and Food Microbiology, Mansinhbhai Institute of Dairy & Food Technology-MIDFT, Dudhsagar Dairy Campus, Mehsana – 384002, Gujarat, India, Tel.: +00-91-9825067311; +00-91-2762243777 (O); Fax: +91-02762-253422; E-mail: amiamipatel@yahoo.co.in

Jashbhai B. Prajapati
Department of Dairy Microbiology, Anand Agricultural University, Anand – 388110, Gujarat, India, Tel.: +91-02692-225851; +91-9879105948; E-mail: jbprajapati@aau.in

Paulo Terumitsu Saito
Department of Food Science and Technology, Londrina State University, Londrina-PR, 86057-970, Brazil, Tel.: +55-43-91069144; E-mail: paulinho_badaui_sephirodark@hotmail.com

Aly Savadogo
Laboratory of Biotechnology and Microbiology (LaBM), Center for Research in Biological, Food and Nutritional Sciences (CRSBAN), Department of Biochemistry-Microbiology (DBM), Training and Research Unit in Life and Earth Sciences (UFR)-SVT, Ouaga University I Pr Joseph KI-ZERBO, 03BP7131 Ouagadougou 03, Ouagadougou (Burkina Faso), Tel.: (+226) 79-38-57-07; E-mail: alysavadogo@gmail.com

Mohamed Eid Shenana
Professor, and Head, Dairy Science Department, Moshtohor, Faculty of Agriculture, University of Benha–13736, Arab Republic of Egypt, Moshtohor, Egypt, Tel.: +201222269505; +2013-2460306; +2013-2468150; Fax: +2-0132467786; E-mails: meshenana@fagr.bu.edu.eg; meshenana1964@hotmail.com

Haziz Sina
Laboratory of Molecular Biology and Typing in Microbiology, Department of Biochemistry and Cellular Biology, Faculty of Science and Technology (FAST), University of Abomey-Calavi; 05BP 1604 Cotonou (Benin), Tel.: (+229) 95-06-44-50; E-mail: sina.haziz@gmail.com

Barkha Singhal
School of Biotechnology, Gautam Buddha University, Greater Noida, Gautam Budh Nagar–201312, U.P., India, Tel.: +91-120-2344290 (O); +91-9871167765; E-mails: barkha@gbu.ac.in; gupta.barkha@gmail.com

Prem Prakash Srivastav
Agricultural and Food Engineering Department, Indian Institute of Technology, Kharagpur–721302, West Bengal, India, Tel.: +91-3222281673; Fax: +91-3222282224; E-mail: pps@agfe.iitkgp.ernet.in

Essodolom Taale
Laboratory of Microbiology and Quality Control of Foodstuffs (LAMICODA), School of Biological and Food Techniques (ESTBA), University of Lomé, BP 1515, Lomé, Togo, Tel.: (+228) 93-34-20-66; E-mail: taaleernest12@hotmail.com

Mamta Thakur
Department of Food Engineering and Technology, Sant Longowal Institute of Engineering and Technology, Longowal – 148106, Punjab, India, Tel.: +91-8219831376; 8352895496; Fax: +91-1672-280057; E-mails: thakurmamtafoodtech@gmail.com; mamta.ft@gmail.com

Alfred S. Traore
Laboratory of Biotechnology and Microbiology (LaBM), Center for Research in Biological, Food and Nutritional Sciences (CRSBAN), Department of Biochemistry-Microbiology (DBM), Training and Research Unit in Life and Earth Sciences (UFR)-SVT, Ouaga University I Pr Joseph KI-ZERBO, 03BP7131 Ouagadougou 03, Ouagadougou (Burkina Faso), Tel.: (+226) 50-33-73-73; E-mail: astraore@gmail.com

Gemilang Lara Utama
Lecturer, Faculty of Agro-Industrial Technology, Universitas Padjadjaran, Sumedang–45363, West Java, Indonesia, Tel.: +62-8122-0272-894; E-mail: lugemilang@gmail.com

Faysa Utba
Faculty of Agro-Industrial Technology, Universitas Padjadjaran, Sumedang–45363, West Java, Indonesia, Tel.: +62-813-1753-0717; E-mail: faysautba@gmail.com

Deepak Kumar Verma
PhD Research Scholar, Agricultural, and Food Engineering Department, Indian Institute of Technology, Kharagpur – 721302, West Bengal, India, Tel.: +91-3222281673; +91-7407170259; Fax: +91-3222282224; E-mails: deepak.verma@agfe.iitkgp.ernet.in; rajadkv@rediffmail.com

Abbreviations

AAB	acetic acid bacteria
ABTS	2,2'-azino-bis(3-etilbenzotiazolin) 6-sulfonic acid
AcDH	acetaldehyde dehydrogenase
ACE	angiotensin-converting-enzyme
ADH	alcohol dehydrogenase
ANVISA	National Health Surveillance Agency Brazil
ATCC	American Type Culture Collection
BLIS	bacteriocin-like inhibitory substances
BOD	biological oxygen demand
bp	base pairs
C	cytosine
C/N ratio	carbon/nitrogen ratio
CA	conventional apple
$CaCO_3$	calcium carbonate
Cas	Crispr associated protein
CC	cheese and curcuma
CDC	Center for Disease control
CDCP	Centers for Disease Control and Prevention
cDNA	complementary DNA
CFU	colony-forming unit
CIP	clean in place
CIRAD	Center for International Cooperation in Economic Research for Development
CNG	compressed natural gas
CO_2	carbon dioxide
COD	chemical oxygen demand
CRISPR	clustered regularly interspaced short palindromic repeats
CSIs	conserved signature indels
Ct	cycle threshold
db	dry basis
DBT	Department of Biotechnology
DDH	DNA-DNA homology
DIV	direct-in-vat inoculations

DNA	deoxyribonucleic acid
DPPH	2,2-difenil-1-picril-hidrazil
DSHEA	Dietary Supplement Health and Education Act
DSM	Deutsche Sammlung von Mikroorganismen und Zellkulturen (German Collection of Microorganisms and Cell Cultures)
DSS	defined strain starter
DVI	direct-in-vat inoculation
DVS	direct vat-set
E. coli	*Escherichia coli*
EBS	extract from black soybean
EBSP	extract from black soybean with pectin
EC	European Commission
EcN	*Escherichia coli* Nissle1917
EFSA	European Food Safety Authority
ELISA	enzyme-linked immunosorbent assays
Ent.	*enterobacter*
EPS	exopolysaccharides
EU	European Union
FAO	Food and Agriculture Organization
FAST	Faculty of Sciences and Techniques
FDA	Food and Drug Administrations
FFA	free fatty acids
FOS	fructooligosaccharides
FOSHU	Food for Specified Health Use
FSANZ	Food Standards Australia and New Zealand
FUFOSE	Functional food science in Europe
g	gram
G	guanine
GAE	gallic acid equivalente
GBF	green banana flour
GHGs	greenhouse gases
GHs	glycoside hydrolases
GIT	gastrointestinal system
GMO	genetically modified organism
GMSC	genetic modified starter culture
GOI	Government of India
GRAS	general recognized as safe

HRP	horseradish peroxidase
HTS	high throughput sequencing
ICMR	Indian Council of Medical Research
IPP	Ile-Pro-Pro
ISO	International Organization for Standardization
ITS	internal transcribed spacer
J	Juçara
Kbp	kilobase pairs
KDa	Kilo Dalton
KH_2PO_4	monopotassium phosphate
LAB	lactic acid bacteria
LCPUFAs	long chain polyunsaturated fatty acids
LOD	limit of detection
LTA	lipoteichoic acid
Mbp	mega base pairs
MC	milk and curcuma
MCR	multi-curve resolution
MF	milk fat
$Mgcl_2$	magnesium chloride
MGE	mobile genetic elements
$MgSO_4$	magnesium sulfate
MHLW	Ministry of Health and Welfare
ML	cells encapsulated in alginate and flaxseed mucilage
MLST	multilocus sequence typing
mm	millimeter
MPS	massive parallel sequencing
MPS	monopersulphate
MQ	cells encapsulated in alginate and okra mucilage
NaCl	sodium chloride
NaOH	sodium hydroxide
NCIMB	National Collection of Industrial and Marine Bacteria
NCTC	National Collection of Type Cultures
NGS	next generation sequencers
NNG	*N*-methyl-*N*'-nitro-*N*-nitrosoguanidine
NRPS	nonribosomal peptide synthetase
NSLAB	non-starter lactic acid bacteria
Nt	nucleotide
OA	organic apple

OECD	Organization for Economic Co-operation and Development
ORF	open reading frame
PAM	protospacer-adjacent motif
PAP	peracetic acid products
PCA	principle component analysis
PC-PLC	phosphatidylcholine-preferring phospholipase C
PCR	polymerase chain reaction
PCR-TTGE	polymerase chain reaction–temporal temperature gradient gel electrophoresis
PedA	pediocin A
PedB	pediocin B
PEP-PTS	phosphoenol-pyruvate phosphotransferase system
PFGE	pulsed field gel electrophoresis
PFU	plaque forming units
PHA	poly-hydroxylalkanoate
PHB	poly-hydroxybutyrate
PIM	phage inhibitory media
PI-PLC	phosphatidylinositol-specific phospholipase C
PLs	polysaccharide hydrolases
PRM	phage resistant media
Pseudo.	*pseudomonas*
RAPD	random amplified polymorphism DNA
RDI	recommended daily intake
rDNA	ribosomal deoxyribonucleic acid
RDP-II	ribosomal database project II
RFLP	restriction fragments length polymorphisms
R-M	restriction-modification
RNA	ribonucleic acid
RNA-Seq	RNA sequencing
rRNA	ribosomal ribonucleic acid
RSM	reconstituted skimmed milk
Scc	somatic cell count
SCFA	short chain fatty acids
SCP	single cell protein
SE	soybean extract
SEG	soybean extract and soybean germ
SFDA	State Food and Drug Administration

Sie	Superinfection exclusion
SKML	Senate Commission on Food Safety of Deutsche Forschungsgemeinschaft
SMRT	single-molecule real time
SMS	single-molecule sequencing
SOPs	standard operating procedures
SPSs	slime producing starters
TDS	total dissolved solids
TRH	TDH-related hemolysin
TSS	total suspended solids
US	United States
UV	ultraviolet
VPP	Val-Pro-Pro
W/V	weight per volume
WASC	West Africa soft cheese
WGS	whole genome sequencing

Symbols

%	percentage
α-la	α-lactalbumin
°C	degree Celsius
μg	microgram
a_w	water activity

Preface

Microorganisms are small entities with enormous potential applications and play a beneficial as well as detrimental role in our life. Recent innovations and advanced technologies have helped us to expand our knowledge and understanding of how microorganisms are related to food quality, food safety, and their ultimate connection to human health. From ancient times, many vital components and ingredients, including organic acids, vitamins, enzymes, antibiotics, bacteriocins, ethanol, biopolymers, and vaccines, have been obtained from food-grade bacteria and fungi at the commercial level. They significantly contribute to the dairy, bakery, food, and beverage industries. Further, several microorganisms have also gained importance in the medical field or pharmaceutical industry. The health beneficial microorganisms, known as probiotics, have gained more attention during the last few decades. Probiotics have been proven to have many health benefits, like for the treatment of diarrhea and related gastrointestinal disturbances; lowering blood cholesterol level; and immunomodulatory as well as immunostimulatory effects in a number of chronic infections and diseases.

This book, *Microbiology for Food and Health: Technological Developments and Advances*, highlights the innovative microbiological approaches and advances made in the field of microbial food industries. The aim of this book is to cover the most recent progress in the field of dairy and food microbiology, emphasizing the current progress, actual challenges, and the success of the latest technologies.

This book is divided into three main parts: Part 1: Technological Advances in Starter Cultures, Part 2: Prospective Application of Food-Grade Microorganisms for Food Preservation and Food Safety, and Part 3: Innovative Microbiological Approaches and Technologies in the Food Industry. Conventionally, starter cultures are employed for biological food fermentation processes. The isolation, screening, and designing functional starters with technological aid advantages are the most fascinating areas of investigation in the food sector. In this context, the chapters in Part 1 discuss types, classification; and systematic use of various starters in addition to probiotics for various commercial fermentation processes. Food safety is one of the frontier area of research in today's world. Thus, Part 2 of the book covers recent breakthroughs in the microbial bioprocessing that can be employed in the food and health industry; such as for an example, prospective antimicrobial application of inherently present fermentative microflora against spoilage and pathogenic type microorganisms and use of potential probiotic LAB biofilms for the control of the formation of pathogenic biofilms by exclusion mechanisms. In of Part 3, advanced analytical and technological approaches such as molecular techniques for the detection of lactic acid bacteria from food matrix as well as foodborne pathogens; environmental friendly approaches for utilization of liquid and solid food waste, have been discussed in three different chapters.

The volume contains seven chapters contributed by a group of international contributors who excel in the frontier field of food microbiology, functional foods, and microbial technology. Together we have produced an outstanding reference book that provides accessible information for a wide audience, especially researchers, teachers, students, and food, nutrition, and health practitioners. It is expected to be a valuable resource for all those working in the dairy, food, and nutraceutical industry.

We extend our sincere thanks to all the contributing authors who have contributed with dedication, persistence, and cooperation in completing their chapters in a timely manner for this book and whose cooperation has made our task as editors a pleasure. We hope that this book will be informative and stimulating to readers.

—Deepak Kumar Verma
Ami R. Patel
Prem Prakash Srivastav
Balaram Mohapatra
Alaa Kareem Niamah

PART I

Technological Advances in Starter Cultures

CHAPTER 1

Starter Culture and Probiotic Bacteria in Dairy Food Products

DEEPAK KUMAR VERMA,[1*] AMI R. PATEL,[2*] JASHBHAI B. PRAJAPATI,[3] MAMTA THAKUR,[4] ALAA JABBAR ABD AL-MANHEL,[5] and PREM PRAKASH SRIVASTAV[6]

[1]*Agricultural, and Food Engineering Department, Indian Institute of Technology, Kharagpur–721302, West Bengal, India,*

[2]*Division of Dairy and Food Microbiology, Mansinhbhai Institute of Dairy & Food Technology-MIDFT, Dudhsagar Dairy Campus, Mehsana–384002, Gujarat, India*

[3]*Department of Dairy Microbiology, Anand Agricultural University, Anand–388110, Gujarat, India*

[4]*Department of Food Engineering and Technology, Sant Longowal Institute of Engineering and Technology, Longowal–148106, Punjab, India,*

[5]*Department of Food Science, College of Agriculture, University of Basrah, Basra City, Iraq*

[6]*Agricultural and Food Engineering Department, Indian Institute of Technology, Kharagpur–721302, West Bengal, India*

[*]*Corresponding author. E-mail: deepak.verma@agfe.iitkgp.ernet.in; rajadkv@rediffmail.com (D. K. Verma); amiamipatel@yahoo.co.in (A. R. Patel)*

1.1 INTRODUCTION

Fermented dairy products (dahi or curd, yogurt, cheese), cereal-legume based products (bread, idli, dosa), vegetables (kimchi, sauerkraut), and meat, as well as fish products, are manufactured using specific microorganisms called as starter cultures at industrial level. Starter cultures may be defined as a group of pure, actively growing microorganisms capable of bringing about desirable changes in the substrate through the process of fermentation. These carefully selected microbial cultures are purposely inoculated in milk to initiate (start) and accomplish the desired fermentation changes during the production of fermented milk products. Starter microorganisms bring about the specific changes in the body-texture, flavor, and appearance of the end product. The inherent microflora of the milk or for instance any substrate may be inefficient, unpredictable, and uncontrollable and generally gets killed by heat treatment given to the substrate. When it is inoculated with a starter culture, which can produce specific characteristics in a controlled and expected manner during fermentation.

Starter cultures mainly produce lactic acid in milk and milk products from milk sugar lactose, which leads or assists to coagulate milk proteins by reducing milk pH. Lactic acid bacteria (LAB) are the main group of food-grade starter culture associated with the development of lactic acid in milk. Apart from acid production, certain starter microorganisms are specifically added for their aptitude to synthesize flavor compounds including diacetyl, acetoin, or acetaldehyde in the fermented product. In addition to the pH effect, these organisms also influence the body-texture of fermented milk products through the breakdown of major milk constituents such as proteins and fats.

In the beginning, this chapter briefly describes the basic microbiological concepts of starter cultures such as their types, classification, different groups, and their presence in diverse fermented milk products of the world. Currently, probiotic cultures also find their way into fermented milk products. These specific microorganisms possess several claimed health benefits for the consumer, e.g., better digestion, improved metabolism of lactose, antihypertensive effect, cholesterol-lowering effect, immunomodulatory, and anti-carcinogenic effect. Probiotic cultures may be directly used in from of starters or as an adjunct with a starter culture of the specific product during the fermentation process.

The details about the probiotics, including their mechanism of action, proven health claims for different probiotic strains and regulatory and legislative aspects have been discussed in the latter part of the chapter.

1.2 HISTORICAL RESUME AND CURRENT STATUS

Fermented milk is of great significance since ancient times due to their attractive organoleptic attributes with a wide diversity of aromas, flavors, and textures apart from preservation. Though our ancestors did not have information about microorganisms' existence or their activities, they gradually acquired the knowledge to use good quality microflora (LAB and related food-grade microorganism) during the fermentation of milk (Prajapati and Nair, 2008). Accidently, when milk was left at room temperature, it was spontaneously fermented and was used as such or after dilution with water. That may be the first fermented milk produced by mankind. Depending on the microflora of different regions, diverse types of traditional fermented milk have arisen globally. Alternatively, draining of whey from fermented milk has resulted in the development of a new product, for instance, the first cheese that could be stored for several months. It is now easy to understand that fermented milk and milk products were developed as a means of preserving milk and organoleptic qualities.

In the area of microbiology, the lack of precise scientific information was the biggest drawback for the failure and undesirable fermentation in traditionally fermented milk products in the earlier times (Prajapati and Nair, 2008). Eventually, traditional milk products were standardized and to some extent, modified to fine-tune their industrial production. Cheese and butter production became industrialized about two hundred years ago. Storch and Conn in 1890 firstly make use of starters for developing flavor and aroma. Probably, in 1895, the first time the starters were used for cheese making in Britain (Prajapati, 1995). Starter cultures were grown in the production units, but the importance of standardized starters increased as the size of the operations grew. In the late 1900s, companies with specialized in the production of starters (Dannish farm of Hansen, currently known as Chr. Hansen) were founded (Prajapati, 1995). In the beginning, the key constraint was the undefined starters, containing a mixture of various unknown strains. Afterward, starters were analyzed and defined in order to get a specific strain or mixture of several strains having preferred properties to develop a standard product (Yadav et al., 1993). Starter culture industry has undergone tremendous development in the last few decades. Currently, the starter culture market is highly competitive with leading players such as Angel Yeast Co. Ltd.; Danisco A/S, LB

Bulgaricum Plc.; Chr. Hansen A/S, SACCO (Italy), Csk Food Enrichment B.V, Lesaffre Group, Dohler Group, Wyeast Laboratories Inc.; Lallemand Inc.; and Lactina Ltd.; which invest in research and development on starter cultures to strengthen their product range.

After the alcoholic beverages, fermented dairy products stand for the second most significant fermentation industry; about 400 diverse products obtained from milk fermentation using specific starter culture are consumed around the world (Chandan, 2006). Along with dairy industry, starter cultures also hold significance in the beverage industry; used in the manufacturing of various alcoholic drinks like beer and wine due to their intense flavor, quality enhancement properties, and other characteristics. According to a recent survey (TMR, 2014) between 2013 and 2018, the market of starter culture is projected to rise at a CAGR of 5.6%. In 2012, the market was dominated by Europe and yeast was the largest segment within the starter industry, both in terms of value and volume. The Asia-Pacific market is anticipated to grow at a rapid growth rate during the forecasting period.

The consumption of fluid milk has decreased during the last decade with a concurrent increase in consumption of fermented milk products, mainly due to changing consumers' preference and health awareness. The comparative consumption of fluid milk, cheese, and butter across the world is shown in Table 1.1. Based on portability, health claims and snack appeal, yogurt, and kefir are booming the market. In case of yogurt, in United States (US) only the sales are expected to raise from 7.3 billion USD in 2012 to 9.3 billion USD by 2017 as per the recent survey (STATISTA, 2016).

In India, fermented milk products like dahi (curd), buttermilk, *lassi* (sweetened fermented mill resembling stirred yogurt), and *shrikhand* (drained/concentrated mass of curd mixed with sugar and flavors) form significant place in a routine diet. Dahi is indigenous fermented milk product of India and nearby countries that has retained its popularity in diet despite modern food habits and changing lifestyles (Prajapati and Nair, 2008).

1.3 STARTER CULTURE IN DAIRY FOOD PRODUCTS

Several starter cultures are used to manufacture diverse dairy products at an industrial level, as shown in Table 1.2. These starters can be grouped

TABLE 1.1 Total Global Consumption of Fluid Milk, Cheese, and Butter from 2013 to 2015

Country Name	Fluid Milk* (Liters per Capita)			Cheese** (Kg per Capita)			Butter*** (Kg per Capita)		
	2013	**2014**	**2015**	**2013**	**2014**	**2015**	**2013**	2014	2015
Argentina	44.1	45.1	48.5	13.0	13.1	12.9	0.9	0.9	0.9
Australia	112.5	114.0	110.3	13.4	13.6	13.4	3.9	4.0	3.9
Austria	79.2	79.7	78.9	19.9	20.9	21.5	5.3	5.3	5.0
Belgium	50.9	52.8	50.9	15.0	15.1	14.9	2.3	2.3	2.3
Brazil	47.7	47.4	46.1	3.7	3.7	3.8	0.4	0.4	0.4
Bulgaria	21.4	20.0	19.7	16.4	15.8	15.6	0.9	0.9	1.0
Canada	75.4	73.3	70.6	12.7	12.6	12.5	2.7	2.9	2.8
Chile	24.0	25.4	25.2	8.8	9.0	9.1	1.2	1.2	1.4
China	17.9	18.4	19.3	-	0.1	0.1	0.1	0.1	0.1
Colombia	37.5	39.2	38.9	1.5	1.4	1.3	0.3	0.2	0.2
Croatia	65.6	69.6	60.1	10.2	11.2	12.3	1.0	1.2	1.6
Cyprus	102.9	104.7	107.8	25.5	19.2	24.2	1.8	1.9	1.9
Czech Republic	62.3	61.9	62.3	16.2	16.6	16.6	5.0	5.1	5.5
Denmark	93.7	92.6	90.4	20.2	24.5	26.2	3.9	4.9	4.9
Ecuador	-	-	-	6.2	6.2	-	1.5	1.5	-
Egypt	18.2	18.6	19.1	4.2	4.3	4.3	0.7	0.8	0.8
Estonia	126.2	122.3	105.6	21.3	21.5	16.3	1.5	2.2	1.6
Finland	135.6	132.7	129.3	24.7	25.6	26.7	3.7	3.2	3.3
France	56.1	54.7	53.1	26.2	26.7	26.8	7.7	8.3	8.0
Germany	55.8	58.0	55.9	23.7	24.1	24.6	5.8	5.7	6.0

TABLE 1.1 *(Continued)*

Country Name	Fluid Milk* (Liters per Capita)			Cheese** (Kg per Capita)			Butter*** (Kg per Capita)		
	2013	2014	2015	2013	2014	2015	2013	2014	2015
Hungary	52.1	49.8	52.1	11.0	11.6	12.9	1.0	1.2	1.4
Iceland	101.9	98.0	96.2	25.5	26.3	26.1	5.2	5.8	5.8
India	44.0	45.4	47.0	-	-	-	3.7	3.8	3.8
Iran	30.1	28.8	28.1	5.0	4.9	4.7	1.1	1.1	1.0
Ireland	122.2	122.2	126.6	10.8	11.2	13.8	2.4	2.4	2.4
Israel	56.1	55.2	54.0	16.3	16.3	16.9	0.9	0.9	0.9
Italy	55.2	51.8	49.3	22.3	22.1	21.5	2.4	2.3	2.4
Japan	31.8	31.4	31.4	2.2	2.2	2.2	0.6	0.6	0.6
Kazakhstan	27.9	29.3	28.7	2.6	2.5	2.6	1.3	1.7	1.3
Kenya	-	-	-	-	-	-	0.4	0.4	0.4
Latvia	38.8	41.4	44.0	18.1	17.6	20.3	2.3	2.9	2.8
Lithuania	33.0	34.8	34.9	20.1	18.6	18.3	2.6	3.0	3.3
Luxembourg	46.5	45.6	46.1	25.6	25.6	22.8	5.4	5.3	4.8
Mexico	33.8	31.7	29.7	3.6	3.8	3.8	0.9	0.7	0.8
Mongolia	9.7	10.1	9.5	0.3	0.2	0.3	0.5	0.5	0.6
Netherlands	49.1	47.0	47.0	20.1	18.2	18.2	3.0	3.0	3.0
New Zealand	104.0	113.6	113.2	8.7	8.7	8.8	4.9	4.9	4.9
Norway	87.8	87.5	85.0	18.6	19.4	19.5	2.9	2.9	3.0
Pakistan	-	-	-	-	-	-	4.8	4.8	4.8
Poland	40.7	39.8	40.1	15.6	15.4	16.1	4.0	4.1	4.3

TABLE 1.1 *(Continued)*

Country Name	Fluid Milk* (Liters per Capita)			Cheese** (Kg per Capita)			Butter*** (Kg per Capita)		
	2013	2014	2015	2013	2014	2015	2013	2014	2015
Russia	36.4	36.0	35.5	6.1	5.8	5.7	2.6	2.4	2.3
Slovakia	51.0	50.1	50.5	11.4	11.5	12.2	3.0	3.2	3.5
South Africa	27.3	28.1	37.0	1.8	1.8	2.0	0.4	0.3	0.5
South Korea	34.6	33.5	33.5	2.1	2.3	2.6	0.2	0.2	0.2
Spain	86.3	82.4	83.1	9.5	9.5	8.9	0.5	0.5	0.4
Sweden	94.7	88.1	87.6	19.7	20.6	20.7	2.2	2.3	2.5
Switzerland	67.8	65.0	61.8	21.3	21.5	21.8	5.5	5.3	5.4
Tanzania	-	-	-	0.3	0.3	-	0.4	0.4	0.4
Turkey	18.0	17.9	18.4	7.6	7.8	8.3	0.8	0.9	0.9
Ukraine	22.2	25.5	22.2	4.6	4.4	4.2	2.3	2.6	2.0
United Kingdom	109.8	108.4	105.6	11.4	11.6	12.1	3.2	2.9	3.2
United States	76.1	73.5	71.9	15.3	15.6	16.0	2.5	2.5	2.6
Uruguay	64.4	65.8	66.3	6.0	7.7	7.9	1.4	1.6	1.5
Zimbabwe	-	2.8	2.7	-	0.6	0.6	-	-	-

Source: Canadian Dairy Information Centre.

*CDIC (2017a),

**CDIC (2017b),

***CDIC (2017c)

-Not Reported.

under different categories based on the composition of microflora, growth temperature, type of products, flavor production, and type of fermentation as discussed in subsections.

1.3.1 TYPES OF STARTER CULTURE

1.3.1.1 BASED ON COMPOSITION OF MICROFLORA

1. **Single-Strain Starters:** These starters contain a pure culture of a single strain. These types of culture are very sensitive and prone to failure by bacteriophage or other conditions.
2. **Paired Compatible Strain:** Two strains of cultures having complementary activities in know proportion are used. This will reduce the chances of culture failures. In the case of bacteriophage attack, only one type of organism will be affected, and the other organism will carry out the fermentation without any problem.
3. **Multi-Strain Starters:** More than two strains are known proportion are used. The quality and behavior of these strains are predictable, and hence they are most popular.
4. **Mixed-Strain Starters:** More than two organisms which may have different characteristics like acid production, flavor production, slime production, etc. in unknown proportion are used. These are natural mixtures and may change their behavior under different conditions.

1.3.1.2 BASED ON GROWTH TEMPERATURE

Based on the growth temperature organisms can be divided into mesophilic and thermophilic:

1. **Mesophilic Starter Cultures:** The optimum growth temperature of these cultures is around 30°C, but they can grow between 20–35°C temperature. The mesophilic starter cultures generally include species of Lactococcus and Leuconostoc, eg. *Lactococcus lactis* subsp. *lactis*, *Lc. lactis* subsp. *cremoris* and *Leuconostoc mesenteroides* subsp. *cremoris*.

TABLE 1.2 Major Fermented Milk Products Originated from Different Countries

Product Name	Starter Culture(s)	Country of Origin
Yogurt	*Streptococcus thermophilus, Lactobacillus delbrueckii* subsp. *bulgaricus*	United States (US)
Cultured Buttermilk	*Lactococcus lactis* subsp. *lactis*, *Lc. lactis* subsp. *diacetylactis* and *Leu. mesenteroides*	
Acidophilus milk	*Lb. acidophilus*	US, Russia
Kefir	Mixed species of mesophilic and thermophilic lactobacilli and Lactococci, and yeasts.	Russia, Central Asia
Koumiss/kumiss/kumys	*Lb. delbrueckii* subsp. *bulgaricus*, *Lb. acidophilus* and yeasts-*Torula* sp.	
Donskaya/Varenetes/Kurugna/ Ryzhenka/Guslyanka	-	Russia
Skyr, surmjolk	*Lb. delbrueckii* subsp. *bulgaricus*, *Str. thermophilus,* lactose-fermenting yeast	Iceland
Filmjolk/Fillbunke/Fillbunk/Surmelk/ Taettemjolk/Tettemelk	*Lc. lactis* subsp. *lactis, Lb. delbrueckii* subsp. *bulgaricus*, *Str. thermophilus,* lactose-fermenting yeast	Sweden, Norway, Scandinavia
Pitkapiima, Viili	*Lc. lactis* subsp. *lactis, Leu. mesenteroides* subsp. *cremoris* and a fungus *Geotrichum candidum.*	Finland
Ymer	*Lc. lactis* subsp. *lactis* biovar. *diacetylactis* and *Leu. mesenteroides* subsp. *cremoris*	Denmark
Lassi, Mattha, Ghol, Chhas, Raita	*Lc.lactis* subsp. *lactis*, *Lc. lactis* subsp. *diacetylactis, Str. thermophilus, Lactobacillus* spp.	India, Pakistan, Bangladesh, Nepal
Dahi/Dudhee/Dahee, Mishti Dahi, Shrikhand	*Lc.lactis* subsp. *lactis*, *Lc. lactis* subsp. *diacetylactis*, or *Leuconostoc* spp. and *Lc. lactis* subsp. *cremoris, Str. thermophilus, Lactobacillus* spp.	India
Shosim/Sho/Thara	-	Nepal
Dogh/Abdoogh/Mast	Yogurt cultures	Afghanistan, Iran

TABLE 1.2 *(Continued)*

Product Name	Starter Culture(s)	Country of Origin
Mazun/Matzoon/Matsun/Matsoni/ Madzoon Tan/Than	*Lb. mazun, Lb. bulgaricus, Str. thermophilus*, spore forming bacillus and lactose fermenting yeast	Armenia
KisselMaleka/Naja/Yaourt/Urgotnic	*Lb. delbrueckii* subsp. *bulgaricus*, *Str. thermophilus,* lactose fermenting yeast	Balkans
Leben/Laban/Laban Rayeb	Thermophilic lactobacilli and mesophilic lactococci- *Lc. lactis* subsp. *lactis*, *Str. thermophilus, L. delbrueckii* subsp. *bulgaricus*	Lebanon, Syria, Jordan
Zabady/Zabade	*Lb. bulgaricus* and *Str. thermophilus*, *Candida krusei*.	Egypt, Sudan
Rob *Gariss/Hameedh*	*Lb. fermentum, Lc. lactis* subsp. *lactis, Lb. helveticus, Candida*, lactococci or streptococci	Sudan
Ergo	*Lc. garvieae, Lc. lactis* subsp. *lactis*, Lactobacillus sp.	Ethiopia
Katyk	-	Transcaucasia
Gioddu	-	Sardinia
Gruzovina	-	Yugoslavia
Tarag, airag, byaslag	-	Mongolia
Tarho/Taho	-	Hungary
Iogurte	-	Brazil, Portugal
Cieddu	-	Italy
Yiaourti	-	Greece
Ayrani	-	Cyprus
Busa	-	Turkestan
Chal	-	Turkmenistan
Matsoni		Georgia
Kurunga	-	Western Asia

2. **Thermophilic Starter Cultures:** They have a growth temperature range of 32–45°C, but the optimum temperature of these cultures is usually 40°C. The good examples of thermophilic starters are *Streptococcus thermophilus, Lactobacillus plantarum, Lb. casei, Lb. delbrueckii* subsp. *bulgaricus* and *Lb. helveticus.*

1.3.1.3 BASED ON TYPE OF FERMENTATION

The starters are classified as homofermentative or heterofermentative based on the end products resulting from glucose metabolism. The homo-fermentative strains chiefly synthesize lactic acid (up to 90%) from sugar metabolism, e.g., *Lactococcus lactis* subsp. *lactis, Str. thermophilus,* whereas heterofermentative strains may produce other organic acids like acetate, propionate together with small amounts of ethanol, and carbon dioxide, e.g., *Leuconostoc dextranicum, Lb. fermentum,* etc.

1.3.1.4 BASED ON FLAVOR PRODUCTION

The starters are grouped into B, D, BD, and N-type based on their ability of flavor production.

1. **B (L) Type:** Leuconostocs as flavor producer (old name is Betacocccus).
2. **D Type:** *Lc. lactis* subsp. *lactis* biovar *diacetylactis.*
3. **BD (LD) Type:** Mixer of both of the above cultures.
4. **N or O Type:** Absence of flavor producing organism.

1.3.1.5 BASED ON PRODUCT MANUFACTURED

Starter cultures can also be classified based on the product in which they are employed such as dahi starters, buttermilk cultures, yogurt starters (*Str. thermophilus, Lb. delbrueckii* subsp. *bulgaricus*), Swiss cheese starter (*Str. thermophilus* and thermophilic lactobacilli such as *Lb. delbrueckii* subsp. *bulgaricus, Lb. helveticus, Lb. casei*), etc.

1.3.1.6 BASED ON PHYSICAL FORM OF STARTERS

Starters can be available in different physical forms such as liquid, frozen concentrated (cultures are concentrated to achieve a higher number of cells after growth and then frozen), Dried cultures (dehydrated freeze-dried cultures), etc. The major fermented milk products of the world, their country of origin, and types of starter cultures involved are summarized in Table 1.2.

1.3.2 ECOLOGY AND CLASSIFICATION OF STARTER CULTURES

In 1981, Tamime proposed an overall classification of LAB while the first classification of LAB was given by Orla-Jensen in 1919 which is still recognized as the most acceptable method and had a great impact on the systematic of LAB (Yadav et al., 1993). Even though it is revised to some extent periodically, the basis of classification remarkably remains unchanged. The classification is mainly based on cell morphology, mode of glucose fermentation, the configuration of the lactic acid produced, growth at diverse temperature and pH scales, ability to grow or withstand high salt percentages, and tolerance to acid or alkali concentrations. Currently, some additional characteristics like fatty acid composition, motility, and antigens present on the cell wall or cell membrane, are used to classify the newly described LAB genera (Ljungh and Wadstrom, 2009).

The LAB group consists of twelve genera viz.; *Lactobacillus*, *Streptococcus*, *Lactococcus*, *Leuconostoc*, *Pediococcus*, *Oenococcus*, *Enterococcus*, *Tetragenococcus*, *Carnobacterium*, *Aerococcus*, *Vagococcus*, and *Weissella*. Out of these twelve, only the first seven genera are employed directly in the fermentation of milk since they exclusively produce lactic acid from carbohydrate fermentation as a chief metabolic end-product (Hutkins, 2006). Earlier, Axelsson (1998) worked to show the phylogenetic relationship of LAB genera and build the unrooted phylogenetic tree of LAB (Figure 1.1). Major microorganisms used routinely in the preparations of fermented milk are discussed in subsections.

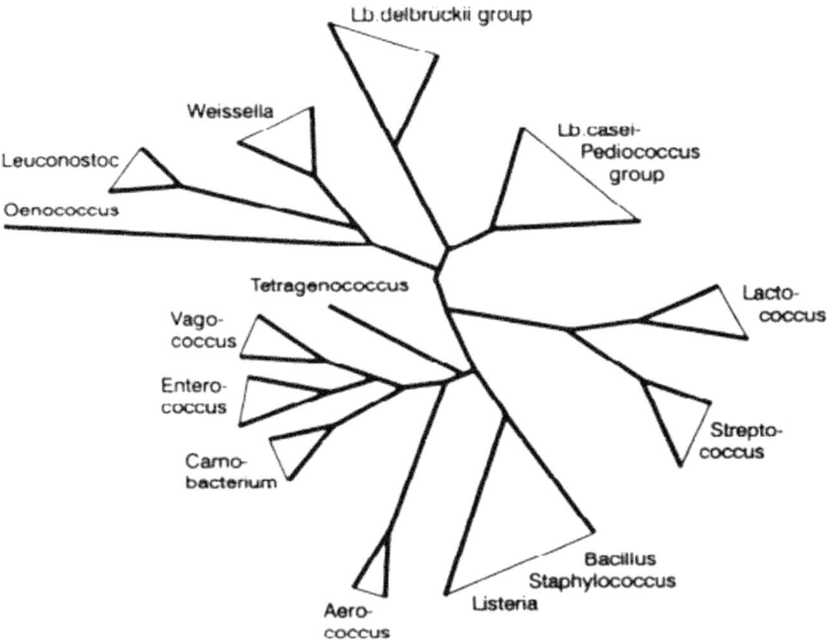

FIGURE 1.1 Unrooted phylogenetic tree of lactic acid bacteria, along with some aerobic and facultative anaerobic gram-positive bacteria of the low G+C subdivision. (Reprinted with permission from Narvhus & Axelsson, 2003. © 2003 Elsevier.)

1.3.2.1 LACTIC ACID BACTERIA (LAB)

1.3.2.1.1 Lactobacillus

The genus *Lactobacillus* (*Lb.*) consists of gram-positive, rod-shaped bacteria, which are facultatively anaerobic or microaerophilic, non-spore former, and non-motile. It is one of the largest genera containing more than 180 species; converts lactose and other sugars into lactic acid chiefly with little amounts of acetic acid, ethanol, and acetaldehyde (Hammes and Hertel, 2009). Based on glucose fermentation, *Lactobacillus* spp. has been divided into three groups, as shown in Table 1.3.

TABLE 1.3 Classification of Lactobacillus Based on Glucose Fermentation

Orla-Jensen Group	Group 1	Group 2	Group 3
	Thermobacterium	**Streptobacterium**	**Betabacterium**
Growth at 15°C	–	+	+
Growth at 45°C	+	–	±
Pentose fermentation	–	+	+
CO_2 from glucose	–	–	+
CO_2 from gluconate	–	+	+
Phosphoketolase	Absent	Inducible by pentose	Present
FDP aldolase	Present	Present	Absent
Example	*Lb. helveticus*, *Lb. acidophilus*	*Lb. plantarum* *Lb. casei*	*Lb. brevis* *Lb. fermentum*

+: growth, –: no growth.

Lactobacillus species are found as normal flora of human and animal body (intestinal tract, colon, and vagina), fruits, vegetables, soil, and naturally fermented foods. They are widely used to prepare diverse variety of fermented milk commercially; specifically *Lb. delbrueckii* subsp. *bulgaricus* is employed for yogurt manufacturing along with *Str. thermophilus*, where these two organisms exhibit a symbiotic relationship (Yadav et al., 1993). *Lactobacillus acidophilus* is the most potential probiotic culture, used for the preparation of acidophilus milk and other probiotic milk products like bifighurt and bioghurt in combination with other probiotic LAB or bifidobacterium species.

1.3.2.1.2 Streptococcus

The genus Streptococcus includes gram-positive homofermentative cocci that usually form pairs or chains. In 1937, Sherman separated the genus into four general groups (Table 1.4) according to physiological and growth characteristics, especially with regards to temperature limitations on growth. This categorization has become somewhat obsolete as relationships between species have been shown to overlap.

The only species used as starter culture is *Str. thermophilus*, a homofermentative facultatively anaerobic strain. It is called yogurt culture, which is thermophilic in nature with an optimum growth temperature of

38–42°C. Some strains of *Str. thermophiles* were explored to produce folic acid at significantly high level (Holasva et al., 2004). Some species of streptococci are pathogenic such as *Str. agalactiae* and *Str. dysgalactiae* are most common isolates in clinical mastitis, an udder disease of milch animals (Abd-Elrahman, 2013).

TABLE 1.4 Groups of Genus Streptococcus and Their Major Characteristics

Lancefield Group	Group	Example(s)	Growth at 10C	Growth at 45 °C
A, B, C, F, H, I	Pyogenic	*Streptococcus agalactiae, Str. pyogenes*	No growth	No growth
-	Viridians	*Str. thermophilus*	No growth	Growth
D	Enterococcus	*E. faecium, E. faecalis, E. durans*	Growth	Growth
N	Lactic cocci	*S. lactis, S. cremoris*	Growth	No growth

1.3.2.1.3 Lactococcus

The genus *Lactococcus* (*Lc.*) belongs to the *Streptococcaceae* family and consists of some important members of LAB. The genus includes total seven species namely *Lc. lactis, Lc. chungangensis, Lc. piscium, Lc. fujiensis, Lc. garvieae, Lc. raffinolactis* and *Lc. plantarum*. Bergey's Manual of Systematic Bacteriology (1986) combined all the mesophilic lactococci with *Lc. lactis* to form a single species since they own (i) high DNA homology, (ii) identical enzyme β-phosphotase and isoprenoid quinines, (iii) identical enzyme lactic dehydrogenase, and (iv) identical G+C% (GC-content). The only characteristic that distinguishes them is presence of plasmid(s). *Lc. lactis* subsp. *lactis* and *Lc. lactis* subsp. *cremoris* both are acid producers but non-flavor producers, while *Lc. lactis* subsp. *lactis* biovar *diacetylactis* is able to produce both acid and flavor producer attributable to the presence of flavor producing plasmid. They are homofermentative and their optimum temperature for growth lies between 25–30°C. In the dairy industry, they are chiefly used in the manufacturing of cultured dairy products such as curd, dahi, buttermilk, and different cheese varieties. Nisin, a bacteriocin obtained from lactococcal strain is legally permitted biopreservative (E234) in dairy and food products.

1.3.2.1.4 Leuconostoc

The genus *Leuconostoc* (*Leu.*) includes gram-positive pleomorphic cocci that are catalase-negative and heterofermentative- capable of producing lactic acid, CO_2 and aromatic compounds (ethanol and acetic acid) from glucose. The genus is placed within *Leuconostocaceae* family, contain 15 species. These organisms are normally used in conjunction with LAB as cheese starters, which produce typical diacetyl flavor. *Leu. mesenteroides, Leu. dextranicum, Leu. citrovorum* and *Leu. cremoris* are the common species of this genus used for manufacturing fermented dairy products at commercial scale. Some leuconostocs found to synthesize exopolysaccharide (EPS) biopolymer; mainly dextran type from carbohydrate fermentation.

1.3.2.1.5 Pediococcus

The genus *Pediococcus* (*P.*) is a member of family *Lactobacillaceae.* They are purely gram-positive homofermentative cocci producing both D and L form of lactic acid and often found in pairs or tetrads. It consists of total nine species and examples of some well-known species are *P. damnosus, P. acidolactici, P. parvulus*, and *P. pentosaceus*. Their aptitude to grow at broad pH range, temperature, and osmotic pressure made them distinct from other lactic cocci and helped to colonize the digestive tract. Certain strains are used as probiotics, and are commonly employed as starters in the preparation of cheeses and yogurts. They are commonly found in fermented vegetables, fermented dairy products (Patel et al., 2012), and meat (Anastasiadou et al., 2008) and considered contaminants of beer and wine.

1.3.2.1.6 Weissella

The genus includes gram-positive coccoid to rod-shaped heterofermentative bacteria. Earlier, the species classified under this genus were classified in either *Lactobacillus* or *Leuconostoc* genera until Collins et al. (1993) proposed that genus *Weissella* species forms a distinct phylogenetic group which can be distinguished from *Lactobacillus and Leuconostoc* members on the basis of 16S and 23S rRNA sequences. Today the genus *Weissella* includes in total nineteen species, reported from diverse fermented foods like sour cream (Van der Meulen et al., 2007), dahi (Patel et al., 2012), soya (Malik et al., 2009), sauerkraut (Plengvidhya et al., 2007; Patel et al., 2012) and fermented cassava (Kostinek et al., 2007), in sourdough (Galle et al., 2010), in meat sausage (Santos et al., 2005) and in traditional Japanese liquor *Shochu* (Endo and Okada, 2005). Also, several species of this genus have been reported from the clinical samples (such as saliva, gut, and vagina) of human (Nam et al., 2007; Lee et al., 2012). Few species were observed to improve the body texture and organoleptic qualities through EPS biosynthesis in diverse fermented foods (Patel and Prajapati, 2016).

1.3.2.2 NON-STARTER LACTIC ACID BACTERIA (NSLAB)

1.3.2.2.1 *Bifidobacterium*

These are gram-positive, non-motile, anaerobic bacteria usually found in the oral cavity, gut, vagina, and intestines of mammals, including various animals and humans, foods, and sewage (Ventura et al., 2007; *Mayo et al., 2010)*. Predominantly, they are rod-shaped bacteria; however, under unfavorable conditions, bifidobacteria show branching and pleomorphism (Leahy et al., 2005). Also, they appear in different shapes like short curved rods, bifurcated Y shaped, and club-shaped rods (Shah, 2006). More than 30 species of genus Bifidobacterium have been isolated so far (Leivers et al., 2011) including some major strains are *Bifidobacterium bifidum, B. dentium, B. longum, B. animalis, B. infantis,* and *B. breve*.

The optimum temperature for bifidobacteria growth lies between 37°C–41°C, and most of them are strict anaerobes. The *Bifidobacterium* species can be classified into four classes, *viz.* O_2-hypersensitive, O_2-sensitive, O_2-tolerant, and microaerophilic on the basis of their growth profiles in different oxygen concentrations. Milk fermented with bifidobacteria has a distinctive vinegar taste due to the production of acetate plus lactate (3:2 ratio) from the carbohydrate metabolism. Thus, they are generally used in combination with LAB for producing fermented milk products of therapeutic significance such as Bioghurt, Biograde, Bifighurt, Cultura 'AB, and Miru-Miru.

1.3.2.2.2 Propionibacterium

The genus *Propionibacterium* consists of gram-positive, catalase-positive, non-motile, and non-spore former bacilli (Stackebrandt et al., 2006). These species produce propionic acid as the key metabolite from carbohydrate fermentation. Some species of this genus are used as starters in the dairy industry, for example, *Propionibacterium freudenreichii* subsp. *shermanii* is used in Swiss cheese or emmental cheese. During ripening/maturation period, it produces large gas holes in the cheese. *P. jensenii, P. thoenii, and P. acidipropionici* are other organisms present in these genera. They are well known for biosynthesis of vitamin B_{12} and tetrapyrrole compounds (Kiatpapan et al., 2002). *P. freudenreichii* found to produce antibacterial compounds (Irina et al., 2012) while some possess probiotic status.

1.3.2.2.3 Brevibacterium

They are gram-positive rods and the sole genus classified in the family *Brevibacteriaceae*. Among different species, *Brevibacterium linens* are used as starter culture in preparation of bacterial surface-ripened cheeses. It imparts characteristic reddish-orange color to the rind of, or forms smear on few cheese varieties like *Brick* cheese and *Camembert* cheese or *Limburger* cheese (Chandan, 2006).

1.3.2.2.4 Acetic Acid Bacteria (AAB)

These, AAB are gram-negative bacilli classified in the *Acetobacteraceae* family. Because of their capability to oxidize ethanol stronger than glucose, many strains are used in the manufacturing of acetic acid (vinegar) at the industrial level. Various AAB are associated with naturally fermented milk like *kefir* (Yadav et al., 1993). Multiple species of AAB are capable of incomplete oxidation of carbohydrates and alcohols to organic acids, ketones, and aldehydes (Matsushita et al., 2003; Deppenmeier et al., 2002).

1.3.2.3 YEASTS

Several yeasts, including Saccharomyces, Candida, and Kluyveromyces are present as natural starter cultures in fermented milk products. Generally, *Candida kefir, Candida maris, Saccharomyces cerevisiae, Saccharomyces kefir, Torulopsis kefir,* and *Kluyveromyces marxianus* are found to be associated with dahi (curd), cheeses, and acid- alcoholic fermented milk products like kefir, kumiss, and yeast-acidophilus milk. Another species *Candida antarctica* is a source of important lipases, an enzyme that can be used to enhance flavor in certain dairy products at industrial level (Choudhury and Bhunia, 2015).

1.3.2.4 MOLDS

Molds are employed in the production of some semi-soft cheese varieties and in some cultured milk products, as shown in Table 1.5. Molds augment the flavor and up to a certain extent, modify the textural properties and appearance of the finished product.

TABLE 1.5 Overview of Applications of Mold in Fermented Dairy Products

Mold Types	Example(s)	Uses(s) as Starter Culture
White mold	*Penicillium camemberti, Penicillium caseicolum, Penicillium candidum*	Manufacturing of surface mold-ripened cheeses like *Camembert* and *Brie* cheeses
Blue mold	*Penicillium roquefortii*	Manufacturing of internal mold-ripened cheeses like *Blue Stilton, Roquefort, Gorgonzola, Danish blue,* and mycelia cheeses
Other molds	*Geotricum candidum*	Used in Villi, a cultured product of Finland where the mold grows on the top of the milk to develop the white velvety layer
	Asperigillus oryzae	In the manufacturing of ripened skim milk cheese (Norway)
	Mucor rasmusen	In the manufacturing of Soya milk cheese (Japan)

1.4 RECENT ADVANCES AND FUTURE PERSPECTIVE TO IMPROVE STARTER PERFORMANCE

In recent years, the microbiological and technological performances of starter cultures have been improved through biotechnological tools. Researchers had successfully improved the rheological properties of the cultured dairy products like yogurt, dahi, cheese, kefir, and many traditional fermented milk products through the application of EPS or slime producing starters (SPSs) (Behare et al., 2009; Ayana and Ibrahim, 2015; Han et al., 2016).

Several products have been enriched with group B vitamins (Beitane and Ciprovica, 2012), bioactive peptides (Pihlanto, 2013; Park and Nam, 2015), and conjugated linoleic acid (Chung et al., 2008; Rodriguez-Alcala et al., 2011). Some peptidases produced by starters (*Str. thermophilus, Lc. lactis* subsp. *cremoris* and species of *Lactobacillus*) improved the body texture and sensory quality of fermented milk while proteolysis and lipolysis enhanced the flavor of most varieties of cheese (Guldfeldt et al., 2001, Gonzalez et al., 2010). Reconstituted skim milk fermented with folate-producing *Bifidobacterium* spp. in conjunction with yogurt cultures suggested that it is possible to increase the folate concentration in fermented milk through appropriate selection of starter bacteria (Crittenden et al., 2003). Another *in-vitro* investigation demonstrated enhanced folic acid production after administration of probiotic *Bifidobacterium* strains in human subjects (Strozzi and Mogna, 2008). Moreover, bacteriocin producing starters have also been employed to enhance the quality and safety aspects of fermented milk products (Arques et al., 2005; Malheiros et al., 2012). Several attempts are being made to genetically improve the starter cultures for better commercial activities. However, till today, no genetically modified starter culture is available in market.

1.5 PROBIOTICS IN DAIRY FOOD PRODUCTS

1.5.1 PROBIOTICS: GENERAL DESCRIPTION

The 'probiotic' word is translated from Greek meaning 'for life.' Fuller (1989) defined probiotics as "a live microbial feed supplement which beneficially affects the host animal by improving its intestinal microbial balance." In 2001, FAO/WHO (Food and Agriculture Organization of the United Nations/World Health Organization) refined the definition as "live microorganisms, which when administered in adequate amounts, confer a health benefit on the host" which is the most commonly used definition and has gained widespread scientific acceptability.

In fact, in the early 1900's the foremost observation concerning health beneficial effect of probiotics was reported by Metchnikoff, now whom we call as the "*Father of Probiotics*." According to him, the consumption of live microorganisms in soured milk may help to improve the balance of the gut microbiota in humans. Afterwards, probiotics have gained much more attention by researchers and health professionals and currently, their use is widely accepted especially to treat gastrointestinal disturbances (Ljungh and Wadstrom, 2009). Fermented milk Yakult added with *Lactobacillus casei* Shirota was the first commercially sold dairy-based probiotic launched in 1935 from Japan. Since then, numerous probiotic food products have arrived in the global market; mainly dominated by diverse milk products, including probiotic yogurts, probiotic ice creams, probiotic cheeses, and kefir. Many non-dairy and unfermented probiotic products have been developed including breakfast cereals, fruit juices and beverages, vegetable products (*kimchi* and *sauerkraut*), snack bars, and as an animal feeds in the last few years. Sales of probiotic foods have a rising trend in recent years, increased by 35% from 23.1 billion US dollars to 31.3 billion US dollars globally (Linares et al., 2016).

Majority of probiotic bacteria belong to *Lactobacillus* and *Bifidobacterium* genera (Table 1.6); however, strains belonging to the *Lactococcus, Streptococcus, Enterococcus, Pediococcus, Propionibacterium* genera and yeast *Saccharomyces* are also considered as probiotics.

TABLE 1.6 Major Groups of Probiotic Microorganisms

Lactobacillus	Bifidobacterium	Other Lactic Acid Bacteria	Non-Lactic Acid Bacteria and Yeasts
Lb. acidophilus	*B. adolescentis*	*Str. thermophilus*	*Propionibacterium freudenreichii*
Lb. rhamnosus L.	*B. animalis*	*Str. salivarius*	*Escherichia coli* Nissle 1917
Lb. casei	*B. bifidum*	*Lactococcus lactis*	*Bacillus cereus* var. *toyoi*
Lb. helveticus	*B. breve*	*Leuconostoc mesenteroides*	*Bacillus clausii*
Lb. johnsonii	*B. infantis*	*Pediococcus acidolactici*	*Bacillus subtilis*
Lb. delbrueckii subsp. bulgaricus	*B. lactis*	*Sporolactobacillus inulinus*	*Saccharomyces boulardii*
Lb. gasseri	*B. longum*	*Enterococcus faecalis*	
Lb. gallinarum		*E. faecium*	
Lb. reuteri			
Lb. plantarum			
Lb. paracasei			
Lb. crispatus			

1.5.2 EFFECTS OF PROBIOTICS IN HEALTH AND MECHANISM OF ACTION

There are plenty of *in-vitro* as well as *in-vivo* studies concerning health benefits of fermented foods and probiotics such as prevention and treatment of diarrhea, atopic dermatitis and food allergy, rheumatoid arthritis, irritable bowel disease, and irritable bowel syndrome; eradication of *Helicobacter pylori* infection; reduction of hypertension, serum cholesterol levels and lactose intolerance; immunomodulatory and immunostimulatory properties; and anticancer activities. Briefly, comprise the details of specific probiotic strain claimed to treat or cure specific health problem(s) by various researchers (*see* Table 1.8 and Table 1.9).

Probiotics may exert a positive influence on the host through modulation of the endogenous ecosystem and stimulation of the immune system as well as maintaining a healthy intestinal microflora. Viability of probiotics would be a reasonable mean to evaluate of their activity; conversely in certain situations improved digestion of lactose, anti-hypertensive effects and immunomodulation activities. Many effects have been linked to dead (non-viable) cells, cell components like proteins, enzymes or fermentation end products (Galdeano et al., 2009).

In order to provide other therapeutic benefits viability of probiotics is a must. Multiple health beneficiary activities of LAB can be associated with diverse mechanisms as depicted in Figure 1.2. In addition to the effects shown in Figure 1.2, eradication of pathogenic bacteria, restoration of a normal balance of intestinal microflora, and the enhanced in humoral immunity are other principle actions exhibited by probiotic strains (Galdeano et al., 2009; Harzallah and Belhadj, 2013). Probiotics also seem to strengthen the non-specific defense mechanisms like phagocytic activity and cytokine synthesis; hence, exerting anti-inflammatory or antimicrobial effects. Natural antimicrobial compounds produced by LAB have been widely used as chemotherapeutic agents that can control the growth of infectious microbes (Liasi et al., 2009).

1.5.3 SELECTION CRITERIA FOR PROBIOTICS

For organisms to achieve probiotic status, they must fulfill several criteria as compiled in Table 1.7. According to Klaenhammer and Kullen (1999), the criteria for selection of probiotics can be categorized fundamentally in four groups *viz.* appropriateness, technological suitability-safety aspects, competitiveness, performance, and functionality. Microbial species or strains that fulfill these criteria can be employed as probiotics to facilitate beneficial effects on health; however, a prospective probiotic strain may not require accomplishing all desired criteria for selection.

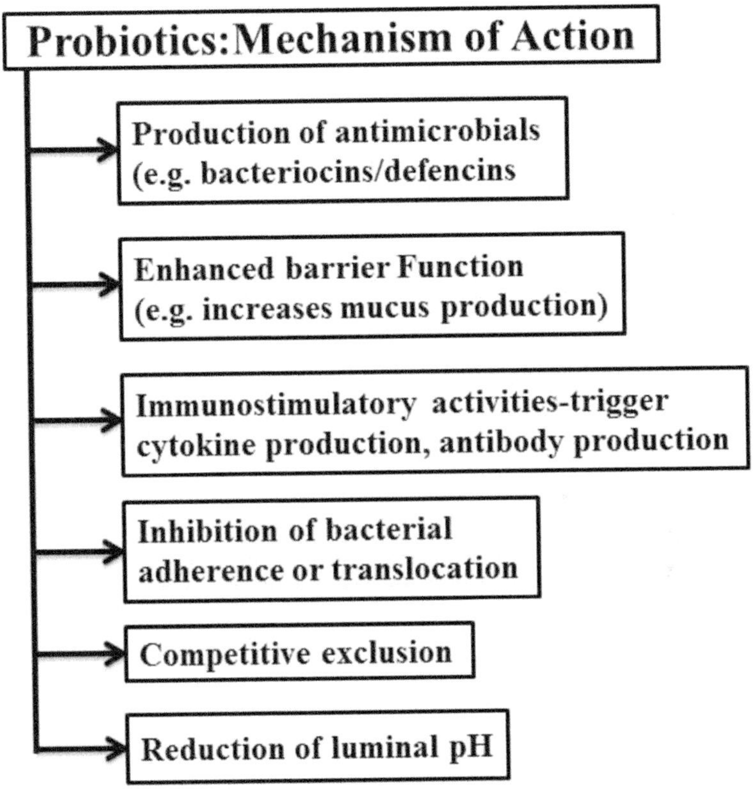

FIGURE 1.2 Overview of probiotic mechanism of action.

During the screening of probable probiotic strains, several *in-vitro* tests are obligatory, and the preliminary step is the determination of its taxonomic classification, which may provide an indication of the origin, habitat, and strain physiology. Particularly, certain LABs are inhabitants of the oral cavity, intestinal tract, or vagina of humans; thus, they may have a positive influence on these ecosystems only. Such characteristics have considerable consequences on the process of selecting novel strains

(Morelli, 2007). Potential probiotic strains should be checked for several important criteria such as phenotypic and genotypic stability (including stability of plasmid, if present); assimilation of carbohydrate and protein patterns; tolerance, survival, and growth in the presence of acid and bile; antibiotic resistance patterns; adhesion to intestinal epithelia; antimicrobial substances production; antimicrobial spectrum against spoilage and pathogenic microorganisms; and immunogenicity.

TABLE 1.7 Principal and Desirable Criteria for Probiotics

S. N.	Properties of Probiotic Strain	Remarks
1.	Human origin for prospective human use	Even though probiotic yeast *Saccharomyces boulardii* is not of human origin, this criterion is important for species dependent health beneficiary activities
2.	Acid and bile tolerance; resistance to digestive enzymes-pepsin, pancreatin, etc.; antimicrobial activity against potentially pathogenic bacteria	Essential criteria for oral consumption of probiotics although it may not be for other applications for survival through the intestine, maintaining adhesiveness and metabolic activity; to inhibit foodborne pathogens within the gut
3.	Adhesion to mucosal surface-gastrointestinal tract	Vital to improve immune system, maintain metabolic activity, compete with pathogens by avoiding their adhesion and colonization on mucosal surfaces
4.	Safe for food and clinical application	Precise taxonomic identification and characterization of strains including tests for virulence factors- toxic effects, metabolic activity and inherent properties like infectivity, pathogenicity, and antibiotic resistance
5.	Clinically validated and documented health effects	For each particular strain, the minimum effective dosage has to be known according to different products. Placebo-controlled, double-blinded, and randomized studies should be conducted
6.	Good technological properties	Desired viability or survival during product processing and storage if viable organisms are obligatory, strain stability, phage resistance, oxygen resistance, culturable at large scales, has no negative influences on product flavor or body-texture

Sources: Modified from Klaenhammer and Kullen, (1999) and Vasiljevic and Shah, (2008).

Among these, adhesion to the intestinal epithelia/mucosa is prerequisite property for the colonization of probiotic strain. Probiotic strain should act as an adjuvant and stimulate the immune system against invading pathogen. It is obvious that a probiotic should be harmless to the host; there must be no local or general allergic, pathogenic, or mutagenic reactions provoked by the microbial cell itself, cell components or its fermentation products.

1.5.4 *MICROBIOLOGICAL CONSIDERATIONS FOR FOOD AND HEALTH CLAIMS*

Probiotic can be marketed either as a food product, nutritional supplement, or pharmaceutical preparation. Establishment as a pharmaceutical product, necessitate significant time to carry out multifaceted and costly investigations (involving clinical trials), and expression of well-defined therapeutic targets (Figure 1.3). Selection of inappropriate strains, poorly regulated quality of the strain, incapability of strain; inefficient handling that impairs the viability of probiotics and difficulties in maintaining new strain in the gut are the few major obstacles in providing probiotic therapy (Tamboli et al., 2003).

1.6 OVERVIEW OF IMPORTANT PROBIOTICS

1.6.1 *PROBIOTIC LACTIC ACID BACTERIA (LAB)*

1.6.1.1 *LACTOBACILLUS SPECIES*

Many species of Lactobacilli are the predominant vital bacteria inherent within small intestine. In the gut, these species collectively ferment carbohydrates and produces lactic acid which creates acidic surroundings in the digestive tract and ward off numerous unwanted bacteria that flourish in alkaline environment. Lactate also enhances mineral absorption such as calcium, iron, magnesium, and copper. Several probiotic species of this genus have been proven for specific health benefits and are presented in Table 1.8.

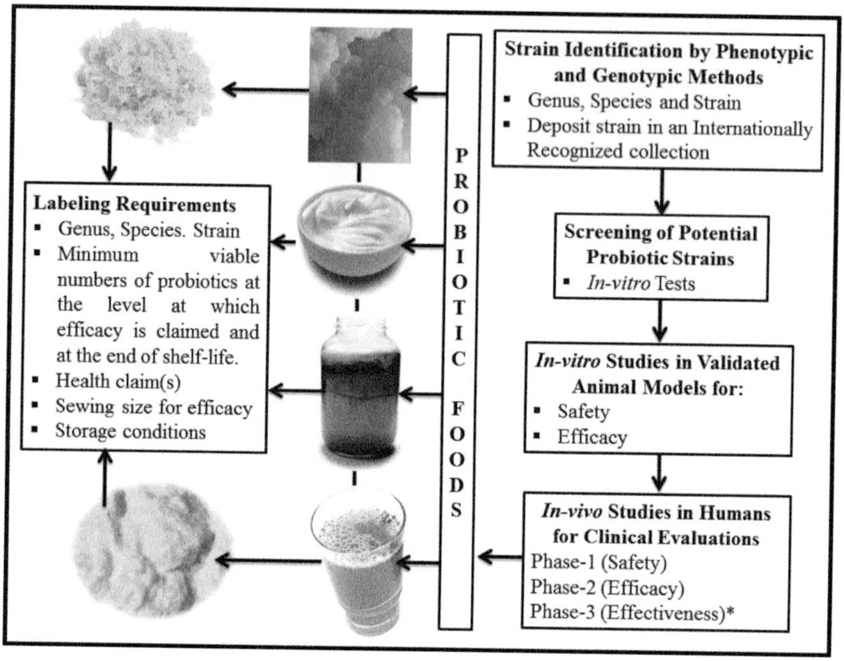

FIGURE 1.3 (See color insert.) Guidelines for evaluation of candidate probiotic strains. *Note:* *required only when a specific health claim is made]. (Reprinted with modification from Ganguly, , et al., 2011.)

1.6.1.2 STREPTOCOCCI AND ENTEROCOCCI SPECIES

Yogurt strain *Str. thermophilis* has also shown some probiotic potential. Another strain closely related to *Str. thermophilus* is *Str. salivarius* K12 which is employed as an oral probiotic strain as it is found in the mucus membranes of oral cavity and well established for their ability to synthesize bacteriocin-like inhibitory substances (BLIS) and to prevent the growth of undesirable bacteria (Stowiks, 2016). According to one study, 10% of the populations who naturally carry BLIS-producing bacterial strains in oral cavity have considerably less sore throats (Williams, 2016).

Enterococcus faecium has been used since long as probiotics, particularly for antibiotic-associated diarrhea prevention. However, some enterococcal strains are opportunistic infectious agents and represent a possible reservoir of virulence and antibiotic resistance genes as demonstrated in few animal

TABLE 1.8 An Overview of Health Effects of Probiotic Lactic Acid Bacteria

Strain	Effect	References
Lb. acidophilus	Diarrhea associated with antibiotic treatment or infection with *Clostridium difficile* and rotavirus	Gao et al., 2010; De Vrese et al., 2011; Lee et al., 2015
	Bacterial vaginosis	Parent et al., 1996
	Cholesterol-lowering effect	Ashar and Prajapati, 1998
	Type 2 diabetes mellitus	Ejtahed et al., 2011
	Anti-*Helicobacter pylori* activity	Canducci et al., 2000; Felley et al., 2001
	Anticarcinogenic effect	Oberreuther-Moschner et al., 2004; Mohania et al., 2013
Lb. helveticus	Cholesterol-lowering effect	Ahire et al., 2012
	Lower blood pressure	Sipola et al., 2001; Seppo et al., 2002
Lb. casei	Diarrhea associated with antibiotic treatments or infection with *Clostridium difficile* and rotavirus	Hickson et al., 2007; Stockenhuber et al., 2008
	Anti-*Helicobacter pylori* activity	Cats et al., 2003; Tursi et al., 2004; Sykora et al., 2005
	Bacterial vaginosis	Kovachev et al., 2013
	Immunomodulatory mechanisms	Shida and Nomoto, 2013
Lb. rhamnosus	Diarrhea associated with antibiotic treatment	Beausoleil et al., 2007
	Treatment of Irritable bowel syndrome	Gawronska et al., 2007
	Bacterial vaginosis	Anukam et al., 2006
	Reduction of food allergy	Wickens et al., 2012; Gruber et al., 2008
	Anticarcinogenic effect	Hatakka et al., 2008; Dominici et al., 2014
	Anti-*Helicobacter pylori* activity	Armuzzi et al., 2001
Lb. plantarum	Cholesterol-lowering effect	Nguyen et al., 2007; Fazeli et al., 2010; Jeun et al., 2010
	Anticarcinogenic effect	Mohania et al., 2013
	Reduce blood pressure and body mass indexes	Sharafedtinov et al., 2013

TABLE 1.8 *(Continued)*

Strain	Effect	References
Lb. reuteri	Cholesterol lowering effect	Taranto et al., 2003; Jones et al., 2012
	Bacterial vaginosis	Anukam et al., 2006
	Anti-*Helicobacter pylori* activity	Imase et al., 2007; Francavilla et al., 2008; Ojetti et al., 2012
Lb. fermentum	Cholesterol-lowering effect	Simons and Amansec, 2006; Pan et al., 2011
Lc. lactis subsp. *Lactis*	Diarrhea associated with antibiotic treatments	Johnston et al., 2011
	Antimicrobial activity against *C. difficile*	Lee et al., 2013
Pd. acidilactici	Anti-*Helicobacter pylori* activity	Kaur et al., 2014
Str. thermophilus	Treatment of irritable bowel syndrome	Wu et al., 2013
Str. salivarius K12 and M18	Oral probiotic inhibits undesirable bacteria in oral cavity	Strowik, 2016
E. faecium	Diarrhea associated with antibiotic treatments	Hempel et al., 2012
E. durans	Efficient animal probiotic	Cao et al., 2013
	Anti-inflammatory activity	Raz et al., 2007

studies (Dirienzo, 2014); and therefore do not possess GRAS status for human use though proven excellent probiotics for animals (Bednorz et al., 2013; Cao et al., 2013). Several strains of lactococci, pediococci, and leuconostocs have also demonstrated positive health claims (Table 1.8).

1.6.2 PROBIOTIC BACTERIA OTHER THAN LACTIC ACID BACTERIA (LAB)

1.6.2.1 BIFIDOBACTERIUM SPECIES

Many species of bifidobacteria are recognized as potential probiotics and claim for various health benefits (Table 1.9). *Bifidobacterium* species usually colonize the epithelial lining of the large intestine (colon) and help discourage invasive pathogens and other microbiota, including yeast (Saavedra et al., 1994; Williams, 2016). *Bifidobacterium* species produce more amount of acetic acid than lactic acid (3:2), which has higher antimicrobial efficiency discourage the growth of other bacteria, plus improving the natural protective barrier in the gut. In addition, low pH surroundings facilitate the minerals absorption (Prasanna et al., 2014; Williams, 2016).

1.6.2.2 BACILLUS SPP.

Bacillus is rod-shaped, spore-bearing bacteria, ubiquitous in nature and thus through air, food, and water get entry in the respiratory and gastrointestinal tracts of healthy people (Cutting, 2011; Sorokulova, 2013). Since they are spore-forming bacteria, they are highly resistant to adverse conditions of stomach such as presence of acid and enzymes, and thus, able to maintain their viability to readily colonize the small intestine (Hoa et al., 2000; Duc et al., 2003). They also inhabited in the body for a longer period than other bacteria and excreted slow gradually (Williams, 2016). There are strong scientific evidences suggesting their credential as probiotics (Table 1.9).

Certain strains of probiotic *Bacillus* have been approved for human use, for instance in Japan *Bacillus subtilis* var. *Natto* OUV23481 was approved as Food for Specified Health Use (FOSHU) (Shimizu, 2003). In Europe, *Bacillus clausii* and *Bacillus cereus,* labeled in products 'Entergermina' and 'Bactisubtil,' respectively were approved for medical use and similarly, in the USA, *Bacillus coagulans* (GanedenBC30) gained

TABLE 1.9 An Overview of Health Effects Observed by Probiotics Other Than Lactic Acid Bacteria

Strain	Effect	References
B. infantis	Reduction and prevention of food allergy	Konieczna et al., 2011
B. longum	Cholesterol lowering effect	Kiessling et al., 2002; Xiao et al., 2003
	Reduction of lactose intolerance	Jiang et al., 1996
B. lactis Bb12	Improvement of gastrointestinal well-being and digestive symptoms	Guyonnet et al., 2009; Waller et al., 2011
	Guyonnet type 2 diabetes mellitus	Ejtahed et al., 2011
	Irritable bowel syndrome	Agrawal et al., 2009
	Reduction and prevention of food allergy	Lindfors et al., 2008
	Anti-carcinogenic effect	Matsumoto and Benno, 2004
	Increase immune responses	Mohan et al., 2008
B. bifidum	Irritable bowel syndrome	Guglielmetti et al., 2011
	Anti-*Helicobacter pylori* activity	Miki et al., 2007
	Prevention of diarrhea	Saavedra et al., 1994
B. animalis	Prevention and treatment of diarrhea	Shah et al., 2006
B. breve and *B. Bifidum*	Treatment for inflammatory bowel disease	Ishikawa et al., 2003; Kato et al., 2004
B. adolescentis and *B. longum*	Immune system stimulation	He et al., 2002
Propionibacterium sp.	Prevention of IgE-associated allergy	Kuitunen et al., 2009
	Anti-carcinogenic properties (binding with carcinogenic toxins)	El-Nezami et al., 2000; Gratz et al., 2005; Zarate, 2009

TABLE 1.9 *(Continued)*

Strain	Effect	References
Escherichia coli-Nisle 1917	Constipation treatment	Chmielewska and Szajewska, 2010
	Gastrointestinal disorders	Xia et al., 2013
	Irritable bowel syndrome	Liebregts et al., 2005; Kruis et al., 2012
	Remission of ulcerative colitis	Kruis et al., 2004
Bacillus coagulans	Treatment of diarrhea, adjunct therapy for rheumatoid arthritis	Jurenka, 2012
	Relieving symptoms of rheumatoid arthritis	Mandel et al., 2010
Bacillus clausii	Treatment of acute diarrhea	Sudha et al., 2013
	Anti-*Helicobacter* pylori therapy	Nista et al., 2004
Bacillus subtilis var. *Natto*	Immunomodulation, stimulation of immune system	Hosoi and Kiuchi, 2004
Saccharomyces boulardii	Anti- *Helicobacter* pylori therapy	Cindoruk et al., 2007
	Diarrhea associated with antibiotic treatment or infection with *Clostridium difficile*	Surawicz et al., 2000; Duman et al., 2005; Can et al., 2006; McFarland, 2010
	Immunomodulation	Martins et al., 2009
	Anti-inflammatory activities	Foligné et al., 2010;
	Irritable bowel disease, Crohn's disease, Ulcerative colitis	Dalmasso et al., 2006; Guslandi et al., 2000, 2003

self-affirmed GRAS status (Sorokulova, 2013). The national authorities of Russia and Ukraine permitted a drug status to *Bacillus subtilis* and *Bacillus licheniformis* (Biosporin) for prophylaxis and acute intestinal infections treatment (Smirnov et al., 1994; Gracheva et al., 1996).

1.6.2.3 ESCHERICHIA COLI NISSLE 1917

The genus *Escherichia* belongs to the *Enterobacteriaceae* family, which mainly recognized for severe virulent serotypes such as *E. coli* O157:H7 (Fijan, 2014). *Escherichia coli* usually reside in lower intestine (colon) of human body. *Escherichia coli* Nissle 1917 (EcN) is one of the strain granted probiotic confirmation, which have been established to treat gastric disorders, constipation, ulcerative colitis, inflammatory bowel disease, and Crohn's disease, either alone or in combination with other probiotic strains (Table 1.9).

1.6.3 PROBIOTIC YEAST

Applications of yeast are not only limited to the food sector but because of diverse biological activities own by some yeast species, makes them competent candidates for several health applications. Probiotic yeasts were used in animal feeds because they improve livestock performance and product quality. In the last few years, interest has increased for human applications, too. *S. boulardii* is the only yeasts to date with the established potential probiotic status. Actually, using multilocus sequence analysis, when the taxonomic position of *S. boulardii* was determined, each locus showed high similarity to the related loci in *S. cerevisiae* (Van Der Aa Kuhle and Jespersen, 2003); thus, *S. cerevisiae* var. *boulardii,* a new denomination has been proposed. However, yet the term *S. boulardii* is widely used in literature (Hatoum et al., 2012).

The cells of *S. boulardii* found to produce a 54-KDa protease that is linked with its beneficial health effects. This protease shows to degrade toxins of *Cl. difficile* (Castagliuolo et al., 1996; Qamar et al., 2001); moreover, inhibit *Cl. difficile* attachment to intestinal cells by modifying cell surface receptors (Tasteyre et al., 2002; Buts, 2009). The yeast cells reported to enhance the immune response against *Cl. difficile* toxins (Buts,

2009), and to stimulate intestinal immunoglobin A secretion (Qamar et al., 2001). Health beneficiary effects of *S. boulardii* are listed in Table 1.9. Yeasts are rarely associated with food-borne illness or outbreaks; many yeast species confirmed Qualified Presumption of Safety status by the European Food Safety Authority (EFSA) (http://www.efsa.europa.eu). Bruno et al. (2009) investigated the effect of feeding probiotic yeasts on dairy cows during summer; resulted in improved milk yield and milk components in heat-stressed cows. Recent data on probiotic properties of yeasts has opened the door for new therapeutic uses of as an immunotherapeutic agent. It is essential to note that *S. boulardii* is linked with fungemia or localized infections in immunocompromised individuals (Thygesen et al., 2012).

1.7 REGULATORY FRAMEWORKS AND LEGISLATION FOR PROBIOTICS

The regulatory framework for probiotics differ a lot from one nation to other nation and sometimes also even within the nation (Reid, 2001; Patel et al., 2008; EFFCA, 2008). In 1991, Japan was the first nation to implement regulatory system for functional foods and nutraceuticals aid with probiotics (Arora and Baldi, 2015); where probiotic products are in distinct category called FOSHU, while to claim about efficacy of probiotic(s), a special permission to the Ministry of Health and Welfare (MHLW) of Japan is must (Amagase, 2008). Afterward, Europe was second to establish a regulatory commission known as functional food science in Europe (FUFOSE) in 1995. Similarly, in US, Food and Drug Administration (FDA) and Dietary Supplement Health and Education Act (DSHEA); in Australia and New Zealand, Food Standards Australia and New Zealand (FSANZ); in Brazil, National Health Surveillance Agency Brazil (ANVISA); while in China, State Food and Drug Administration (SFDA) is involved with regulations of probiotics in food and pharmaceutical products.

In common, these regulations mainly address the identification of the proper strain, screening for probiotic features through *in-vitro* tests, subsequently followed by *in-vivo* animal and human clinical studies to establish efficiency, labeling of probiotic products with specification of strain(s), storage conditions and viable cell count at the end of shelf-life in order to facilitate the consumers to uphold their awareness. During the last decade, the influx of probiotic dairy and food products in the market has been increase tremendously in many developing countries, including India. Department of Biotechnology (DBT) and Indian Council of Medical Research (ICMR) in concert with Government of India (GOI) have formulated a guideline that defines a set of parameters requisite for a strain or product to called as 'probiotic' for the regulation of probiotic products in the country (Ganguly et al., 2011). Canada and Malaysia have very complex and unsatisfactory legislative systems (Arora and Baldi, 2015) while many other developing countries of the world lack such regulations or legislation. A regulatory framework and guidelines must be created at the national and international level to better address various issues of probiotics such as dose, efficacy, safety, claims, and labeling across the globe.

1.8 CONCLUSION

Starter cultures are those microorganisms that are purposefully employed for the production of fermented food products to bring out desirable change(s). Biotechnological tools can be helpful to enhance the technological performances of starter cultures as well as probiotic strains. Careful selection of bacteriocin producing starters would help to enhance the quality and safety aspects of fermented milk products while incorporation of bioactive compound (like vitamins, peptides, conjugated linoleic acid, enzymes, etc.) producing strains may serve to develop value-added food products. With regard to probiotics, the effective doses of probiotics vary from strain to strain, the food matrixes, and the host like the age, gender, healthy or immunocompromised person. Also, few strains are effective when used alone, while some show efficacy in combinations. Thus, to address such issues and validate the specific health claims of specific probiotic strains, there is a need for multicentric approach for human clinical trials. Formulation of even Standard Operating Procedures (SOPs) and analytical methods to assess the efficacy and safety of probiotics is necessary. Though there exist guidelines and regulatory standards in different countries, it varies in different nations. The guidelines and regulations warrant harmonization at the global level that would ensure better quality and safety of probiotic foods for effective utilization.

KEYWORDS

- **acetic acid bacteria**
- **antibiotic resistance**
- **antihypertensive effect**
- **antimicrobial compound**
- **bioactive peptides**
- **bioghurt**
- **carbohydrate fermentation**
- **dairy-based probiotic**
- **fermented milk**
- **food for specified health use (FOSHU)**
- **immunostimulatory properties**
- **starter culture**
- **yeast-acidophilus milk**
- **β-phosphatase**

REFERENCES

Abd-Elrahman, A. H., (2013). Mastitis in housed dairy buffaloes: Incidence, etiology, clinical finding, antimicrobial sensitivity and different medical treatment against *E. coli* mastitis. *Life Science Journal, 10*(1), 531–538.

Abou, A. I. A. A., & Ibrahim, A. E., (2015). Attributes of low-fat yogurt and kareish cheese made using exopolysaccharides producing lactic acid bacteria. *American Journal of Food Technology, 10*, 48–57.

Agrawal, A., Houghton, L. A., Morris, J., Reilly, B., Guyonnet, D., Goupil, F. N., Schlumberger, A., Jakob, S., & Whorwell, P. J., (2009). Clinical trial: The effects of a fermented milk product containing *Bifidobacterium lactis* DN-173 010 on abdominal distension and gastrointestinal transit in irritable bowel syndrome with constipation. *Alimentary Pharmacology & Therapeutics, 29*, 104–114.

Ahire, J. J., Bhat, A. A., Thakare, J. M., Pawar, P. B., Zope, D. G., Jain, R. M., & Chaudhari, B. L., (2012). Cholesterol assimilation and biotransformation by *Lactobacillus helveticus*. *Biotechnology Letters, 34*(1), 103–107.

Amagase, H., (2008). Current marketplace for probiotics: A Japanese perspective. *Clinical Infectious Diseases, 4*, 73–75.

Anastasiadou, S., Papagianni, M., Filiousis, G., Ambrosiadis, I., & Koidis, P., (2008). Growth and metabolism of a meat isolated strain of *Pediococcus pentosaceus* in submerged fermentation: Purification, characterization and properties of the produced pediocin SM-1. *Enzyme and Microbial Technology*, *43*(6), 448–454.

Anukam, K. C., Osazuwa, E., Osemene, G. I., Ehigiagbe, F., Bruce, A. W., & Reid, G., (2006). Clinical study comparing probiotic *Lactobacillus GR-1* and RC-14 with metronidazole vaginal gel to treat symptomatic bacterial vaginosis. *Microbes Infection, 8*, 2772–2776.

Armuzzi, A., Cremonini, F., Ojetti, V., Bartolozzi, F., Canducci, F., Candelli, M., et al. (2001). Effect of *Lactobacillus* GG supplementation on antibiotic-associated gastrointestinal side effects during *Helicobacter pylori* eradication therapy: A pilot study. *Digestion, 63*, 1–7.

Arora, M., & Baldi, A., (2015). Regulatory categories of probiotics across the globe: A review representing existing and recommended categorization. *Indian Journal of Medical Microbiology, 33*(S1), 2–10.

Arques, J. L., Rodríguez, E., Gaya, P., Medina, M., & Nuñez, M., (2005). Effect of combinations of high-pressure treatment and bacteriocin-producing lactic acid bacteria on the survival of *Listeria monocytogenes* in raw milk cheese. *International Dairy Journal, 15*, 893–900.

Ashar, M. N., & Prajapati, J. B., (1998). Bile tolerance, bile deconjugation and cholesterol-reducing properties of dietary lactobacilli. *Indian Journal of Microbiology, 38*(3), 145–148.

Axelsson, L., (1998). Lactic acid bacteria: Classification and physiology. In: Salminen, S., & Von Wright, A., (eds.), *Lactic Acid Bacteria: Microbiology and Functional Aspects* (2nd edn., pp. 1–72). Marcel Dekker, Inc., New York.

Beausoleil, M., Fortier, N., Guenette, S., Lecuyer, A., Savoie, M., Franco, M., et al. (2007). Effect of a fermented milk combining *Lactobacillus acidophilus* Cl1285 and *Lactobacillus casei* in the prevention of antibiotic-associated diarrhea: A randomized, double-blind, placebo-controlled trial. *Canadian Journal of Gastroenterology, 21*, 732–736.

Bednorz, C., Guenther, S., Oelgeschläger, K., Kinnemann, B., Pieper, R., Hartmann, S., et al. (2013). Feeding the probiotic *Enterococcus faecium* strain NCIMB 10415 to piglets specifically reduces the number of *Escherichia coli* pathotypes that adhere to the gut mucosa. *Applied and Environmental Microbiology, 79*, 7896–7904.

Behare, P., Singh, R.,&Singh, R. P., (2009). Exopolysaccharide-producing mesophilic lactic cultures for preparation of fat-free Dahi–Indian fermented milk. *Journal of Dairy Research, 76*(1), 90–97.

Beitane, I., & Ciprovica, I., (2012). The study of added prebiotics on0 b group vitamins concentration during milk fermentation. *Romanian Biotechnology Letters, 16*, 92–96.

Bruno, R. G. S., Rutigliano, H. M., Cerri, R. L., Robinson, P. H., & Santos, J. E. P., (2009). Effect of feeding *Saccharomyces cerevisiae* on performance of dairy cows during summer heat stress. *Animal Feed Science and Technology, 150*(3), 175–186.

Buts, J. P., (2009). Twenty-five years of research on *Saccharomyces boulardii* trophic effects: Updates and perspectives. *Digestive Disease Science, 54*, 15–18.

Can, M., Besirbellioglu, B. A., Avci, I. Y., Beker, C. M.,&Pahsa, A., (2006). Prophylactic *Saccharomyces boulardii* in the prevention of antibiotic-associated diarrhea: A prospective study. *Medical Science Monitor, 12*, PI19–PI22.

Canducci, F., Cremonini, F., Armuzzi, A., Di Caro, S., Gabrielli, M., Santarelli, L., et al. (2000). A lyophilized and inactivated culture of *Lactobacillus acidophilus* increases *Helicobacter pylori* eradication rates. *Aliment Pharmacol. Therapy, 14*, 1625–1629.

Cao, G. T., Zeng, X. F., Chen, A. G., Zhou, L., Zhang, L., Xiao, Y. P., & Yang, C. M., (2013). Effects of a probiotic, *Enterococcus faecium*, on growth performance, intestinal morphology, immune response, and cecal microflora in broiler chickens challenged with *Escherichia coli* K88. *Poultry Science, 92*, 2949–2955.

Castagliuolo, I., LaMont, J. T., Niku-lasson, S. T., & Pothoulakis, C., (1996). *Saccharomyces boulardii* protease inhibits *Clostridium difficile* toxin A effects in the rat ileum. *Infection and Immunology, 64*, 5225–5232.

Cats, A., Kuipers, E. J., Bosschaert, M. A., Pot, R. G., Vandenbroucke-Grauls, C. M., & Kusters, J. G., (2003). Effect of frequent consumption of a *Lactobacillus casei*-containing milk drink in *Helicobacter pylori*-colonized subjects. *Alimentary Pharmacology& Therapeutics, 17*, 429–435.

CDIC (Canadian Dairy Information Centre), (2017a). Global milk consumption (liters per capita). In: *Global Consumption of Dairy Products by Country (Annual)*. URL: http://www.dairyinfo.gc.ca/index_e.php?s1=dff-fcil&s2=cons&s3=consglo&s4=tm-lt (accessed on 6 September 2019).

CDIC (Canadian Dairy Information Centre), (2017b). Global milk consumption (liters per capita). In: *Global Consumption of Dairy Products by Country (Annual)*. URL: http://www.dairyinfo.gc.ca/index_e.php?s1=dff-fcil&s2=cons&s3=consglo&s4=tm-lt (accessed on 6 September 2019).

CDIC (Canadian Dairy Information Centre), (2017c). Global milk consumption (liters per capita). In: *Global Consumption of Dairy Products by Country (Annual)*. URL: http://www.dairyinfo.gc.ca/index_e.php?s1=dff-fcil&s2=cons&s3=consglo&s4=tm-lt (accessed on 6 September 2019).

Chandan, R. C., (2006). History and consumption trends. In: Chandan, R. C., (ed.), *Manufacturing Yogurt and Fermented Milks* (pp. 3–17). Blackwell Publishing.

Chmielewska, A., & Szajewska, H., (2010). Systematic review of randomized controlled trials: Probiotics for functional constipation. *World Journal of Gastroenterology, 16*, 69–75.

Choudhury, P., & Bhunia, B., (2015). Industrial application of lipase: A review. *Biopharma Journal, 1*(2), 41–47.

Chung, S. H., Kim, I. H., Park, H. G., Kang, H. S., Yoon, C. S., Jeong, H. Y., et al. (2008). Synthesis of conjugated linoleic acid by human-derived *Bifidobacterium breve* LMC 017: Utilization as a functional starter culture for milk fermentation. *Journal of Agriculture and Food Chemistry, 14, 56*(9), 3311–3316.

Cindoruk., M, Erkan, G., Karakan, T., Dursun, A., & Unal, S., (2007). Efficacy and safety of *Saccharomyces boulardii* in the 14-day triple anti-*Helicobacter pylori* therapy: A prospective randomized placebo-controlled double-blind study. *Helicobacter, 12*, 309–316.

Collins, M. D., Samelis, J., Metaxopoulos, J., & Wallbanks, S., (1993). Taxonomic studies on some leuconostoc-like organisms from fermented sausages - description of a new genus *Weissella* for the *Leuconostoc paramesenteroides* group of species. *Journal of Applied Bacteriology, 75*, 595–603.

Cutting, S. M., (2011). Bacillus probiotics. *Food Microbiology, 28*, 214–220.

Dalmasso, G., Cottrez, F., Imbert, V., Lagadec, P., Peyron, J. F., Rampal, P., Czerucka, D., Groux, H., Foussat, A., & Brun, V., (2006). *Saccharomyces boulardii* inhibits inflammatory bowel disease by trapping T cells in mesenteric lymph nodes. *Gastroenterology, 131*, 1812–1825.

De Vrese, M., Kristen, H., Rautenberg, P., Laue, C., & Schrezenmeir, J., (2011). Probiotic *lactobacilli* and bifidobacteria in a fermented milk product with added fruit preparation

reduce antibiotic associated diarrhea and *Helicobacter pylori* activity. *Journal of Dairy Research, 78*, 396–403.
Deppenmeier, U., Hoffmeister, M., & Prust, C., (2002). Biochemistry and biotechnological applications of *Gluconobacter* strains. *Applied Microbiology Biotechnology, 60*, 233–242.
Desai, A., (2008). *Strain Identification, Viability and Probiotics Properties of Lactobacillus Casei* (p. 228). PhD Thesis Victoria University, Victoria, Australia.
Dirienzo, D. B., (2014). Effect of probiotics on biomarkers of cardiovascular disease: Implications for heart-healthy diets. *Nutrition Review, 72,* 18–29.
Dominici, L., Villarini, M., Trotta, F., Federici, E., Cenci, G., et al. (2014). Protective effects of probiotic *lactobacillus rhamnosus* IMC501 in mice treated with PhIP. *Journal of Microbiology and Biotechnology, 24*, 371–378.
Duc, L. H., Hong, H. A., & Cutting, S. M., (2003). Germination of the spore in the gastrointestinal tract provides a novel route for heterologous antigen delivery. *Vaccine, 21*, 4215–4224.
Duman, D. G., Bor, S., Ozütemiz, O., Sahin, T., Oguz, D., Istan, F., et al. (2005). Efficacy and safety of *Saccharomyces boulardii* in prevention of antibiotic-associated diarrhea due to *Helicobacter pylori* eradication. *European Journal of Gastroenterology and Hepatology, 17*, 1357–1361.
Ejtahed, H. S., Mohtadi-Nia, J., Homayouni-Rad, A., Niafar, M., Asghari-Jafarabadi, M., Mofid, V., & Akbarian-Moghari, A., (2011). Effect of probiotic yogurt containing *Lactobacillus acidophilus* and *Bifidobacterium lactis* on lipid profile in individuals with type 2 diabetes mellitus. *Journal of Dairy Science, 94*(7), 3288–3294.
El-Nezami, H., Mykkanen, H., Kankaanpaa. P., Salminen, S., & Ahokas, J., (2000). Ability of *lactobacillus* and *propionibacterium* strains to remove aflatoxin B-1 from the chicken duodenum. *Journal of Food Protection, 63*, 549–552.
Endo, A., & Okada, S., (2005). Monitoring the lactic acid bacterial diversity during shochu fermentation by PCR-denaturing gradient gel electrophoresis. *Journal of Bioscience and Bioengineering, 99*, 216–221.
European Food and Feed Culture Association, (2008). *EFFCA Guidelines for Probiotics in Food and Dietary Supplements*. http://bbs.bio668.com/simple/index.php?t35919.html (accessed on 6 September 2019).
FAO/WHO, (2002). Guidelines for the evaluation of probiotics in food–Joint Food and Agricultural Organization of the United Nations and World Health Organization Working Group Meeting Report, London, Ontario, Canada.
Fazeli, H., Moshtaghian, J., Mirlohi, M., & Shirzadi, M., (2010). Reduction in serum lipid parameters by incorporation of a native strain of *L. plantarum* A7 in mice. *Iranian Journal of Diabetes & Lipid Disorders, 9*, 1–7.
Felley, C. P., Corthesy-Theulaz, I., Rivero, J. L., Sipponen, P., Kaufmann, M., Bauerfeind, P., et al. (2001). Favourable effect of an acidified milk (LC-1) on *Helicobacter pylori* gastritis in man. *European Journal of Gastroenterology and Hepatology, 13*, 25–29.
Fijan, S., (2014). Microorganisms with claimed probiotic properties: An overview of recent literature. *International Journal of Environmental Research and Public Health, 11*, 4745–4767.
Foligne, B., Dewulf, J., Vandekerckove, P., Pignède, G., & Pot, B., (2010). Probiotic yeasts: Anti-inflammatory potential of various non-pathogenic strains in experimental colitis in mice. *World Journal of Gastroenterology, 16*(17), 2134–2145.

Francavilla, R., Lionetti, E., & Castellaneta, S. P., (2008). Inhibition of *Helicobacter pylori* infection in humans by *Lactobacillus reuteri* ATCC 55730 and effect on eradication therapy: A pilot study. *Helicobacter, 13*, 127–134.

Galdeano, C. M., De Moreno, De LeBlanc, A., Carmuega, E., Weill, R., & Perdigón, G., (2009). Mechanisms involved in the immunostimulation by probiotic fermented milk. *Journal of Dairy Research, 76*(4), 446–454.

Galle, S., Schwab, C., Arendt, E., & Ganzle, M., (2010). Exopolysaccharide-forming *Weissella* strains as starter cultures for sorghum and wheat sourdoughs. *Journal of Agricultural Food Chemistry, 58*, 5834–5841.

Ganguly, N. K., Bhattacharya, S. K., Sesikeran, B., Nair, G. B., Ramakrishna, B. S., Sachdev, H. P. S., Batish, V. K., Kanagasabapathy, A. S., Muthuswamy, V., Kathuria, S. C., & Katoch, V. M., (2011). ICMR-DBT guidelines for evaluation of probiotics in food. *Indian Journal of Medical Research, 134*(1), 22–25.

Gao, X. W., Mubasher, M., Fang, C. Y., Reifer, C., & Miller, L. E., (2010). Dose–response efficacy of a proprietary probiotic formula of Lactobacillus acidophilus CL1285 and *Lactobacillus casei* LBC80R for antibiotic-associated diarrhea and Clostridium difficile-associated diarrhea prophylaxis in adult patients. *The American Journal of Gastroenterology, 105*(7), 1636–1641.

Gawronska, A., Dziechciarz, P., Horvath, A., & Szajewska, H., (2007). A randomized double blind placebo-controlled trial of *Lactobacillus* GG for abdominal pain disorders in children. *Alimentary Pharmacology & Therapeutics, 25*, 177–184.

Gonzalez, L., Sacristan, N., Arenas, R., Fresno, J. M., & Tornadijo, M. E., (2010). Enzymatic activity of lactic acid bacteria (with antimicrobial properties) isolated from a traditional Spanish cheese. *Food Microbiology, 27*, 592–597.

Gracheva, N. M., Gavrilov, A. F., Soloveva, A. I., Smirnov, V. V., Sorokulova. I. B., et al. (1996). The efficacy of the new bacterial preparation biosporin in treating acute intestinal infections. *Journal of Microbiology, Epidemiology, and Immunobiology*, 75–77.

Gratz, S., Mykkanen, H., & El-Nezami, H., (2005). Aflatoxin B-1 binding by a mixture of *Lactobacillus* and *Propionibacterium*: *In vitro* versus *ex vivo*. *Journal of Food Protection, 68*, 2470–2474.

Gruber, C., Keil, T., Kulig, M., Roll, S., & Wahn, U., (2008). Randomized, placebo-controlled trial of *Lactobacillus rhamnosus* GG as treatment of atopic dermatitis in infancy. *Pediatric Allergy and Immunology, 19*(6), 505–512.

Guglielmetti, S., Mora, D., Gschwender, M., & Popp, K., (2011). Randomized clinical trial: *Bifidobacterium bifidum* MIMBb75 significantly alleviates irritable bowel syndrome and improves quality of life–a double-blind, placebo-controlled study. *Alimentary Pharmacology & Therapeutics, 33*, 1123–1132.

Guldfeldt, L. U., Sorensen, K. I., Stroman, P., Behrndt, H., Williams, D., & Johansen, E., (2001). Effect of starter cultures with a genetically modified peptidolytic or lytic system on cheddar cheese ripening. *International Dairy Journal, 11*, 373–382.

Guslandi, M., Giollo, P., & Testoni, P. A., (2003). A pilot trial of *Saccharomyces boulardii* in ulcerative colitis. *European Journal of Gastroenterology and Hepatology, 15*, 697–698.

Guslandi, M., Mezzi, G., Sorghi, M., & Testoni, P. A., (2000). *Saccharomyces boulardii* in the maintenance treatment of Crohn's disease. *Digestive Diseases and Science, 45*, 1462–1464.

Guyonnet, D., Schlumberger, A., Mhamdi, L., Jakob, S., & Chassany, O., (2009). Fermented milk containing *Bifidobacterium lactis* DN-173 010 improves gastrointestinal well-being and digestive symptoms in women reporting minor digestive symptoms: a randomized, double-blind, parallel, controlled study. *British Journal of Nutrition, 102*, 1654–1662.

Hammes, W. P., & Hertel, C., (2009). Genus I. *Lactobacillus* Beijerink 1901. In: De Vos, P., Garrity, G. M., Jones, D., Krieg, N. R., Ludwig, W., Rainey, F. A., Schleifer, K. H., & Whitman, W. B., (eds.), *Bergey's Manual of Systematic Bacteriology* (2nd edn., Vol. 3, pp. 465–510). Berlin: Springer.

Han, X., Yang, Z., Jing, X., Yu, P., Zhang, Y., Yi, H., & Zhang, L., (2016). Improvement of the texture of yogurt by use of exopolysaccharide producing lactic acid bacteria. *BioMed. Research International* (p. 6), Article ID 7945675.

Harzallah, D., & Belhadj, H., (2013). *Lactic Acid Bacteria as Probiotics: Characteristics, Selection Criteria and Role in Immunomodulation of Human GI Muccosal Barrier* (pp. 197–216). Kongo M.

Hatakka, K., Holma, R., El-Nezami, H., Suomalainen, T., Kuisma, M., et al. (2008). The influence of *Lactobacillus rhamnosus* LC705 together with *Propionibacterium freudenreichii* ssp. *shermanii* JS on potentially carcinogenic bacterial activity in human colon. *International Journal of Food Microbiology, 128*, 406–410.

Hatoum, R., Labrie, S., & Fliss, I., (2012). Antimicrobial and probiotic properties of yeasts: From fundamental to novel applications. *Frontiers in Microbiology*, 3.

He, F., Morita, H., Ouwehand, A. C., Hosoda, M., Hiramatsu, M., Kurisaki, J., et al. (2002). Stimulation of the secretion of pro-inflammatory cytokines by *Bifidobacterium* strains. *Microbiology and Immunology, 46*(11), 781–785.

Hempel, S., Newberry, S. J., Maher, A. R., Wang, Z., Miles, J. N., Shanman, R., Johnsen, B., & Shekelle, P. G., (2012). Probiotics for the prevention and treatment of antibiotic-associated diarrhea: A systematic review and meta-analysis. *JAMA, 307,* 1959–1969.

Hickson, M., D'Souza, A. L., Muthu, N., Rogers, T. R., Want, S., Rajkumar, C., et al. (2007). Use of probiotic *Lactobacillus* preparation to prevent diarrhea associated with antibiotics: Randomized double-blind placebo-controlled trial. *British Medical Journal, 335*, 380.

Hoa, T. T., Duc, L. H., Isticato, R., Baccigalupi, L., Ricca, E., et al. (2001). Fate and dissemination of *Bacillus subtilis* spores in a murine model. *Applied and Environmental Microbiology, 67*, 3819–3823.

Holasova, M., Fiedlerova, V., Roubal, P., & Pechacova, M., (2004). Biosynthesis of folates by lactic acid bacteria and propionibacteria in fermented milk. *Czech Journal of Food Science, 22*, 5, 175–181.

Hosoi, T., Kiuchi, K., Ricca, E., Henriques, A. O., & Cutting, S. M., (2004). Production and probiotic effects of natto. *Bacterial Spore Formers: Probiotics and Emerging Applications*, 143–153.

Hutkins, R. W., (2006). *Meat Fermentation: Microbiology and Technology of Fermented Foods* (pp. 207–229). Blackwell Publishing.

Imase, K., Tanaka, A., Tokunaga, K., Sugano, H., Ishida, H., & Takahashi, S., (2007). *Lactobacillus reuteri* tablets suppress *Helicobacter pylori* infection a double-blind randomised placebo-controlled crossover clinical study. *Kansenshogaku Zasshi, 81*, 387–393.

Irina, V. D., Lee, H., Tourova, T. P., Ryzhkova, E. P., & Netrusov, A. I., (2012). *Propionibacterium freudenreichii* strains as antibacterial agents at neutral ph and their production on food-grade media fermented by some lactobacilli. *Journal of Food Safety, 32*, 48–58.

Ishikawa, H., Akedo, I., Umesaki, Y., Tanaka, R., Imaoka, A. I., & Otani, T., (2003). Randomized controlled trial of the effect of bifidobacteria-fermented milk on ulcerative colitis. *Journal of the American College of Nutrition, 22*(1), 56–63.

Jeun, J, Kim, S., Cho, S. Y., Jun, H. J., Park, H. J., Seo, J. G., Chung, M. J., & Lee, S. J., (2010). Hypocholesterolemic effects of *Lactobacillus plantarum* KCT3928 by increased bile excretion in C57BL/6 mice. *Nutrition, 26*, 321–330.

Jiang, T., Mustapha, A., & Savaiano, D. A., (1996). Improvement of lactose digestion in humans by ingestion of unfermented milk containing bifidobacterium longum. *Journal of Dairy Science, 79*(5), 750–757.

Johnston, B. C., Goldenberg, J. Z., Vandvik, P. O., Sun, X., & Guyatt, G. H., (2011). Probiotics for the prevention of pediatric antibiotic-associated diarrhea. *Cochrane Database Systematic Review, 11*, CD004827, doi: 10.1002/14651858.CD004827.pub2.

Jones, M. L., Martoni, C. J., Parent, M., & Prakash, S., (2012). Cholesterol-lowering efficacy of a microencapsulated bile salt hydrolase-active *Lactobacillus reuteri* NCIMB 30242 yogurt formulation in hypercholesterolaemic adults. *British Journal of Nutrition, 107*(10), 1505–1513.

Jurenka, J. S., (2012). *Bacillus coagulans. Alternative Medicine Review, 17*, 76–81.

Kato, K., Mizuno, S., Umesaki, Y., Ishii, Y., Sugitani, M., Imaoka, A., et al. (2004). Randomized placebo-controlled trial assessing the effect of bifidobacteria fermented milk on active ulcerative colitis. *Alimentary Pharmacology & Therapeutics, 20*(10), 1133–1141.

Kaur, B., Garg, N., Sachdev, A., & Kumar, B., (2014). Effect of the oral intake of probiotic *Pediococcus acidilactici* BA28 on *Helicobacter pylori* causing peptic ulcer in C57BL/6 mice models. *Applied Biochemistry and Biotechnology, 172*, 973–983.

Kiatpapan, P., & Murooka, Y., (2002). Genetic manipulation system in propionibacteria. *Journal of Bioscience and Bioengineering, 93*, 1–8.

Kiessling, G., Schneider, J., & Jahreis, G., (2002). Long-term consumption of fermented dairy products over 6 months increases HDL cholesterol. *European Journal of Clinical Nutrition, 56*(9), 843–849.

Klaenhammer, T. R., & Kullen, M. J., (1999). Selection and design of probiotics. *International Journal of Food Microbiology, 50*(1), 45–57.

Konieczna, P., Groeger, D., Ziegler, M., Frei, R., Ferstl, R., Shanahan, F., & O'Mahony, L., (2011). *Bifidobacterium infantis* 35624 administration induces Foxp3 T regulatory cells in human peripheral blood: Potential role for myeloid and plasmacytoid dendritic cells. *Gut, Gutjnl. 61*(3), 354-366

Kostinek, M., Specht, I., Edward, V. A., Pinto, C., Egounlety, M., Sossa, C., Mbugua, S., Dortu, C., Thonart, P., Taljaard, L., & Mengu, M., (2007). Characterization and biochemical properties of predominant lactic acid bacteria from fermenting cassava for selection as starter cultures. *International Journal of Food Microbiology, 114*(3), 342–351.

Kovachev, S., & Dobrevski-Vacheva, R., (2013). Effect of *Lactobacillus casei* var. *rhamnosus* (Gynophilus) in restoring the vaginal flora by female patients with bacterial vaginosis—randomized, open clinical trial. *Obstetrics and Gynecology (Sofia), 52*(1), 48–53.

Kruis, W., Chrubasik, S., Boehm, S., Stange, C., & Schulze, J., (2012). A double-blind placebo-controlled trial to study therapeutic effects of probiotic *Escherichia coli* Nissle

1917 in subgroups of patients with irritable bowel syndrome. *International Journal of Colorectal Disease, 27*, 467–474.

Kruis, W., Fric, P., Pokrotnieks, J., Lukas, M., Fixa, B., Kascak, M., et al. (2004). Maintaining remission of ulcerative colitis with the probiotic *Escherichia coli* Nissle 1917 is as effective as with standard mesalazine. *Gut, 53*, 1617–1623.

Kuitunen, M., Kukkonen, K., Juntunen-Backman, K., Korpela, R., Poussa, T., Tuure, T., Haahtela, T., & Savilahti, E., (2009). Probiotics prevent IgE-associated allergy until age 5 years in cesarean-delivered children but not in the total cohort. *Journal of Allergy & Clinical Immunology, 123*, 335–341.

Leahy, S. C., Higgins, D. G., Fitzgerald, G., & Sinderen, D., (2005). Getting better with bifidobacteria. *Journal of Applied Microbiology, 98*(6), 1303–1315.

Lee, D. K., Park, J. E., Kim, M. J., Seo, J. G., Lee, J. H., & Ha, N. J., (2015). Probiotic bacteria, *B. longum* and *L. acidophilus* inhibit infection by rotavirus *in vitro* and decrease the duration of diarrhea in pediatric patients. *Clinics and Research in Hepatology and Gastroenterology, 39*(2), 237–244.

Lee, J. S., Chung, M. J., & Seo, J. G., (2013). *In vitro* evaluation of antimicrobial activity of lactic acid bacteria against *Clostridium difficile*. *Toxicology Research, 29*, 99–106.

Lee, K. W., Park, J. Y., Jeong, H. R., Heo, H. J., Han, N. S. J., & Kim, H., (2012). Probiotic properties of *Weissella* strains isolated from human faeces. *Anaerobes, 18*, 96–102.

Leivers, S., Hidalgo-Cantabrana, C., Robinson, G., Margolles, A., Ruas-Madiedo, P., & Laws, A. P., (2011). Structure of the high molecular weight exopolysaccharide produced by *Bifidobacterium animalis* subsp. *lactis* IPLA-R1 and sequence analysis of its putative EPS cluster. *Carbohydrate Research, 346*(17), 2710–2717.

Liasi, S. A., Azmi, T. I., Hassan, M. D., Shuhaimi, M., Rosfarizan, M., & Ariff, A. B., (2009). Antimicrobial activity and antibiotic sensitivity of three isolates of lactic acid bacteria from fermented fish product, Budu. *Malaysian Journal of Microbiology, 5*(1), 33–37.

Liebregts, T., Adam, B., Bertel, A., Jones, S., Schulze, J., Enders, C., Sonnenborn, U., Lackner, K., & Holtmann, G., (2005). Effect of *E. coli* Nissle 1917 on post-inflammatory visceral sensory function in a rat model. *Neurogastroenterology & Motility, 17*, 410–414.

Linares, D. M., Ross, P., & Stantona, C., (2016). Beneficial microbes: The pharmacy in the gut. *Bioengineered, 7*(1), 11–20. doi: 10.1080/21655979.2015.1126015.

Lindfors, K. T., Blomqvist, K., Juuti-Uusitalo, Stenman, S., Venalainen, J., Maki, M., & Kaukinen, K., (2008). Live probiotic *Bifidobacterium lactis* bacteria inhibit the toxic effects induced by wheat gliadin in epithelial cell culture. *Clinical and Experimental Immunology, 152*, 552–558.

Ljungh, A., & Wadström, T., (2009). *Lactobacillus Molecular Biology: From Genomics to Probiotics* (pp. 2–44), Horizon Scientific Press.

Malheiros, P. S., Santanna, V., Barbosa, M. S., Brandelli, A., & Franco, B. D., (2012). Effect of liposome-encapsulated nisin and bacteriocin-like substance P34 on *Listeria monocytogenes* growth in Minas frescal cheese. *International Journal of Food Microbiology, 156*(3), 272–277.

Malik, A., Radji, M., Kralj, S., & Dijkhuizen, L., (2009). Screening of lactic acid bacteria from Indonesia reveals glucansucrase and fructansucrase genes in two different *Weissella confusa* strains from soya. *FEMS Microbiology Letters, 300*, 131–138.

Mandel, D. R., Eichas, K., & Holmes, J., (2010). *Bacillus coagulans*: A viable adjunct therapy for relieving symptoms of rheumatoid arthritis according to a randomized, controlled trial. *BMC Complementary and Alternative Medicine, 10*(1), 1.

Martins, F. S., Silva, A. A., Vieira, A. T., Barbosa, F. H., Arantes, R. M., Teixeira, M. M., & Nicoli, J. R., (2009). Comparative study of *Bifidobacterium animalis, Escherichia coli, Lactobacillus casei* and *Saccharomyces boulardii* probiotic properties. *Archives of Microbiology, 191*(8), 623–630.

Matsumoto, M., & Benno, Y., (2004). Consumption of *Bifidobacterium lactis* LKM512 yogurt reduces gut mutagenicity by increasing gut polyamine contents in healthy adult subjects. *Mutation Research, 568*, 147–153.

Matsushita, K., Fujii, Y., Ano, Y., Toyama, H., Shinjoh, M., Tomiyama, N., Miyazaki, T., Sugisawa, T., Hoshino, T., & Adachi, O., (2003). 5-Keto-D-gluconate production is catalyzed by a quinoprotein glycerol dehydrogenase, major polyol dehydrogenase in *Gluconobacter* species. *Applied and Environmental Microbiology, 69*, 1959–1966.

Mayo, B., & Van Sinderen, D., (2010). *Bifidobacteria: Genomics and Molecular Aspects. Caister* (pp. 17–44). Academic Press. ISBN 978–1–904455–68–4.

McFarland, L. V., (2010). Systematic review and meta-analysis of *Saccharomyces boulardii* in adult patients. *World Journal of Gastroenterology, 16*, 2202–2222.

Miki, K., Urita, Y., Ishikawa, F., Iino, T., Shibahara-Sone, H., Akahoshi, R., et al. (2007). Effect of *Bifidobacterium bifidum* fermented milk on *Helicobacter pylori* and serum pepsinogen levels in humans. *Journal of Dairy Science, 90*, 2630–2640.

Mohan, R., Koebnick, C., Schildt, J., Mueller, M., Radke, M., & Blaut, M., (2008). Effects of *Bifidobacterium lactis* Bb12 supplementation on body weight, fecal pH, acetate, lactate, calprotectin, and IgA in preterm infants. *Pediatric Research, 64*(4), 418–422.

Mohania, D., Kansal, V., Sagwal, R., & Shah, D., (2013). Anticarcinogenic effect of probiotic dahi and piroxicam on DMH-induced colorectal carcinogenesis in wistar rats. *American Journals of Cancer Therapy and Pharmacology, 1*, 17.

Morelli, L., (2007). *In vitro* assessment of probiotic bacteria: From survival to functionality. *International Dairy Journal, 17*, 1278–1283.

Nam, H., Whang, K., & Lee, Y., (2007). Analysis of vaginal lactic acid producing bacteria in healthy women. *Journal of Microbiology, 45*, 515–520.

Narvhus, J. A., & Axelsson, L. (2003). Lactic Acid Bacteria, in Caballero, B., Finglas, B., and Toldra, F. (eds.) in *Encyclopedia of Food Sciences and Nutrition* (Second Edition, pp. 3465–3472). Elsevier Science Publishing Co. Inc.

Nguyen, T. D. T., Kang, J. H., & Lee, M. S., (2007). Characterization of *Lactobacillus plantarum* PH04, a potential probiotic bacterium with cholesterol-lowering effects. *International Journal of Food Microbiology, 113*, 358–361.

Nista, E. C., Candelli, M., Cremonini, F., et al. (2004). *Bacillus clausii* therapy to reduce side-effects of anti-*Helicobacter pylori* treatment: Randomized, double-blind, placebo-controlled trial. *Alimentary Pharmacology & Therapeutics, 20*, 1181–1188.

Oberreuther-Moschner, D. L., Jahreis, G., Rechkemmer, G., & Pool-Zobel, B. L., (2004). Dietary intervention with the probiotics *Lactobacillus acidophilus* 145 and *Bifidobacterium longum* 913 modulates the potential of human faecal water to induce damage in HT29clone19A cells. *British Journal of Nutrition, 91*, 925–932.

Ojetti, V., Bruno, G., Ainora, M. E., Gigante, G., Rizzo, G., Roccarina, D., et al. (2012). Impact of *Lactobacillus reuteri* supplementation on anti-*Helicobacter pylori* levofloxacin-based second-line therapy. *Gastroenterology Research and Practice,* 740381.

Orla-Jensen, S., (1919). The lactic acid bacteria. In: *The Royal Danish Scientific Society and Writings (Danish)* (pp. 184–196). Frederik Host & Son, Kongligge Hof-Boghandel, Kobenhaven, Denmark.

Pan, D. D., Zeng, X. Q., & Yan, Y. T., (2011). Characterization of *Lactobacillus fermentum* SM-7 isolated from Koumiss, a potential probiotic bacterium with cholesterol-lowering effects. *Journal of Science of Food and Agriculture, 91*(3), 512–518.

Parent, D. B., Bayot, D., Kirkpatrick, C., Graf, F., Wilkinson, F. E., & Kaiser, R. R., (1996). Therapy of bacterial vaginosis using exogenously-applied *Lactobacillus acidophilus* and a low dose of estriol: A placebo-controlled multicentric clinical trial. *Drug Discovery, 46,* 68–73.

Park, Y. W., & Nam, M. S., (2015). Bioactive peptides in milk and dairy products: A review. *Korean Journal of Food Science of Animal Resources, 35,* (6).

Patel, A. R., Shah, N. P., & Prajapati, J. B., (2012). Antibiotic resistance profile of lactic acid bacteria and their implications in food chain. *World Journal of Dairy & Food Science, 7*(2), 202–211.

Patel, A., & Prajapati, J. B., (2016). Partial characterization of exopolysaccharides obtained from novel isolates of *Weissella* Spp. *Indian Journal of Dairy Science, 69*(3), 310–315.

Patel, A., Lindström, C., Patel, A., Prajapati, J. B., & Holst, O., (2012). Probiotic properties of exopolysaccharide producing lactic acid bacteria isolated from vegetables and traditional Indian fermented foods. *International Journal of Fermented Foods, 1*(1), 87–101.

Patel, A., Prajapati, J. B., & Nair, B. M., (2012). Methods for isolation, characterization and identification of probiotic bacteria to be used in functional foods. *International Journal of Fermented Foods, 1*(1), 1–13.

Patel, D., Dufour, Y., & Domigan, N., (2008). Functional food and nutraceutical registration process in Japan 25 and China: Similarities and differences. *The Journal of Pharmacy and Pharmaceutical Sciences, 11*(4), 1–11.

Pihlanto, A., (2013). *Lactic Fermentation and Bioactive Peptides* (pp. 282–309). INTECH Open Access Publisher.

Plengvidhya, V., Breidt, F., Lu, Z., & Fleming, H. P., (2007). DNA fingerprinting of lactic acid bacteria in sauerkraut fermentations. *Applied and Environmental Microbiology, 73,* 7697–7702.

Prajapati, J. B., & Nair, B., (2008). The history of fermented foods. In: Farnworth, E. D., (eds.), *Handbook of Functional Fermented Foods* (pp. 1–24). CRC Press.

Prajapati, J. B., (1995). In: Akta, P., (ed.), *Fundamentals of Dairy Microbiology* (pp. 48–120). Gujarat, India.

Prasanna, P. H. P., Grandison, A. S., & Charalampopoulos, D., (2014). Bifidobacteria in milk products: An overview of physiological and biochemical properties, exopolysaccharide production, selection criteria of milk products and health benefits. *Food Research International, 55,* 247–262.

Qamar, A., Aboudola, S., Warny, M., Michetti, P., Pothoulakis, C., LaMont, J. T., et al. (2001). *Saccharomyces boulardii* stimulates intestinal immunoglobulin-A immune response to *Clostridium difficile* toxin-A in mice. *Infection and Immunology, 69,* 2762–2765.

Raz, I., Gollop, N., Polak-Charcon, S., & Schwartz, B., (2007). Isolation and characterization of new putative probiotic bacteria from human colonic flora. *British Journal of Nutrition, 97*, 725–734.
Reid, G., (2001). *Regulatory and Clinical Aspects of Dairy Probiotics*. Background paper for FAO/WHO 23. Expert consultation on evaluation of health and nutritional properties of probiotics in food including powder milk and live lactic acid bacteria. Cordoba, Argentina.
Rodríguez-Alcala, L. M., Braga, T., Malcata, X. F., Gomes, A., & Fontecha, J., (2011). Quantitative and qualitative determination of CLA produced by Bifidobacterium and lactic acid bacteria by combining spectrophotometric and Ag +-HPLC techniques. *Food Chemistry, 25*(4), 1373–1378.
Saavedra, J. M., Bauman, N. A., Perman, J. A., Yolken, R. H., & Oung, I., (1994). Feeding of *Bifidobacterium bifidum* and *Streptococcus thermophilus* to infants in hospital for prevention of diarrhea and shedding of rotavirus. *The Lancet, 344*(8929), 1046–1049.
Santos, E. M., Jaime, I., Rovira, J., Lyhs, U., Korkeala, H., & Bjorkroth, J., (2005). Characterization and identification of lactic acid bacteria in "morcilla de Burgos." *Int. J. Food Microbio., 97*, 285–296.
Seppo, L., Kerojoki, O., Suomalainen, T., & Korpela, R., (2002). The effect of a *Lactobacillus helveticus* lbk-16 h fermented milk on hypertension: A pilot study on humans. *Milchwissenschaft, 57*(3), 124–127.
Shah, N. P., (2006). In: Chandan, R. C., (ed.), *Health Benefits of Yogurt and Fermented Milk* (pp. 327–340). Oxford: Blackwell Publishing Ltd.
Sharafedtinov, K. K., Plotnikova, O. A., Alexeeva, R. I., et al. (2013). Hypocaloric diet supplemented with probiotic cheese improves body mass index and blood pressure indices of obese hypertensive patients–a randomized double-blind placebo-controlled pilot study. *Nutrition Journal, 12*, 138.
Shida, K., & Nomoto, K., (2013). Probiotics as efficient immunopotentiators: Translational role in cancer prevention. *Indian Journal of Medical Research, 138*, 808–814.
Shimizu, T., (2003). Health claims on functional foods: The Japanese regulations and an international comparison. *Nutrition Research Reviews, 16*, 241–252.
Simons, L. A., & Amansec, S. G., (2006). Effect of *Lactobacillus fermentum* on serum lipids in subjects with elevated serum cholesterol. *Nutrition Metabolism and Cardiovascular Diseases, 16*, 531–535.
Sipola, M., Finckenberg, P., Santisteban, J., Korpela, R., Vapaatalo, H., & Nurminen, M. L., (2001). Long-term intake of milk peptides attenuates development of hypertension in spontaneously hypertensive rats. *Journal of Physiology and Pharmacology, 52*, 745–754.
Smirnov, V. V., Reznik, S. R., & Sorokulova, I. B., (1994). Highly effective probiotic Biosporin. *Likarska Sprava*, 133–137.
Sorokulova, I., (2013). Modern status and perspectives of *bacillus bacteria* as probiotics. *Journal of Probiotics & Health, 1*, e106. doi: 10.4172/2329–8901.1000e106.
Stackebrandt, E., Cummins, C. S., & Johnson, J. L., (2006). Family propionibacteriacea: The genus *propionibacterium*. In: Dworkin, M., Falkow, S., Rosemberg, E., Schleifer, K. H., & Stackebrandt, E., (eds.), *The Prokaryotes: A Handbook on the Biology of Bacteria* (pp. 400–418). Singapur: Springer.
STATISTA, (2016). *Statistics and Facts on the Yogurt Market in the U.S.* URL: https://www.statista.com/topics/1739/yogurt/ (accessed on 6 September 2019).

Stockenhuber, A., Kamhuber, C., Leeb, G., Adelmann, K., Prager, E., Mach, K., et al. (2008). Preventing diarrhea associated with antibiotics using a probiotic *Lactobacillus casei* preparation. *Gut, 57*(2), A20.

Stowik, T. A., (2016). Contribution of probiotics *Streptococcus salivarius* strains K12 and M18 to oral health in humans: A review. *Honors Scholar Theses,* p. 488. http://digitalcommons.uconn.edu/srhonors_theses/488 (accessed on 6 September 2019).

Strozzi, G. P., & Mogna, L., (2008). Quantification of folic acid in human faeces after administration of *Bifidobacterium* probiotic strains. *Journal of Clinical Gastroenterology, 42*, 179–184.

Sudha, M. R., Bhonagiri, S., & Kumar, M. A., (2013). Efficacy of *Bacillus clausii* strain UBBC-07 in the treatment of patients suffering from acute diarrhea. *Beneficial Microbes, 4*, 211–216.

Surawicz, C. M., McFarland, L. V., Greenberg, R. N., Rubin, M., Fekety, R., Mulligan, M. E., et al. (2000). *The Search for a Better Treatment for Recurrent Clostridium Difficile Disease: Use of High-Dose Vancomycin Combined with Saccharomyces Boulardii.* Oxford University Press. Available at: http://www.ncbi.nlm.nih.gov/pubmed/11049785 (accessed on 6 September 2019).

Sykora, J., Valeckova, K., Amlerova, J., Siala, K., Decek, P., & Watkins, S., (2005). Effects of a specially designed fermented milk product containing probiotic *Lactobacilli casei* DN-114 001 and the eradication of H. pylori in children. *Journal of Clinical Gastroenterology, 39*, 692–698.

Tamboli, C. P., Caucheteux, C., Cortot, A., Colombel, J. F., & Desreumaux, P., (2003). Probiotics in inflammatory bowel disease: A critical review. *Best Practice & Research Clinical Gastroenterology, 17*(5), 805–820.

Tamime, A. Y., & Robinson, R. K., (2007). *Yoghurt: Science and Technology* (3rd edn., pp. 348–725). CRC Press, Cambridge, UK.

Taranto, M. P., Medici, M., Perdigon, G., Ruiz, H. A. P., & Valdez, G. F., (2003). Effect of *Lactobacillus/reuteri* on the prevention of hypercholesterolemia in mice. *Journal of Dairy Science, 83*, 401–403.

Tasteyre, A., Barc, M. C., Karjalainen, T., Bourlioux, P., & Collignon, A., (2002). Inhibition of *in vitro* cell adherence of *Clostridium difficile* by *Saccharomyces boulardii. Microbial Pathogens, 32*, 219–225.

Thygesen, J. B., Glerup, H., & Tarp, B., (2012). *Saccharomyces boulardii* fungemia caused by treatment with a probioticum. *BMJ Case Reports* (pp. 1–3). doi: 10.1136/bcr.06.2011.4412.

TMR (Top Market Reports), (2014). *Starter Culture Market by Type (Yeast, Bacteria, Molds), Application [Alcoholic Beverages (Beer, Wine, Whisky), Non-Alcoholic Beverages (Dairy-Based, Cereal-Based, Kombucha)] & Geography-Global Trends & Forecast to 2018.* URL: http://www.marketsandmarkets.com/Market-Reports/starter-culture-market-213083494.html (accessed on 6 September 2019).

Tursi, A., Brandimarte, G., Giorgetti, G. M., & Modeo, M. E., (2004). Effect of *Lactobacillus casei* \ supplementation on the effectiveness and tolerability of a new second-line 10-day quadruple therapy after failure of a first attempt to cure *Helicobacter pylori* infection. *Medical Science Monitor, 10*, 662–666.

Van Der Aa Kuhle, A., & Jespersen, L., (2003). The taxonomic position of *Saccharomyces boulardii* as evaluated by sequence analysis of the D1/D2 domain of 26S rDNA, the

ITS1–5.8S rDNA-ITS2 region and the mitochondrial cytochrome-c oxidase II gene. *Systematic and Applied Microbiology, 26*, 564–571.

Van der Meulen, R., Grosu-Tudor, S., Mozzi, F., Vaningelgem, F., Zamfir, M., De Valdez, G. F., & De Vuyst, L., (2007). Screening of lactic acid bacteria isolates from dairy and cereal products for exopolysaccharide production and genes involved. *Int. J. Food Microbio., 118*, 250–258.

Vasiljevic, T., & Shah, N. P., (2008). Review: Probiotics—from Metchnikoff to bioactives. *International Dairy Journal, 18*, 714–728.

Ventura, M., O'Connell-Motherway, M., Leahy, S., Moreno-Munoz, J. A., Fitzgerald, G. F., & Van Sinderen, D., (2007). From bacterial genome to functionality, case bifidobacteria. *International Journal of Food Microbiology, 120*(1/2), 2–12.

Waller, P. A., Gopal, P. K., Leyer, G. J., et al. (2011). Dose-response effect of *Bifidobacterium lactis* HN019 on whole gut transit time and functional gastrointestinal symptoms in adults. *Scandinavian Journal of Gastroenterology, 46*, 1057–1064.

Wickens, K., Black, P., Stanley, T. V., Mitchell, E., Barthow, C., Fitzharris, P., Purdie, G., & Crane, J., (2012). A protective effect of *Lactobacillus rhamnosus* HN001 against eczema in the first 2 years of life persists to age 4 years. *Clinical & Experimental Allergy, 42*(7), 1071–1079.

Williams, D., (2016). *Probiotic Species and Strains: What Are Their Differences?* URL: http://www.drdavidwilliams.com/probiotic-strains/ (accessed on 6 September 2019).

Wu, Z. J., Du, X., & Zheng, J., (2013). Role of *Lactobacillus* in the prevention of *Clostridium difficile*-associated diarrhea: A meta-analysis of randomized controlled trials. *Chinese Medical Journal, 126*, 4154–4161.

Xia, P., Zhu, J., & Zhu, G., (2013). *Escherichia coli* Nissle 1917 as safe vehicles for intestinal immune targeted therapy—A review. *Acta Microbiologica Sinica, 53*, 538–544.

Xiao, J. Z., Kondo, S., Takahashi, N., Miyaji, K., Oshida, K., Hiramatsu, A., et al. (2003). Effects of milk products fermented by *Bifidobacterium longum* on blood lipids in rats and healthy adult male volunteers. *Journal of Dairy Science, 86*, 2452–2461.

Yadav, J. S., Grover, S., & Batish, V. K. A., (1993). *Comprehensive Dairy Microbiology* (pp. 463–524). Metropolitan Publisher, New Delhi, India.

Zarate, G., & Perez, C. A., (2009). Dairy bacteria remove *in vitro* dietary lectins with toxic effects on colonic cells. *Journal of Applied Microbiology, 106*, 1050–1057.

CHAPTER 2

Starter Cultures: Classification, Traditional Production Technology and Potential Role in the Cheese Manufacturing Industry

MOHAMED EID SHENANA[1*] and AMI R. PATEL[2]

[1]*Dairy Science Department, Moshtohor, Faculty of Agriculture, University of Benha-13736, Arab Republic of Egypt, Moshtohor, Egypt*

[2]*Division of Dairy and Food Microbiology, Mansinhbhai Institute of Dairy & Food Technology-MIDFT, Dudhsagar Dairy Campus, Mehsana-384002, Gujarat, India*

[*]*Corresponding author. E-mail: meshenana@fagr.bu.edu.eg; meshenana1964@hotmail.com*

2.1 INTRODUCTION

Cheese is generally a fermented milk-based product manufactured in a large number of flavors and forms in different countries. Still time, more than 1000 types of cheese varieties have been recognized throughout the worldwide; however, there is no trustworthy assessment of the total number of cheese types generated. Cheese is the classic handiness food type; indeed 'there is a cheese for every taste preference and a taste preference for every cheese.' Cheese manufacturing is an essentially method, known from about 6000–7000 BC for conserving the nutritional value of milk through microbial fermentation process together with removal of a part of moisture and salt addition, that serve as a taste material and preservatives.

Biologically and bio-chemically, cheeses are dynamic and hence unstable food. Very often cheese is stated as "alive" product since it contains microbial ecosystem, which considerably changes during maturation (ripening) attributable to a series of biochemical conversions. The balanced biochemical transformations lead to the synthesis of numerous kinds of flavor substances and ultimately results in the most desirable flavors and aromas of diverse cheese varieties. A number of flavors can be synthesized from the same vital components *viz.* milk, rennet, starter cultures; often salt is a fascinating fact of the cheese manufacture process. While cheese-making is an ancient art, modern industrial cheese manufacturing processes greatly relies on different scientific disciplines applied to cheese manufacturing process including enzymology, microbiology, biochemistry, flavor chemistry, molecular genetics; rheology (the science of the flow and deformation of the cheese) to ensure a consistently high-quality product. Cheese industry utilizes a large quantity of enzymes, and technologically cheese fermentation is more complex than many other industrial fermentation processes.

Generally, the species of four genera from lactic acid bacteria (LAB) including *Lactococcus, Streptococcus, Lactobacillus,* and *Leuconostoc* are the most largely used in cheese making; however, the use of genus, *Pediococcus,* have been proposed recently to be used in cheese making but still under study. So, the main aim of this chapter is to discuss the starter culture bacteria and their role either in cheese making or cheese repining.

2.2 CHEESE: CLASSIFICATION AND TYPES

Numerous varieties of cheese can be categorized into two major families based on the manner the milk is coagulated during cheese-making. In the first family of cheeses, while manufacturing cheese usually milk is coagulated by adding rennet, a milk coagulating enzyme which converts milk into a gel. Cheddar, Swiss, Gouda, Camembert, Blue, Mozzarella, etc. are well-known examples of rennet-coagulated cheeses. In the second family of cheeses, instead of enzymes generally milk is coagulated by acidification process using food grade organic acids (acetic acid, lactic acid); these acid-coagulated varieties include Cottage cheese and Quark cheese.

Nowadays, cheese is mostly consumed or utilized in the form of processed cheese. Processed cheese is produced from young and mature cheeses; shredding and melting cheeses with emulsifying salts into a smooth molten mass which is subsequently cooled and molded into the desired shape. During the manufacture of processed cheese, some optional ingredients may also be added including milk powders, vegetable oils, extra salts, flavors, colors, stabilizers, and preservative agents. Majority of processed cheese is manufactured in the form of blocks or alternatively molded into large sheets which are further cut into ribbons, interleaved with film, stacked, and cut to form convenient slices.

Cheese analog (cheese alternatives), also called imitation cheese is another type of commercially available processed cheese that do not meet the criteria as true cheeses. They are cheese-like products formulated from ingredients such as casein powder (rennet casein); butterfat or vegetable oils as a source of fat. Cheese analogs are chiefly used in consumer-ready to eat meals or for feeding special people such as vegan consumers. The melting points of imitation cheese and their lower manufacturing costs make them attractive for businesses.

2.3 CHEESE MANUFACTURE

Manufacturing of most cheeses involves two different stages, manufacturing phase and ripening phase. Generally, manufacturing of fresh cheese curd takes approximately 5–24 hours; subsequently, most cheese types are then kept for ripening (matured) for the development of flavor and body-texture according to the cheese variety.

At the industrial level, selection, and pre-treatment of milk are the first steps for the manufacture of cheese curd. Afterward, the milk is heat-treated (pasteurized) to destroy microbial pathogens, if any present; however, some cheese types are still made from the raw (unpasteurized) milk may be at a small scale. Manufacturing of cheese is basically a kind of dehydration processes where the milk fat and milk casein are concentrated between six to twelve folds, based on the cheese type. In the milk, the ratio of fat to protein is often adjusted to provide the preferred composition of the final cheese; also in certain cheese varieties color may be added into milk for stable product color.

Cheese is a fermented milk product whereby the action of starter bacteria, lactose (milk sugar) is converted to lactic acid. Strains of LAB which are used to acidify milk during cheese-making are generally carefully chosen and deliberately incorporated into the milk as a starter (since it "starts" acid production) culture. Fundamentally, during cheese-making moisture is removed from milk through its biochemical conversion into a gel and subsequent contraction of the gel in a process called 'syneresis.' For most of the cheese varieties, the milk is coagulated enzymatically using rennet enzyme, but for some specific cheese types, milk is gelled by acidification. Afterward, the gel is cut (broken) into pieces that help to expel whey- much of the liquid that was trapped inside the gel. At this stage of manufacture, the solid material suspended in the whey is known as cheese curd. Syneresis is encouraged by acidification of milk and is further facilitated by heating of the curds together with whey mixture in the process popularly known as "cooking" and "scalding." Eventually, this process separates the curds from the whey which is a valuable by-product of cheese- manufacturing industry and the drained curds are then transferred to molds to own a typical shape and size characteristic of the cheese variety. For low moisture varieties, the curd is pressed to form a continuous mass while to prepare high moisture cheeses, and curds are fuse together under their own weight gravity. In certain types, like Cheddar and related cheeses, curds are "cheddared" before molding which develops more acid. On the other hand, the acidified curds are kneaded and stretched in hot water to characterize before molding step for other cheese types like Mozzarella.

During the manufacturing of almost all cheese varieties salt (NaCl) is added not only for taste but also as a preservative and flavoring agent. After molding, cheeses are salted usually by immersing in a brine solution. Conversely, the curds for Cheddar cheese and related cheese types are cut (called "milled") into small pieces after an additional step known as 'cheddaring' and then dry salted before being molded and pressed. Also, usually dry salt is rubbed to the cheese surface after molding for Blue cheeses.

2.4 THE GENERAL ROLE OF STARTER MICROORGANISMS IN CHEESES MANUFACTURE

In modern cheese-making practice, a group of food-grade bacteria commonly referred as LAB are inoculated in milk as starter cultures; the key function of these starters is to form lactic acid through fermentation of milk sugar (lactose). Apart from this, lactate is accountable for the fresh clean acidic taste and flavor of the un-ripened cheeses and in general for fermented milks, too. It is also responsible for the formation and texturizing of the curd during cheese-making. Starter cultures also have other vital roles to play such as the synthesis of volatile flavor compounds like diacetyl, acetoin, and aldehydes; and of lipolytic and proteolytic enzymes which would be taking part in the ripening of some cheese types together with the inhibition of spoilage type and pathogenic microflora. Four genera of LAB including *Lactococcus* sp; *Streptococcus* sp; *Lactobacillus* sp.; *Leuconostoc* sp. are the most extensively employed in cheese making; however, the application of genus *Pediococcus* has been anticipated in recent times to be used in cheese making (Caldwel et al., 1996; Verachia, 2005).

2.5 TECHNOLOGICALLY IMPORTANT CHARACTERISTICS OF LAB

2.5.1 METABOLISM OF LACTOSE

The process for the transport and metabolism of lactose differs in the different starter cultures. An instance, in *Lactococcus* sp. transport of lactose into the bacterial cell involves the phosphoenol-pyruvate phosphotransferase system (PEP-PTS) and the concurrent formation of lactose-phosphate. This lactose-phosphate is then hydrolyzed to glucose and galactose-6-phosphate by phospho-ß-galactosidase that is further metabolized via the glycolysis and galactose metabolism pathways, respectively.

In case of *Lactobacillus* sp; *Leuconostoc* sp.; *Streptococcus thermophilus,* lactose enters in the intact form in the cell via permease system and is later get hydrolyzed by *β*-galactosidase to glucose and galactose. Afterward, glucose is utilized through the glycolysis in *Lactobacillus* sp. and *Str. thermophilus* while in *Leuconostoc* it follows the phosphoketolase pathway. Usually, galactose is metabolized via conversion into glucose-1-phosphate. Few strains of *Lb. delbrueckii* sub sp. *lactis, Lb. delbrueckii* sub sp. *bulgaricus; Str. thermophilus* are not able to metabolize galactose; thus excreted out in growth media.

2.5.2 PROTEIN METABOLISM

Milk contains insufficient free amino acids to support the luxuriant growth of LAB, and thus, starter cultures must own proteolytic activity. In LAB, the proteinases are present in the cell wall, and according to the bacterial strain types, enzymes differ in their specificity, location, as well as in number. Proteinases carried out hydrolysis of milk proteins into larger peptides or oligopeptides, which are further hydrolyzed into small peptides and amino acids by a cell wall and cell membrane-bound peptidases. Furthermore, proteolytic enzymes of starter microflora also contribute to the cheese ripening.

2.5.3 CITRATE METABOLISM

Citrate gets metabolize by oxaloacetate and pyruvate pathways. The only starter strains competent to metabolize citrate is *Lactococcus lactis* sub sp. *lactis* biovar *diacetylactis* and *Leuconostoc* sp. Citrate metabolism is an imperative trait since end products of this pathway, including acetoin, diacetyl, and acetate are accountable for the development of flavor in un-ripened and to a certain extent in ripened cheeses. The CO_2 produced during citrate metabolism contribute to desirable 'eye formation' character in some cheeses like Swiss cheeses.

2.6 STARTER CULTURES-CLASSIFICATION

Cheese starters may be classified in several different ways. As an instance, starters may be classified according to their optimal growth temperatures into two types, mesophilic, and thermophilic cultures. *Lactococcus* and *Leuconostoc* falls in mesophilic starter category have an optimal growth temperature of ~30°C, while *Lactobacillus* species and *Str. thermophilus* fall under thermophilic type starters having an optimal growth temperature of about 40–45°C.

Mixed mesophilic cultures may be also categorized on the basis of flavor production attribute, either as flavor-producing or non-flavor-producing. Most species of the genus *Leuconostocs* and lactococci strain *Lc. lactis* biovar *diacetylactis* are the only starters fall under the category of flavor producing starters in cheese. Starters may consist of single, paired or multiple strains. At large-scale industrial usage, single or multiple strains mainly containing defined thermophilic starters in known proportion have chiefly replaced the traditionally employed mesophilic mixed cultures. It is because actually in traditionally used mixed strain starters the proportions of different strains are unknown and their growth is also unpredictable, in contrast to that of multiple strain starters. Moreover, in multiple cultures it is feasible to reduce the total number of strains (e.g., maybe from originally used six to three or four) devoid of any undesirable consequence on the performance of other combine used starter cultures.

2.7 TRADITIONAL PRODUCTION TECHNOLOGY OF STARTER CULTURES

Conventionally, starter cultures were maintained in liquid forms as stock cultures (usually glycerol stocks) in dairy plants or cheese factories. These stock cultures were regularly sub-cultured systematic manner like mother culture followed by intermediate or feeder culture to prepare adequate large bulk inoculums for cheese production. There are several problems associated with this methodology such as the high risk of starter contamination; sometimes alterations in the inherent characteristics of starter bacteria due to frequent sub-culturing and very lengthy laborious technique which necessitate a skilled person to deal with microorganisms aseptically. Afterward, with advances in technologies, from last many decades now, it is easier to maintain the starter cultures in the convenient form including lyophilized dried powders, liquids, or as a frozen concentrate.

Commercially many companies maintain and supplied starter cultures; though such cultures may be received as stocks their propagation as mother culture and bulk (feeder) culture is more straightforward and convenient than conventional cultures. Recently, several starters are available in ready to use form, which is directly required to be inoculated into the bulk tank, or even into the production tank, called 'direct vat-set' (DVS) or direct-in-vat inoculation' (DVI) cultures.

While using the conventional method to maintain starters, stringent precautions must be taken to prevent and control bacteriophage infection of LAB strains. Air filtration, radiations (ultraviolet (UV) light and gamma rays) and air sanitization with disinfectants are further means of preventing airborne bacteriophages. Application of phage inhibitory media (PIM) or phage resistant media (PRM) also helps to control phage infection.

Several cheese industries utilize mechanically protected systems in that the mother culture, feeder culture as well as bulk culture medium, cheese milk or reconstituted skimmed milk (RSM) powder is inoculated under aseptic conditions. Further, it is protected inside the bulk tank through maintaining slight positive pressure by sterile air from filters.

DVS or DIV cultures offer a highly convenient process by a less threat of bacteriophage attack and consistent performance of starter cultures, and most advantageously minimize the culture handling. Such starter cultures are grown in specific media, concentrated, made free from metabolites and ultimately preserved either in lyophilized (freeze-dried) form or under liquid nitrogen. The convenient DVS/DIV culture technology finds wider application in cheese making, particularly in small-scale level. Such starters may also be tailored for making specific cheese varieties with definite features matched to the need of consumers or market.

2.7.1 MICROBIOLOGICAL EXAMINATION OF STARTER CULTURES

Critical care and precautions regarding microbiological control are required while making different types of cheese except direct-in-vat cultures are employed. Starter cultures are required o regularly examined for activity tests and purity tests. If starter cultures are propagated from commercially available mother cultures, this may be restricted to performing only activity tests; however, at large scale production starters should be subjected for various recommended tests, chiefly for investigating for the tile presence of phage(s).

Rate of acid production by a simple acidification test is the usual activity test performed for determining the starter. Secondly, it is easier to observe the pH value of cheese milk inoculated with the specific starter strain at the proper temperature during incubation. Dye reduction test and flavor production are also effective tests to judge the starter activity. Although it does not directly associate with either total viable cell counts or production of lactic acid, satisfactorily indicates or confers activeness of cheese starters.

Starter cultures performance may get reduced by contamination with other microbial types. Determining the presence of contaminants by direct microscopic examination (Gram's staining) and by streaking onto non-selective media are common methods to check the starters for purity. It is advisable to use both carbohydrates containing and non-carbohydrate containing microbiological media and further, the identified and observed colonies should be subsequently subjected for microscopic assessment and also through determining some simple biochemical characteristics like catalase test. Inoculation in broth media can also indicate the presence of contaminants; undesirable odors, excessive gas production, uneven growth pattern, pellicle or sediment formation are few indications of contaminated starter culture.

2.7.2 CHARACTERISTICS OF STARTER CULTURES BACTERIA USED IN CHEESE AND FERMENTED DAIRY PRODUCTS

The bacteria commonly used in the production of cheese like fermented milk products belong to the group of LAB; however, some other bacteria such as *Bifidobacterium* sp; *Brevibacterium linen*; *Propionibacterium freudenreichii* sub sp. *shermanii* are not LAB used for manufacturing of certain cheese varieties. Along with lactic acid, these strains may produce acetic acid or propionic acid and hence, responsible for specific flavor in some cheeses or may contribute to surface ripening of cheeses, i.e., Limburger cheese, brick cheese. Some molds like *Penicillium* sp. are employed as starter for manufacture of Roquefort, Camembert, Stilton, and Gorgonzola cheeses.

Currently, sixteen different genera of LAB are used in cheese-like fermented dairy products. Among all genera, species of *Lactococcus*, *Streptococcus*, *Enterococcus*, *Lactobacillus*, *Leuconostoc*, *Pediococcus*, *Oenococcus*; *Tetragenococcus* are chiefly employed for making different cheese varieties. The important characteristics of these major genera that play a vital role in cheese making and dairy fermentation are described in Table 2.1. While Table 2.2 comprises details of other non-lactic starter microorganisms.

TABLE 2.1 Important Characteristics of Few Major Genera of LAB Used in Cheese Making and Dairy Fermentations

Genus	Properties
Enterococcus	They are Gram-positive, cocci shaped, catalase-negative bacteria that form chains of varying length and superficially similar to lactococci. They are often considered as indicators of fecal contamination since are found as normal inhabitants of the intestinal tract of human and animals, and few species are pathogenic. They are able to grow in high concentrations of salt, at high temperature- up to 45°C, at alkaline pH (9.6), in the presence of high concentrations of bile salts; are resistant or insensitivity to a number of antimicrobial compounds. Previously, *Enterococcus* sp. employed as starters were classified as fecal streptococci and Group D streptococci. Some of the species are associated with cheese and other fermented milk products, especially that use a high temperature during their processing, i.e., 45°C.

TABLE 2.1 *(Continued)*

Genus	Properties
Streptococcus.	Only a single species of this genus *Str. thermophilus* is used as starter culture in the dairy industry. It is a thermophilic organism, optimum growth at 42–45°C and higher and employed for yogurt manufacturing, Mozzarella, Cheddar, and other cheeses types. In combination with lactococci, it leads to rapid lactic acid production during scalding, and further may confer supplementary measure protection against bacteriophage.
	They appear in long chains, homofermentative, and produce L-lactic acid only. It is sensitive to low levels of salts, specifically to NaCl (sensitive to ~2% conc.). High levels of streptococci may occasionally found in cheese as *Str. thermophilus* and *Str. thermophilus*-like organisms may proliferate in the regeneration section of pasteurizers. In addition, most strains are unable to metabolize galactose that may support the growth of Non-Starter culture LAB (NSLAB) for gas production due to the availability of significant concentrations of a fermentable carbohydrate. Also, galactose and other reducing sugars can react with amino acids in the Maillard reaction; hence it is advisable to select only galactose-utilizing strains or in mixed with other lactic culture to diminish the probability of undesirable color changes.
Lactobacillus	This is the largest genera consisting about 150 distinct species of a rod-shaped Gram-positive and catalase-negative bacteria. It includes both homofermentative and heterofermentative species; few may produce chiefly L-lactate from glucose, while few produce D-lactate. Also, several lactobacillus strains exhibit significant racemase (an isomerase enzyme) activity that leads to the production of D/L lactic acid. Few strains may also exhibit coccoid - coccobacilli type morphology that can create confusion which resembles to like leuconostocs and even lactococci, too.
	Lactobacilli are widely employed in the production of fermented milk like yogurt and Mozzarella cheese as starter cultures. Some strains may find use as an adjunct starter in Cheddar and related cheeses to promote faster ripening so as to decrease bitterness. Many lactobacilli have potential probiotics and thus commercially used in such preparations.
	Usually homofermentative *Lb. delbrueckii* subsp. *bulgaricus* is used as a starter in combination with *Str. thermophilus* for yogurt formation. This strain has an optimum temperature of ~42°C or higher, sensitive to bile salts and produces almost 2% w/v lactic acid in milk.
	Lb. acidophilus, generally found in the small intestine is widely used as a probiotic bacterium in combination with other LAB for manufacturing of probiotic cheese or other fermented food products. It is a homofermentative bacterium, has an optimum temperature of 37°C and produces high amounts of D-lactic acid in milk. Compared with

TABLE 2.1 *(Continued)*

Genus	Properties
	Bifidobacterium sp. that is often used in conjunction with this organism, *Lb. acidophilus* is relatively tolerant to oxygen. Also, since bacterium is synthesizes D-lactate there have been some concerns about its application in infant formula and nutrition part of D&L lactic acid.
	Another lactobacillus *Lb. casei* is also inherent to the small intestine, used as a probiotic and is commonly associated as a member of NSLAB in Cheddar cheese. It chiefly produces L-lactate from lactose, but some strains also form small concentrations of D-lactate related to weak racemase activity.
	Lb. helveticus is homofermentative thermophilic lactobacilli (grow at 45°C) used as starter culture in Mozzarella and Emmental type cheeses. *Lb. helveticus* is synthesizes high amount of D/L lactic acid and able to metabolize galactose and thus, can be advantageous if the finished product is required to be free of reducing sugars (Mullan, 2001). Many strains of this bacterium have shown to possess proline-iminopeptidase-like activity, which has been successfully employed to manufacture modified *Cheddar type cheese* characterized with "sweetness" of Swiss cheeses. Several designated strains have also been incorporated as starter adjuncts in some cheese varieties to accelerate ripening, to reduce bitterness;/or to improve flavor in cheeses. It reduces bitterness because of the action of peptidase on starter-derived hydrophobic peptides.
Lactococcus	These are Gram-positive, cocci shaped homofermentative mesophilic (25–30°C) bacteria extensively involved in the manufacturing of varieties of cheese. Originally, bacteria in this genus were classified along with other species of *Streptococcus* genus and were categorized as 'lactic streptococci.' By their specific reaction with Group N antiserum and their tolerance to temperature and salt, they can be easily differentiated from other streptococci. Nisin, a bacteriocin obtained from lacotcocci strains is legally permitted biopreservative (E234) in dairy and food products.

TABLE 2.2 Non-Lactic Starter Microorganism Used in Cheese Manufacturing

Microbial Group	Microbial Species	Type of Cheese
Bacteria	*Lactobacillus* sp; *Pediococcus* sp; *Propionibacterium* sp; *Micrococcus* sp; Corynebacteria (*Bre. linens*)	Hard and semi-hard cheese types such as Swiss-type, Surface ripened cheeses
Yeast	*Saccharomyces* sp; *Candida* sp; *Kluyveromyces* sp; *Debaryomyces* sp.	Mold ripened cheeses like blue cheese; surface-ripened cheeses, washed rind
Mold	*Penicillium* sp; *Geotrichum* sp.	Soft mold-ripened cheeses like Camembert, Roquefort, Brie, Stilton

2.7.3 DIFFERENTIATION OF LACTOCOCCI UP TO SPECIES LEVEL

Majority of fermented dairy products are manufactured using different lactococci species either alone or in combination with other LAB. Due to the presence of some pathogens found in the lactococci group, it is important to make species-level differentiation to these. There are several differentials, but not selective microbiological media existing that can be helpful for strain identification and quality control. Bromocresol purple agar or Reddys' differential agar contains the differential ingredients lactose, L-arginine, calcium citrate; bromocresol purple, a pH indicator dye that produces yellow and blue to purple color colonies under respective acidic and alkaline conditions.

Among different species, *Lc. lactis* subsp. *lactis* and *Lc. lactis* subsp. *cremoris* develops yellow colonies because of lactic acid formation from lactose metabolism. Besides, another strain *Lc. lactis* subsp. *lactis* synthesizes lactic acid plus ammonia from arginine production of that neutralizes the acid formed and eventually develops blue/purple colored colonies due to alkaline conditions. A biovarient, *Lc. lactis* subsp. *lactis* biovar *diacetylactis* as well shows blue-purple colony. Apart from this, *Lc. lactis* subsp. *lactis* and *Lc. lactis* subsp. *cremoris* both are acid producers but non-flavor producers, while *Lc. lactis* subsp. *lactis* biovar *diacetylactis* is able to develop both acid and flavor producer attributable to the presence of flavor producing plasmid.

2.7.4 TYPES OF LAB STARTER CULTURES

There are many types of starter cultures can be used, but commercially three major types of cheese starter employed in overall the world as follows:

1. **Single-Strain Starters:** It always contains a single microorganism, e.g., the strain of *Lc. lactis* subsp. *cremoris* or *Lc. lactis* subsp. *lactis* for the preparation of cheese. In several fermentation plants these two may also be used as pair starters.
2. **Multiple Strain Starters:** Such preparations possess defined numbers or mixtures of three or more single strains like *lactis* subsp. *cremoris* and/or *Lc. lactis* subsp. *lactis, Lc. lactis* subsp. *lactis* biovar *diacetylactis; Leuconostoc* sp. The main advantage of this type is predictable quality and behavior used strains.

3. **Mixed-Strain Starters:** There is a mixture of two or more starters in unknown proportion in this type of preparations. The quality and behavior of these strains are always unpredictable because their composition may vary on sub-culturing.

2.7.5 PRACTICAL USE OF STARTER CULTURES

At the commercial level, starter cultures are prepared mainly in two forms, either as Bulk starter or Starter concentrates. The starter medium should be sterilized, protected from contamination during ripening and cooling; avoided from phage-contamination during incubation. Also, maintaining the pH value during starter preparations and during the keeping conditions are important criteria.

2.7.6 ANTAGONISTIC PROPERTIES OF STARTER CULTURES BACTERIA

Lactic acid produced by starter cultures reduces the pH of the medium, and in turn, play contributes to suppressing the growth and in some cases, even survival of undesirable spoilage bacteria. Diacetyl, produced as a flavor compound, also possess antimicrobial activity against few Gram-negative bacteria, but not towards Gram-positive. It is well known that certain strains used as a starter may produce specific antimicrobial agents which may inhibit the growth of other microflora. As an instance, *Lc. lactis* subsp. *lactis* produces 'nisin,' the polypeptide bacteriocin that possesses a broad spectrum of inhibitory effect against a number of Gram-positive bacteria, including spore formers (Servin, 2004). It is legally permitted biopreservative in specified foods like processed cheese.

Recently, there has been substantial interest in starter microorganisms which develop antimicrobial effect along with a key role in the cheese fermentation process. The use of antagonists to control specific microbial pathogens like *Staphylococcus* sp. or *Listeria* sp. is of particularly interesting along with controlling the spoilage microorganisms in different cheeses. Bacteriocinogenic cultures are the viable alternative for such effects with their long-lasting antagonistic effect even in finished product. Nisin producing lactococci strains effectively restricted the growth of *S. aureus*, *L. monocytogenes*, and *Clostridium* sp. in various cheese types; however, poor technological performance (low acidification and proteolytic activity) recommends their use as adjunct cultures (Deegan et al., 2006; Galvez et al., 2008). Cottage cheese-related spoilages can be diminished by employing inhibitor producing *Lc. lactis* subsp. *lactis* strain (Dal Bello et al., 2011). Modified starters producing lacticin 3147 successfully inhibited the growth of *L. monocytogenes* in various cheese types (O'Sullivan et al., 2006).

2.8 STARTER CULTURE FAILURE

Starter culture failure is unpleasant event that occurs several times during manufacturing of fermented milk and cheese. If the starter bacteria growth ceases or leads to significantly slow acid production than the accepted normal cases, the resultant product is of poor quality. Failure of starter culture is also an indicative of possible serious public health implications due to either the proliferation or survival of enterotoxigenic *Staphylococcus* or *Salmonella* strains during fermentation and then sequent growth during maturation and storage of cheese.

Modern cheese-making processes almost eliminates the possibility of starter failure because of the inadequate or inactive starter strains use. However, possible contamination with bacteriophage of more significant as the main reason for starter culture failure in recent fermentation processes besides other extrinsic factors. Bacteriophage specific to different LAB starter strains has been isolated in particular for widely employed strains of the genus *Lactococcus*.

Similar to other bacteria, in case of LAB also there are two basic types of phage-host relationships observed, i.e., lytic cycle and lysogenic cycle. Of these two, infection with lytic phages is the main cause of starter culture failure, thus of larger industrial significance. In the case of lytic phage, infection occurs in major four sequential stages including phage adsorption, phage (genetic material) injection, and intracellular proliferation and bursting of the host cell for release of phage particles. About 200 phages per bacterium can be released following a little latent period infection; it takes nearly ~60 min between the infection and release of the phage particle. Hence, infected starter cultures may quickly lose their activity.

Temperate phage or lysogenic phage is generally get integrated into the host bacterial chromosome (prophage) or else maintained as a plasmid. Accordingly, in both cases, the phage particle replicates along with the host bacterial cell. Such prophage containing bacteria also becomes resistant from infection with other virulent phages either of the same or closely connected lysogenic phage.

Even though infection with bacteriophage is one of the major significant causes of starter failure, several other factors may also affect the activity of starter bacteria. For example, agglutinins, a normal protein inherently present in milk can cause aggregation of starter bacteria with casein micelles. It leads to uneven starter cells distribution with eventual slow acid production in addition to poor quality curd having "grainy" texture. Troubles related to agglutinin activity were noticed during the manufacturing of cottage cheese only, but it may perhaps affect the production of other cheeses.

Failure of starter cultures due to the occurrence of traces of antibiotics in milk is also very common. Nowadays it is largely abolished by rapid milk testing systems, and financial penalties levied on the antibiotic-incorporating milk producers. Continuous rinsing of the equipments and pipelines with sanitizer as well as detergents during or after cleaning (CIP) may subsequently affect the starter cultures. The sensitivity of different starters varies largely with sanitizers or detergents depending to the species and strain; however, presence of iodine and chlorine at a low level have a minute consequence on the performance of major starter cultures as compare to that of quaternary ammonium compounds, having potentially higher problems. Generally, late lactation milk has superior lactoperoxidase activity, a normal antimicrobial system of milk that can result into failure of starter growth. Though there is a substantial dissimilarity in the sensitivity of different starters, only resistant strains of bacteria should be preferred as starters for making cheese.

2.9 RIPENING OF CHEESE AND RELATED PHENOMENON

Majority of other cheese varieties involves a ripening step ranging from few weeks (like in Mozzarella cheese) to few months and even few years (e.g., Parmigiano-Reggiano or extra-mature Cheddar cheeses). Few rennet-coagulated and major acid-coagulated cheese types are generally utilized fresh without ripening after manufacturing. Ripening or maturation of cheese leads to the development of a typical pre-determined flavor, aroma, and other organoleptic changes in the finished product. Ripening of cheese is characterized by a series of biochemical and microbiological changes associated mainly with the enzymatic action of inherent milk enzymes or enzymes produced by natural microflora of milk or starter cultures as well as residual rennet.

Although starter bacteria die due to several processing steps during the cheese manufacture process, the bursting of cells liberates specific proteolytic, lipolytic, and other enzymes within the food matrixes that contribute in the maturation of cheese. During ripening, in addition to LAB, other NSLAB may also take part in long ripened cheese types; may eventually show a higher population. Secondary microflora brings about several apparent microbial changes take place in specific cheese types such as the distinct eyes formation/holes formation in Swiss cheese or Emmental cheese is due to the conversion of residual lactose or produced lactic acid to other metabolic end-products including CO_2 gas, which partially get trapped in the cheese to form bubbles. During the ripening of Blue cheese, *Penicillium roquefortii* dominates other microbiota, contribute greatly to pungent aroma of the cheese and develop typical blue-green veins on the surface of product through their growth in fissures within the cheese matrixes. Similarly, other strain *P. camembertii* contribute to the development of typical surface growth and specific flavor to other cheese varieties like Camembert cheese and Brie cheese. Overall, few smear-ripened cheese varieties, including Tilsit, Limburger; Munster exhibits the most complex microbiological and biochemical changes due to the growth of different microorganism on the surface. Initially there is a predominant growth of yeast species, and later on, growth of mixed strains of bacteria gives typical red-orange color texture and strong aroma to these cheeses.

During ripening, the conversion of organic acids (in particular of lactic acid), milk protein and milk fat into an exceptionally wide range of flavor compounds lead most significant biochemical changes in the cheese varieties. A well-ripened cheese has numerous uncountable flavor/aroma compounds. The distinctive flavor of different cheese varieties are often linked with the unique chemical compounds at different proportions and levels. Biochemical conversion of lactic acid into other organic acids and CO_2 is highly significance in manufacturing of Swiss cheese, which leads eye formation; while in white mold cheese varieties like Brie and Camembert, it causes texture softening. Inherently, very little amount of citrate is present in milk, some of which get trapped in the curd during cheese-making and later during ripening stage transform into flavor compounds and low level of CO_2 generating small eyes typical that of few Dutch types cheese (e.g., Gouda, and Edam).

Proteolysis means the breakdown of protein is the foremost essential incident takes places during ripening of major cheese varieties. Casein and whey proteins are the main component of milk proteins. Milk caseins are initially broken down by milk clotting enzyme rennet and inherent milk enzyme plasmin, to a number of bigger and medium-sized peptides. These peptides are further attacked by microbial enzymes and lead to free amino acids formation; free amino acids serve as precursors for several key flavor compounds formed by a series of complex biochemical reactions. Some small peptides may contribute to off-flavors also like bitterness.

A triglyceride molecule that contains three fatty acids linked with glycerol in the backbone is the main component of milkfat. Unlike other fats, milk fat is comprised of a higher amount of short-chain fatty acids (SCFA). Lipolysis is largely favored when these SCFA released from milk triglycerides by enzymatic action. Free fatty acids (FFA) are advantageous at low levels, but very high levels are not desirable as it can lead rancidity in cheeses. During the preparation of some hard Italian varieties and also in Blue cheeses lipolysis is particularly of more significant than other cheese types. Lipolysis has a direct impact on the flavor of cheese; even during ripening certain fatty acids as well act as precursors of other flavor compounds.

Cheese ripening is a dynamic procedure which involves complex metabolic reactions such as lactose, protein; lipid (Fox et al., 1998; Upadhyay, 2003). These biochemical changes are carried out by means of microbial cells and their enzymes, indigenous milk enzymes (proteases and lipases); residual rennet or any other milk clotting enzyme used in the cheese manufacturing process. Although the influence of starter LAB and many other intrinsic, as well as extrinsic factors on various biochemical alterations, have been deeply analyzed and reviewed but still the complete significance of NALAB in the cheese ripening process remains indecisive. Many researchers mentioned that the presence of NSLAB either which survive the heating or as post pasteurization contaminant is adventitious in few cheeses (Turner et al., 1986). In such cases, the inhabitant flora in the dairy plant equipments and also of raw milk itself is the major source of contamination. The viable of NSLAB may vary from low numbers in the beginning to 10^7–10^8cfu per g of the product during the initial weeks of ripening stage and they may found as dominate flora after the death of LAB (Fox et al., 1998). In cheese varieties made with pasteurized milk, chiefly mesophilic LAB, especially strains of lactobacilli have been observed as NSLAB (McSweeney et al., 1993). Eventually, the strains of lactic starters employed in cheese making usually influence the growth rate and proportion of NSLAB during the cheese ripening process. Autolytic starter bacteria could accelerate the growth of NSLAB during ripening as suggested by several researchers (Thomas, 1987; Martley and Crow, 1993; Crow et al., 1995) though so far many experimental evidences have not confirmed or supported the results (Lane et al., 1997; Hynes et al., 2000, 2001).

Ripening of cheese has been the entity of substantial scientific research over many years. Although the common mechanisms of major cheese types ripen are now well recognized, it remains impracticable to pledge the finest quality cheese. Thus, the ripening of cheese still remains a thrust area of research and the art of cheese-making.

2.9.1 RIPENING OF CHEESES-HARD, SEMI-HARD, AND SEMI-SOFT VARIETIES

Before consumption, all hard, semi-hard, and semi-soft cheese varieties are matured through the ripening process; the ripening period may vary from two weeks to 2 years. Compared to LAB, some NSLAB including *Lactobacillus* and *Pediococcus* species play imperative role in bring about desirable changes during cheese ripening. Conversely, cheese with regular organoleptic properties can be made with starter cultures alone. Also, in certain cheese varieties synthesis of biogenic amines and defects like calcium lactate crystals production through some NSLAB suggest strict observation of related strains for commercial use as secondary flora in cheese.

Presence of *Propi. freudenreichii* subsp. *freudenreichii*, as main starter bacteria strongly influenced the basic pattern of ripening of Swiss-type cheese as compared to other hard cheese. The strain metabolizes lactic acid and converts it into propionic acid, acetic acid, and CO_2; the organic acids are responsible for distinctive acidic flavor of Swiss cheese while CO_2 is responsible for formation of eyes like structure in cheese.

In the ripening of semi-hard cheese types, surface ripening performed by NSLAB is crucial process. In Brick cheese, *Brevibacterium linens* is of most significance and is accountable for the characteristic brownish-red color of the mature cheese. In several cheese types, presence of salt-tolerant microorganisms such as *Micrococci, Enterococci, Caseobacter*, and *Geotrichum candidum,* may contribute to the ultimate characteristics of the cheese. *Geotrichum candidum* metabolizes lactic acid and raise the pH value of cheese matrixes to be optimum for the growth of *Bre. linens*.

Depending on the cheese types, the extent of ripening by *Bre. linens* may vary a lot. For instance, in compare to Brick cheese, *Bre. linens* growth on the Monterey cheese and related cheese types is less due to relatively dry curd which results in a firm texture and a low flavor intensity of mature cheeses. Over ripening is one of the serious problems in case of the surface-ripened cheeses. Lowering of the temperature, physical removal of surface growth or restricting the area of the cheese exposed during maturation period can help to control over-ripening of different cheeses types.

Ripening of blue vein cheeses like Roquefort, Stilton, and Gorgonzola is initiated by rennet and perhaps more from enzymes released by starter bacteria; however, the foremost changes bring about from the proliferation of *P. roquefortii*. In such cheese, during ripening air is introduced to the cheese interior through needles that trigger growth of mold into the lines of piercing. Stilton cheese is pierced after five to six weeks of maturation in which growth usually reaching a maximum after 3 months while in other cheeses like Gorgonzola and Roquefort cheeses are pierced after second or third week of ripening. In the surface-salted varieties like Gorgonzola and Roquefort, growth of fungi is maximum in the mid-zone where the salt concentration is optimum (about 1–3 g $100g^{-1}$) of the cheese. Furthermore, during ripening excessive growth of *P. roquefortii* or any other mold is undesirable in many blue vein cheese types. Thus, once the mold has grown on the surface up to the desired level, the cheese is subsequently wrapped in tin foil to limit the further mold growth without any unfavorable effect on the continuing enzyme activity.

Development of a smooth, impermeable outer layer (called rind), which can safeguard the interior from aerial contamination is a key feature of the blue-vein cheese-making process. Surface microbiota plays an imperative role in the rind formation and then die depending on the amount of moisture in the cheese. Subsequently, the salt-tolerant and lactate-degrading yeasts emerge first on it that raises the pH to a value which permits the growth of other microflora. In French type cheeses which contain high moisture, usual growth of *Bre. linens* dominates, but in dry cheese varieties like Stilton different microorganisms like yeasts, *Geotrichum, Micrococcus, Lactobacillus;* sometimes, *Bacillus* species may developed during ripening.

Cheese ripening always involves specific requirements like storage under controlled humidity and temperature. For this reason, mechanical refrigeration is generally prerequisite; however, some cheese varieties allowed maturing in limestone caverns and in naturally cooled cellars (storerooms). More often, the employed temperature is adjusted in a way to compromise between two important aspects, first the fast ripening of cheese; secondly, prevention of growth of undesirable microflora that can develop atypical flavors in the end product. According to cheese type, mostly the relative humidity is kept to ~ 85–90% while the ripening temperatures are usually adjusted from 10–15°C. In specific cheese types like Emmental cheese, in the first stage of ripening temperature is kept about 18–22°C to promote the growth of main starter *Propi. freudenreichii* subsp. *freudenreichii,* on the other hand, the relative humidity on blue-vein cheese is as high as more than 90% to support the growth of *P. roquefortii.*

The time-span and extent of ripening period greatly depend on the manufacturing cheese type. The degree of cheese ripening is also varied in some varieties such as Cheshire, Cheddar; Parmesan cheeses on the basis of the preferences of consumers and the final application of cheese. Cheese ripening is very expensive practice, and accelerating the ripening process is one of the common practices at a commercial level to reduce the cost. Much attention has been received for accelerated ripening of hard cheese varieties like Cheddar, semi-soft as well as soft blue-vein cheeses.

Ripening process can be accelerated by rising the maturation temperature. However, a higher temperature may also potentially enhance the proliferation of spoilage types and few pathogenic microorganisms. To overcome such limitations, a number of new additional approaches have been employed. For instance, incorporation of few proteolytic enzymes associated with ripening; currently, it is the most sophisticated approach used at commercial scale for the production of enzyme-modified cheese. For hard cheeses, proteinases, and peptidases obtained from various sources such as mammalian tissues, *Penicillium, Aspergillus, Pseudomonas fluorescens, Micrococcus* sp; *Bacillus subtilis; Lb. casei* subsp. *casei* have been successfully utilized for accelerated cheese ripening.

Mixing bacterial or mold derived endopeptidases with exopeptidases that metabolize the short, bitter peptides have partly helped to overcome the former problems associated with flavor and textural defects. Introduction of a novel technique like 'Microencapsulation' for the introduction of ripening enzymes have greatly reduced difficulties related with separation of enzymes once the product is made. Microencapsulation is obtained by using liposomes or artificial vesicles with an aqueous, enzyme-containing core enclosed by concentric lipid lamellae layers. At the end of the cheddaring process, about 90% of enzyme encapsulated liposomes remain in intact position within the curd matrix but during the cheese maturing it gets disrupted.

Accelerated cheese ripening is also possible by employing a higher concentration of starter cultures, but it may lead to inherent problems caused by excess acid production in cheese. In order to overcome such technical problems, in-vat neutralization with NaOH solution is one of the alternative approaches. Heat-shock treatment to starters is another attractive approach which may selectively inactivate their acid production ability while retaining proteinase and peptidase activity. However, the organoleptic feature of cheese manufactured using heat-shocked starter cultures is extremely variable, and also it is not an economical alternate. Ripening using cheese slurries is a more satisfactory approach where *Lactococcus* strains are used as starters to get the maximum required viable cells exclusive of excess acid production.

2.9.2 SOFT CHEESE RIPENING

In case of soft cheese, non-starter microflora are used as ripening agents; otherwise, it involves similar basic steps as that of hard varieties. As an instance, during ripening of camembert cheese the growth of main mold *P. camembertii* is preceded by proliferation of yeasts and other molds like *Geotrichum* sp. though there is disagreement over the role of these microorganisms in creating favorable condition for *P. camembertii*. Once the pH increase because of lactate metabolism by species of yeast, the mold *P. camembertii* begins to grow profoundly, rapidly spread on the cheese surface. Apparently, within 10 to 12 days the maximum growth is achieved essential for softening of the cheese body mass after which the product is wrapped and boxed for marketing. During the later phases of ripening period, pH value of the cheese matrix raises up to certain extent; it contributes to softening of cheese body. Higher pH also allows the proliferation of non-acid tolerant microflora which may either positively or negatively affect the flavor of the cheese.

Use of pasteurized or heat-treated milk in conjunction with enhanced general standards of hygiene in recent times had significantly reduced the numbers of adventitious microorganisms in the cheese-making process. This, in turn, leads to the production of a bland cheese, which lacks typical cheese flavor. To overcome such problem, commercially a preparation known as ripening cultures, containing adventitious microorganism are available. Ripening cultures usually contain mixtures of microbial cells, including species of *Bre. linens, Micrococcus, and* yeast sp.

In mold-ripened soft cheeses, throughout ripening process control of humidity and storage temperature is importance, particularly during the initial 10–20 days. *P. camembertii* can actively grow at 11–14°C with a relative humidity of about 90%. A very high humidity and temperatures keep the cheese surface wet which permits proliferation of both advantageous and unwanted contaminating microbiota. On the other hand, low humidity and temperature could delay ripening process and even cheese may get dehydrate in extreme conditions without finishing ripening period.

2.9.3 BIO-CHEMICAL CHANGES DURING CHEESE RIPENING

2.9.3.1 ROLE OF METHANETHIOL AS FLAVORING COMPOUND OF CHEDDAR CHEESE AND RELATED VARIETIES

Ripened cheese contains numerous flavor compounds which influence the flavor and aroma of cheese either positively or negatively. Such flavor components may comprise diacetyl, acetaldehyde, alcohols (ethanol and methanol), FFA, methyl ketones and esters of fatty acids, peptides, and amino acids; sulfur-containing substances like methanethiol, hydrogen sulfide and dimethyl sulfide. According to previously used "component balance" theory to describe organoleptic characteristics, though many of these components have either limited or almost no impact on flavor they significantly contribute to background cheese flavors. Among all these, methanethiol, a volatile sulfur compound plays a vital role in completing the flavor of Cheddar cheese and related varieties. It has been also likely to associate the characteristics of Cheddar cheese with the charisma of methanethiol.

The definite function of methanethiol is not known. It is believed that this compound does not directly contribute to flavor; the ultimate interactions of methanethiol with other compounds results in development of flavoring compounds in cheese. Non-enzymatic reactions between casein or methionine and H_2S lead to production of methanethiol. Up to certain extent, starter cultures are involved with providing reducing conditions for methanethiol production and stability.

Among different starters, *Bre.linens* is well established for ability to produce methanethiol in smear-ripened cheese varieties. The breakdown of the side chains of some amino acids like methionine and cysteine results into the methanethiol formation. Further, probably an interaction between *Bre. linens* and non-starter bacteria like micrococci is involved in the conversion of a part of methanethiol to thioesters.

2.9.3.2 PROTEOLYSIS

In the cheese ripening, proteolysis is the most significant which affects the flavor and overall body-texture of the finished product. It is highly indispensable for the flavor and aroma formation in Cheddar cheese. Residual rennet or any other milk coagulating enzyme used; inherent milk enzyme plasmin; and starter plus non-starter microbial cells are the main three sources of proteolytic enzymes. The key pathway of proteolysis involves breakdown of para-casein by residual rennet or related enzyme to generate polypeptides. These polypeptides are then subsequently acted upon by microbial proteinases and peptidase to yield peptides and amino acids. Though rennin has very limited action towards paracasein, it has a significant extended activity during ripening phase.

The combined activity of rennin like coagulant, plasmin, and microbial enzymes are involved with degradation of β-casein. Plasmin has a protease activity and believed play a major role in para-casein degradation and further continues polypeptide hydrolysis. However, the activity of plasmin is a subject of controversy among researchers. Its activity is more pH-dependent and hence, plasmin is considered to play an essential role in the ripening of cheese having high pHs such as Gouda, Emmental, and related types, but not in Cheddar cheese or related cheese types having low pH (Ismail and Nielsen, 2010).

Proteolysis leads to weakening of the casein network of cheese matrix; and immediately affect the texture of cheeses, especially of hard and semi-hard types causing softening like an Edam cheese. In general, proteolysis rate is higher and rapid in the middle part of cheese due to comparatively less salt concentration as compared to other parts. The interstitial water in hard cheeses like Cheddar is bound by means of ionic groups and thus, at higher proteolysis cheese becomes crumbly.

As compared to hard and semi-hard cheese types, the proteolysis rate is higher in soft surface-ripened cheese varieties with complex textural changes. Until residual lactose is used via glycolysis by microorganisms, a significant level of proteolysis cannot take place in surface mold-ripened cheeses, as it will lead to a rise in pH necessary to stimulates the activity of milk plasmin. It will result in further enhanced proteolysis and thus, weakening of soft cheese matrix. Additionally, at high pH, surface precipitation of calcium phosphate cause more softening of cheese structure, and subsequently, the calcium gradient gets established to the center of cheese from the surface.

In case of bacterial smear-ripened cheese, apparently, proteolysis occurs in a similar way however the ability of surface microorganisms to do proteolysis in cheese is comparatively limited as the diffusion of proteolytic enzymes in cheese is possible for a small distance only. The pH of the cheese raises because of surface proteolysis as the hydrolyzed metabolites diffuse to great distances. Intracellular proteinases of few microorganisms like *P. roquefortii* are more significant to that of extracellular ones; the former contributes in overall metabolism of proteins. Due to the lysis of mycelia or breakage in microbial cell, level of intracellular proteinases increased remarkably between 10 to 16 weeks of cheese ripening.

The peptides produced by protein metabolism have been categorized into five major kinds *viz*. sweet, salty, sour, bitter; umami, a range of taste qualities. Depending on the prevalence of different types, the contribution of peptides may be pleasant or unpleasant. Rather than acting as a flavor component, many a time peptides may be associated with the overall taste of the product. Mainly the water-soluble fraction of the peptide contributes to the flavor of various types of cheeses.

Formation of bitter peptides usually leads to defect in cheese since it results in a bitter flavor. Polypeptides containing a very high content of hydrophilic residues commonly cause bitterness in a product. Both plasmin and chymosin are considered in the synthesis of bitter peptides. Apart from these, the cell wall-bound proteinases of several LAB have also been strongly implicated for production of bitter peptides. Often there is a large variation in the bitter peptides quantities formed by different *Lactococcus* species and strains. In many circumstances, if produce at low a level, bitter peptides as well as related astringent flavors may contribute positively to the organoleptic quality of the ripened cheese.

Apart from peptides, amino acids like proline are also significantly determines the flavor of some cheeses. For instance, *Propionibacterium* sp. is the main source of proline-releasing peptidases in Swiss-type cheeses. Other characteristic flavor compounds are formed by several lactobacilli like *Lb. helveticus*. The patterns amino acids have a specific connection with particular cheese types; in Greek type and related cheeses, alanine, valine, leucine; α-amino butyric acid is regarded as the dominating amino acids.

However, during ripening process most amino acids are metabolized by microorganisms. For example, in Limburger cheese *Bre. linens* mediated amino acid catabolism is responsible for the characteristic flavors including 3-methylthiopropanol, phenylethanol, and 3-methyl-1-butanol derived respectively from methionine, phenylalanine, and leucine. In mold-ripened cheese varieties, excessive breakdown of amino acid produces aldehydes, ammonia, amines, and acids like volatile compounds.

On the other hand, proteolysis produces bioactive peptides which are considered to be one of the major functional peptides in dairy and dairy products, especially that fermented with different lactic microorganisms. The bioactive peptides play an important role in treating a different disease like hypertension without any side effects. These bioactive peptides can be found in most ripened cheese types.

2.9.3.3 LIPOLYSIS

In the majority of cheese types, lipolysis is considered to play a secondary role in cheese ripening. Primarily the fat content and the extent of lipid degradation ultimately determine when to encourage lipolysis, either during cheese making or cheese ripening. During cheese making, sometimes rennet preparations containing a high concentration of lipase activity, lipases from microbial origin or pre-gastric enzymes may lead to more amount of FFA formation that may perhaps dominate the flavor spectrum, also.

The total FFA value can be very high (up to 0.05%) in most of soft cheeses which contain SCFA (C_4 to C_8). Oxidation of FFAs may further lead to synthesis of other flavor compounds like methyl ketones. Among molds, *P. roquefortii* possess very high lipolytic activity resulting into formation of methyl ketones, FFA; other flavoring substances.

In several cheese types, FFA have been responsible with flavor defects like "cowy" taints in cheese made from low quality milk. Fungal produced compound phenyl ethanol is primarily associated with almost all mold-ripened cheeses while oct-1-en-3-ol is involved with aroma of other cheeses.

2.9.3.4 GLYCOLYSIS AND ASSOCIATED REACTIONS

Degradation of the remaining lactose by LAB results in numerous changes during cheese ripening process such as the racemization of L-lactate to D-lactate. Metabolism of lactate produces CO_2 and H_2O in case of surface molds ripened cheeses; increased the value of pH with simultaneous stimulation of proteolysis and *Bre. linens* growth, both.

In the case of Swiss-type cheeses, the degradation of L-lactic acid to propionic acid, acetic acid and CO_2 by species of *Propionibacterium* is imperative for the production of characteristic flavor and eyes formation. Lactose metabolism comprises oxidation of lactic to pyruvic acid, which is partially oxidized to acetyl-CoA and CO_2. Subsequently, the Acetyl-CoA is transformed to acetate, a one of flavor component of cheese. Propionic acid biosynthesis involves a reductive randomizing pathway, balancing the oxidative synthesis of acetate and CO_2. Lactic acid may be converted to acetate through non-starter microflora like *Pediococcus* in Cheddar cheese and related types. Very high amounts of acetic acid are valuable in some cheese like Dutch cheeses, but to be a fault in some cheese like Cheddar cheeses.

2.10 CHEESE MATRIX AS AN ENVIRONMENT FOR MICROBIAL GROWTH

2.10.1 HARD AND SEMI-HARD CHEESES

A high concentration of lactic acid, low pH, and variable salt (NaCl) content in the aqueous phase of hard and semi-hard cheeses play a significant role to restrict the growth of various microorganisms. In high moisture cheese varieties like Gouda cheese, usually, the water activity (a_w) varies from 0.97 to 0.94 while in other cheeses may have a_w value of 0.9 or below. In matured Cheddar cheese may have this value below 0.85. Thus, naturally hard and semi-hard cheese varieties restrict the growth of a broad range of microflora.

2.10.2 SOFT CHEESES

Soft cheeses such as cottage have high pH values plus high moisture content which support the proliferation of a number of microorganisms. In particular, throughout the early stages of cheese storage, the conditions are adequate enough to inhibit as well as delay the commencement of growth. Soft un-ripened high acid cheeses are normally inhibitory to nearly all microorganisms. In young surface-ripened soft cheeses also the condition is also comparatively same, however during maturation the increased pH value lessen environmental stresses and allows the multiplication of a diverse number of microorganisms. Also, the presence of NaCl is sufficient to hamper the proliferation of almost all gram-negative bacterial species in certain cheese varieties.

2.11 CHEESE AND FOOD-BORNE DISEASE

Generally, cheese is believed to be one of the low-risk foods. The ability of some pathogenic microorganisms including *Listeria monocytogenes* and few strains of *Escherichia coli* to proliferate in several soft cheeses can be found, but not either in semi-hard or hard cheese varieties. The major risk factors in such cases are:

1. The use of unpasteurized milk;
2. Insufficient growth of starter microorganisms;
3. Post-pasteurization contamination.

2.11.1 HARD AND SEMI-HARD CHEESES

A hard and semi-hard cheese variety has been concerned because of several *Salmonella* poisoning outbreaks. Unpasteurized milk is generally served as the main source of this bacterium is many cases. *Salmonella* can survive under cheese manufacturing conditions and still being competent of limited growth in the low-acid curd cheese (pH > 4.95). During the ripening period and 60 days at low temperature (~ 4.4°C) storage, these pathogenic bacteria get disappears, which was sufficient to ensure the safety of cheese made from raw milk. Conversely, some *Salmonella* strains have been establishing to have an acid adaptation that helps to survive and resist low pH, and even few cells can survive at refrigeration temperature for longer periods (> 60 days).

Consumption of hard and semi-hard cheeses has been accountable for several staphylococcal food poisoning outbreaks. Human handlers and raw milk, especially mastitic milk is considered to be the potential sources of *Staph. aureus*. Primarily if the starter culture is slow acid producer, it favors the growth of pathogen and subsequent production of enterotoxin. Further, *Staph. aureus* proliferation and production of enterotoxin can continue in the cheese curd at high temperature and high salt concentration. Although the pathogen is usually destroyed during ripening enterotoxin produced by bacterium is heat stable.

With enhanced concern and improved hygiene levels outbreaks related to *Staph. aureus* are generally less common in recent years. However, it is more common in small-scale production if proper care is not taken.

2.11.2 SOFT CHEESE

Some of the soft cheese has been associated with *Salmonella* and staphylococcal food poisoning. In such cases, contamination of the cheese curd or finished product appears to be a major risk factor, particularly at small scale production.

E. coil has been responsible for many outbreaks of food poisoning incidences in Brie and similar cheeses. Poor hygienic conditions and uncontrolled temperature are the main cause of *E. coli* poisoning as this organism is able to grow during cheese ripening. Usually, *E. coli* are present in higher numbers in soft cheeses, and they are mostly non-pathogenic. Several enterotoxigenic *E. coil* has been isolated in some cases from soft cheeses.

L. monocytogenes has been establish to grow during ripening in most soft cheeses, although this appears to be restricted to outer layer only. These listeriosis causing bacteria are usually gain entry in cheese though the use of unpasteurized milk and further their presence in several pasteurized milk samples have generated more concern which is related to heat resistance of *L. monocytogenes*.

Milk products have been barely implicated with anaerobic spores; cheese is the exceptional milk product in which proliferation of *Clostridium botulinum* cells noticed to cause botulism. Processed cheese spreads also considered as a vehicle of this anaerobic bacilli involved with latest outbreak of *Cl. botulinum* type A.

2.12 CHEESE RIPENING AND PRODUCTION OF BIOGENIC AMINES

Presence of biogenic amines like histamine and tyramine in ripened cheese may be linked with decarboxylation of respective histidine and tyrosine amino acids. These biogenic amines are responsible for a significant elevation of blood pressure in conjunction with headaches, flushing, and rashes and occasionally gastrointestinal disturbances with or without diarrhea incidences in sensitive individuals.

In most cases, non-starter microorganism emerges to cause decarboxylation of histidine and production of histamine. Cheese isolates *Lb. helveticus, Lb. casei, Lb. fermentum, Lc. lactis* and *Enterococcus faecium* have been established to produce histamine in little amount (Burdychova and Komprda, 2007). In case of LAB, histidine decarboxylate positive strains are relatively very less; if present then also considerable quantities of biogenic amines are synthesized only under definite growth conditions like the low concentration of NaCl or high temperature. Generally, cheeses contain very less number of free histidine molecules. Proteolysis is believed as the primary step in formation of histamine. As a result, problems of histamine toxicity arise mainly in cheeses that undergo extensive proteolysis during ripening process, for instance in Roquefort, Stilton, Swiss-type cheese and Gorgonzola. Thus, practically it is impossible to rule out histidine decarboxylating strains from cheese, but maturation at low temperatures such as 7°C or below seems to be the most effective to restrict the formation of biogenic amines in cheeses.

2.13 SPOILAGE OF CHEESES

2.13.1 HARD AND SEMI-HARD CHEESES

There observe two main types of spoilage in hard and semi-hard cheeses: 1) microbial growth on surface, frequently by molds; 2) gas production owing to microflora present inside cheese. Except certain species of mold which are used for ripening specific cheese types, mold growth on surface of cheese is undesirable. Surface growth of mold on cheese produces visible spoilage which is usually accompanied by extensive lipolysis and proteolysis in severe cases. *Penicillium* and *Aspergillus* are the main fungi responsible for spoilage of 60–80% of these cases. Mold present at the packaging time more often cause visible defect/spoilage in packed cheeses. Mold spoilage is more frequent problem in pre-packaged cheese and during packaging careful safety measures should be implemented including regular UV light and/or chemical disinfection of handling surfaces, air sanitization or air filtration; perhaps coating of anti-fungal agents on packaging material. Treatment of packaging materials and working surface with and ionizing radiation are effective means to control mold growth. Application of vacuum and modified atmosphere packaging techniques in modern dairy plant has helped to reduce mold spoilage incidences.

The harmful molds may perhaps grow alone or in alliance with some yeast species in wrinkles and folds of plastic film. As consequence, there may be development of green, brown, black or dark threads or spots associated with cheese water or whey. Various species of *Alternaria, Cladosporium, Aspergillus; Penicillium* are the most frequent molds while *Candida* sp. are the chief yeast associated with such spoilage. Mold growth not only spoils the appearance of cheese but also produces off-flavors and further lead to concern over the likelihood of mycotoxin formation in the product.

If the surface of hard cheese is moist, it may also allow the growth of bacteria and yeasts. "Rind rots" formation and discoloration attributable to the proteolysis are main spoilages associated with the multiplication of molds, yeasts; bacteria. It may be associated with softening and off-flavors of cheese. Usually, discoloration on the surface is due to pigment-producing microorganisms, for instance, *Bre. linens* and *Aureobacterium liquefaciens.* A rusty spot in cheddar and Cheshire varieties of cheese is caused by pigmented variants of *Lb. plantarum* and *Lb. brevis* subsp. *rudensis.* While Swiss cheese colored spots are produced by pigmented species of Propionibacteria (*Propi. rubrum*) (Yadav et al., 1993). Typical pink discoloration predominantly by *Micrococcus* in nitrate-containing cheese has been noticed. It is because of reduction of nitrate to nitrite by the bacteria originally present in storage shelves and further, this nitrite reacts with the annatto dye to generate pink color on the cheese surface between 5.2 to 6.7 pH.

Gas production is another common defect of hard cheeses. It which may occur either in the cheese curd or green or raw cheese known as "early blowing" of cheese and if it occurs during maturation, it is called "late blowing" of cheese. The members of *Enterobacteriaceae* family are the common causative agents of early blowing of cheese. Several species of genus *Bacillus* and lactose fermenting yeasts like *Candida* and *Kluyveromyces* are also responsible for gassiness in cheese. Proper care and precautions together with improved general hygiene at production level may effectively reduce the early blowing of cheese.

Late blowing of cheese is generally associated with gas production by anaerobic sporeformers, i.e., Clostridia, which are able to convert lactose to organic acids like butyric acid and acetic acid, hydrogen, and CO_2. *Cl. butyricum, Cl. sporogenes*, and *Cl. tyrobutyricum* are the major species associated with late blowing in Edam, Emmental; Gouda cheeses. Generally, endospores of these bacteria are present in the milk employed for cheese making and once favorable environment appears, vegetative cells formed and spoil the product during storage.

Certain yeasts like *Candida* has been observed to produce high amounts of ethanol ethyl acetate and ethyl butyrate in Cheddar cheese; which impart characteristic yeasty flavor to cheese. High moisture and low NaCI content may favor the yeast growth within six months of cheese ripening.

Curd flatting is another problem associated with cheeses if raw milk is heavily loaded with coliforms. Uncontrolled temperature during cottage cheese manufacturing may favor growth of coliforms as well as of starter culture *Leuconostocs* which may produce CO_2 during their normal metabolism and contribute in curd floating defect. *Leuconostocs* species also contributed to a specific defect in Cheddar cheese called 'excessive openness' if ripening temperature is above 10°C.

Most NSLAB has been associated with several microbial spoilages, and they may contribute to visual and textural defects like calcium lactate crystal precipitation. Some *Leuconostoc, Lactococcus; Pediococcus* strains which can do racemization of L - (+) lactate to D - (-) lactate are also accountable for formation of calcium lactate crystal, particularly it occurs at high temperatures usually kept during accelerated ripening of cheese.

The role of starter microorganisms is producing health-improving bioactive peptides is well known, but continue metabolic activity of cultures during ripening can lead to flavor and aroma defects. Few strains such as *Lc. lactis* sub sp. *lactis* are able to produce esters including ethyl butyrate and ethyl hexanoate via reactions between ethanol and butyric or hexanoic acid which lead to generate fruity flavor in the cheese. Furthermore, transformation of several amino acids like leucine to the related aldehydes 3-methylbutanol through transaminase and the decarboxylase of *Lc. lactis* subsp. *lactis* lead to development of "malty" flavors in cheese.

2.13.2 SOFT CHEESE

Few Gram-negative bacteria from the genus *Pseudomonas* (*Ps. fluorescens, Ps. Putida*), *Enterococcus, and Enterobacter* (*E. agglomerans*) can be found in wash water or added ingredients are the main widespread spoilage bacteria in soft cheeses. The growth of *Pseudomonas* sp. may be managed by acidification to 4.5 pH, but *Enterobacter* sp. are able to multiply at low pH (<3.8). Yeasts and molds are normally considered to be spoilage organisms at high acid cheeses "pH below 5." A large ranges of yeasts including *Candida, Cryptococcus, Kluyveromyces, Pichia, and Torulopsis* species have been involved in spoilage of cottage cheeses and other un-ripened cheese varieties like Quarg cheese; where they usually cause gassiness, Flavor, and aroma defects; show visible growth, i.e., discoloration.

Apparently, in the surface-ripened soft cheeses, spoilage is frequently a result of ripening stage in that the normal ripening microflora may be dominated by unwanted microbes. Mainly molds, but occasionally yeast and bacteria are also involved as such contaminants in cheese spoilage.

2.14 GENETICS OF STARTER MICROORGANISMS

2.14.1 BACKGROUND

According to the new technology strategies for improvement, the production of foods either from the characteristics or the yield of the foods some modifications can be done to make desired traits in animals, plants, or microorganisms used for making food began few decades ago. These alterations together with natural evolutionary transformations have consequence out in ordinary food species that are genetically dissimilar from their ancestors nowadays. There are a number of beneficial effects of genetic transformations, including enhanced food manufacture and yields; food with better taste and nutritional values; and reduced losses owing to diverse abiotic and biotic factors. Such positive outcomes trigger further development of newer genetic transformation techniques for screening, selecting, identifying; evaluating each microbial cell that acquires genetically improved features for increasing the desired properties.

2.14.2 GENETIC IMPROVEMENT OF STARTER MICROORGANISMS

Fermentation is a highly accepted process for food preservation around the world. Nearly from ancient times, every civilization has a wide variety of fermented milk (cheeses, buttermilk, curd, kefir, koumiss, etc.), vegetable, fruit, meat, fish or cereal products as a part of their regular diet. Fermenting microflora produce a huge range of metabolic end products that may provide distinctive attribute(s) to the food, i.e., as a flavoring, coloring texturizers, stabilizers; preservative agents. Several different traditional and non-traditional techniques have been used to improve such fermentation related metabolic properties of related microflora. It mainly comprises mutation; natural gene transfer methods like transformation, transduction, and conjugation; and recombinant DNA technology- genetic engineering. These techniques have been briefly discussed with major emphasis on the merits and demerits of each method for improvement of fermenting microorganisms used in the dairy and food industry.

2.14.3 GENETIC IMPROVEMENT STRATEGIES OF MICROORGANISMS

The importance of starter culture bacteria known from 10,000 years ago and has been used either for preserving food or dairy products also, has been spreads for supporting the human survival. Most of the people at different countries used different fermented products at their diet, i.e., cheeses, fermented food, fermented milk, etc. Different organisms for example, LAB, some yeast; some molds, these microorganisms can produce preservatives, texturizing, and stabilizing agent and at the same time can produce flavoring and coloring organic materials from their metabolism.

So, many methods have been used for improving the metabolism of such microorganisms used in the fermentation process. The mutation and selection transfer of the natural gene transformation and genetic engineering. All of these methods can improve the characteristics of microorganisms used for food and dairy fermentation and in the following some white spot on these methods:

2.14.3.1 MUTATION

The mutations consider to be one of the methods in which changes of the microorganism chromosome can happen at every low rate in the microorganisms generation, i.e., one mutational each (10^6 to 10^7) cell. Most of mutation occurs randomly in the normal metabolic pathways of food and dairy fermentations.

Nowadays, the scientists can increase the mutations of the microorganisms using some mutagenic materials, for instance, UV light or some chemicals which makes a change in the structure of deoxyribonucleic acid (DNA) of the selected microorganism cells. Most of the mutations attempt increased to be event every 10^1 or 10^2 cells per generations. A method of selection considered to be the main critical stage which needs evaluations of several individual isolates to find one improved property of the microorganisms.

Mutation can be used to improve the characteristics of starter microflora used for dairy and food fermentations. The mutation of the microorganisms happened randomly, so the specificity is not possible in this technique. On the other hand, the traditional mutation procedure high in their cost and consume much time. The mutation and selection method also consider being one of the most important methods for improving the important microorganisms used in food and dairy industry. The improved microorganisms can produce metabolic materials more than the normal production level, which would be better for the industry. For example, in the dairy industry, the mutation can be used for LAB to improve the certain characteristic of fermented end products such as Flavor, structure, nutritional value, phage resistance, lactose metabolism and increased the proteolytic activity of the modified microorganisms.

2.14.3.2 TRANSDUCTION

Transduction considers to be also one of the controlled genetic modification of DNA methods. This method involves the exchange of genetic material mediated by virus (bacteriophage). The virus acquires a portion of the plasmid or chromosomes from the starter bacterium (the host) and transfers it to another bacterial cell (the recipient) during interval viral infection.

Transduction has been adapted as very efficient gene transfer methods in case of gram-negative bacteria like *E. coli,* there is no recommendations for using this methodology for improving the microorganisms used for dairy and food fermentation processes. So, there are constraints for using this technique for improving the important strains. Also, the most of starter culture bacteriophages are not characterized up tell now.

2.14.3.3 CONJUGATION

Conjugation is one of the controlled methods for innate gene transfer system which requires close contact between two partner types (called donor and recipient) for the dissemination of the segment of DNA or plasmid. There are many types of bacteria have plasmid DNA. During the conjugation there are physical contacts between the donors and the recipients and once the plasmid DNA introduced into the new strain, the plasmid-encoded properties can be expressed in the recipient bacterial cell.

Generally, LAB holds more than 10 discrete plasmids, which consider to be the most important trails from metabolism side of view, including lactose fermenting ability, bacteriocin production, and phage resistance. So, the conjugation method can be used for transfer these plasmids into the recipient strains in order to improve the important characteristics of starter culture used in the dairy industry.

From the applications side, there are limitations for using this technique in case of improving the useful bacterial strains for different reasons, i.e., there is no complete understanding for the plasmid biology, the plasmid-encoded genes are insufficiently identified; further not all the strains able to serve as recipients during conjugation procedure. So, there are no clear applications of this technique on the industrial scale.

2.14.3.4 TRANSFORMATION

Some microorganisms can be selected for the removal of undesirable components from different food materials, i.e., LAB used in making dairy products can ferment the lactose partially, one of the fermentation by-products is galactose which considered to be harmful to some people who suffer from galactosemia. The transformation technique can be used for improvement the fermentation characteristics. This method has been developed to remove the bacterial cell wall to develop the protoplasts. By adding of polyethylene glycol, DNA can uptake by the protoplasts under standardized conditions.

The transformed protoplasts of LAB generate cell walls which express the transformed DNA for modification some characteristics. This methodology is often having low efficiency, highly variable, time-consuming, and represents limited applications. In the case of using this method, the technique parameters must be optimized for each strain.

2.14.3.5 ELECTROPORATION

This method is used for gene transfer by application of high-voltage electrical pulses for short times. This method depends on making pores in the bacterial cell wall and cell membranes. Further, under special conditions, DNA can be present in the media contain the treated bacteria by high-voltage electrical pulses for a short time and maybe enter through these pores. This technique used for special strains and easy controlled conditions but needs more studies.

2.14.3.6 GENETIC ENGINEERING OF LAB

Genetic engineering of LAB is considered to be a specific and alternative technique for a different methodology for improving the characteristics of LAB. This technique provides isolation and transfers a single gene of LAB under specific and controlled conditions. The transferred gene can be coded for a specific and desirable trait. The improved LAB can be used for increasing food and dairy fermentation impacts such as improving the Flavor, texture, functionality of the fermented food and dairy, etc. Also, most of the genetically engineered bacteria, yeast, and molds can be employed for producing different products including food and dairy ingredients, food additives, some processing tools such as different enzymes and pharmaceutical components.

2.15 LIMITATIONS OF USING GENETIC MODIFIED STARTER CULTURES

Several issues must be resolved before employing genetically engineered starter cultures in the dairy and food industry. Genetically modified organisms (GMO) will primarily need to be approved by appropriate regulatory agencies for specific use and need to be established as generally recognized as safe (GRAS) status by the specific regulatory bodies of respective countries like Food and Drug Administration (FDA) of United States (US).

The biotechnology of LAB must be guaranteed that products containing these GMO are safe for consumer use; otherwise, these new technologies will be rejected for use by the consumers. The use of the genetics improved microbial starter culture still have a limitation; especially, this technique will not improve all the required characteristics of either the microorganisms or the end products. More studies still need to clear the clawed of this new technology in order to satisfy the consumer.

2.16 SUMMARY AND CONCLUSION

Cheese manufacturing, known from 7000 BC is biochemically and microbiologically dynamic process. Diverse types of microflora involving LAB, other non-starter LAB, yeast, and molds are involved with the manufacturing of wide varieties of cheese at different countries. Nevertheless, very few selective strains are employed as a starter during initial acid development, during ripening a number of different microorganisms play a significant role in the development of specific cheese varieties. Ripening of cheese is characterized by extensive metabolism of residual lactose and other milk components through glycolysis, proteolysis; lipolysis. It leads to the specific development of body, texture, and sensory qualities of the particular cheese variety. Ripening is associated with biosynthesis of vitamins, enzymes, bioactive peptides; distinct flavors components.

Although genetically engineered strains have shown immense potential for their probable application as a starter in cheese making, no genetically modified microorganisms have been approved as GRAS yet by regulatory bodies of the world. So, natural strain improvement processes are of vital choices to enhance the acid production, flavor production, bacteriophage resistance, stress resistance, and controlled proteolytic and lipolytic activities of normal starter cultures. Further, the use of health beneficiary bacteria, called probiotics, could enhance the therapeutic and health significance of cheese.

KEYWORDS

- **cheese flavor**
- **cheese ripening**
- **green cheese**
- **hard cheese**
- **lipolysis**
- **nonstarter LAB**
- **proteolysis**
- **rennet**
- **ripened cheese**
- **semi-hard cheese**
- **soft cheese**
- **starter cultures**

REFERENCES

Ammor, M. S., Florez, A. B., & Mayo, B., (2007). Antibiotic resistance in non- enterococcal LAB and bifidobacteria. *Food Microbiology, 24*, 559–570.

Aquilanti, L., Dell'Aquila, L., Zannini, E., Zocchetti, A., & Clementi, F., (2006). Resident LAB in raw milk Canestrato Pugliese cheese. *Letters in Applied Microbiology, 43*, 161–167.

Ayad, E. H. E., Verheul, A., Wouters, J. T. M., & Smit, G., (2002). Antimicrobial-producing wild *Lactococci* isolated from artisanal and non-dairy origins. *International Dairy Journal, 12*, 145–150.

Aymerich, T., Martín, B., Garriga, M., Vidal-Carou, M. C., Bover-Cid, S., & Hugas, M., (2006). Safety properties and molecular strain typing of LAB from slightly fermented sausages. *Journal of Applied Microbiology, 100*, 40–49.

Batt, C. A., Erlandson, K., & Bsat, N., (1995). Design and implementation of a strategy to reduce bacteriophage infection of dairy starter cultures. *International Dairy Journal, 5*, 949–962.

Beresford, T. P., Fitzsimons, N. A., Brennan, N. L., & Cogan, T. M., (2001). Recent advances in cheese microbiology. *International Dairy Journal, 11*, 259–274.

Bernardeau, M., Vernoux, J. P., Henri-Dubernet, S., & Guéguen, M., (2008). Safety assessment of dairy microorganisms: The *Lactobacillus* genus. *International Journal of Food Microbiology, 126*, 278–285.

Bierkland, N. K., & Holo, H., (1966). Transduction of a plasmid carrying the cohesive end region from *Lactococcus lactis* bacteriophage LC3. *Applied and Environmental Microbiology, 59*, 1966–1968.

Briggiler-Marcó, M., Capra, M. L., Quiberoni, A., Vinderola, G., Reinheimer, J., & Hynes, E., (2007). Nonstarter *Lactobacillus* strains as adjunct cultures for cheese making: *In vitro* characterization and performance in two model cheeses. *Journal of Dairy Science, 90*, 4532–4542.

Broome, M. C., Powel, I. B., & Limsowtin, G. K. Y., (2003). Starter cultures: Specific properties. In: Regisnki, H., Fuquay, J. W., & Fox, P. F., (eds.), *Encyclopedia of Dairy Sciences* (Vol. I, pp. 269–275). London: Academic Press.

Buckenhiiskes, H. J., (1993). Selection criteria for LAB to be used as starter cultures for various food commodities. *FEMS Microbiology Reviews, 12*, 253–272.

Burdychova, R., & Komprda, T., (2007). Biogenic amine-forming microbial communities in cheese. *FEMS Microbiology Letter, 276*(2), 149–155.

Caldwell, S. L., McMahon, D. J., Oberg, C. J., & Broadbent, J. R., (1996). Development and characterization of lactose-positive pediococcus species for milk fermentation. *Applied and Environmental Microbiology, 62*, 936–941.

Caplice, E., & Fitzgerald, G. F., (1999). Food fermentations: Role of microorganisms in food production and preservation. *International Journal of Food Microbiology, 50*, 131–149.

Carminati, D., Giraffa, G., Quiberoni, A., Binetti, A., Suarez, V., & Reinhemer, J., (2010). Advances and trends in starter cultures for dairy fermentation, Chapter 10. In: Mozi, F., Raya, R. R., & Vignolo, G. M., (eds.), *Biotechnology of LAB: Novel Applications* (pp. 177–192), Wiley Blackwell Publisher.

Chapot-Chartier, M. P., Deniel, C., Rousseau, M., Vassal, L., & Gripon, J. C., (1994). Autolysis of two strains of *Lactococcus lactis* during cheese ripening. *International Dairy Journal, 4*, 251.

Cocolin, L., Rantsiou, K., Iacumin, L., Urso, R., Cantoni, C., & Comi, G., (2004). Study of the ecology of fresh sausages and characterization of populations of LAB by molecular methods. *Applied and Environmental Microbiology, 4*, 1883–1894.

Coppola, R., Succi, M., Tremonte, P., Reale, A., Salzano, G., & Sorrentino, E., (2005). Antibiotic susceptibility of *Lactobacillus rhamnosus* strains isolated from Parmigiano Reggiano cheese. *Lait, 85*, 193–204.

Crow, V. L., Coolbear, T., Goparl, P. K., Martley, F. G., McKay, L. L., & Riepe, H., (1995). The role of autolysis of LAB in the ripening of cheese. *International Dairy Journal, 5*, 855.

Dal, B. B., Cocolin, L., Zeppa, G., Field, D., Cotter, P. D., & Hill, C., (2011). Technological characterization of bacteriocin producing *Lactococcus lactis* strains employed to control *Listeria monocytogenes* in cottage cheese. *International Journal of Food Microbiology, 153*, 58–65.

Deegan, L. H., Cotter, P. D., Hill, C., & Ross, P., (2006). Bacteriocins: Biological tools for bio-preservation and shelf-life extension. *International Dairy Journal, 16*, 1058–1071.

Dekkers, J. C., & Hospital, F., (2002). The use of molecular genetics in the improvement of agricultural populations. *Nature Review Genetics, 3*, 22–32.

Delcour, J., De Vuyst, L., & Shortt, C., (1999). Recombinant dairy starters, probiotics, prebiotics: Scientific, technological, regulatory challenges. *International Dairy Journal, 9*, 3–80.

FAO/IAEA (Food and Agriculture Organization of the United Nations/International Atomic Energy Agency), (2001). *FAO/IAEA Mutant Varieties Database*. Available at http://www-infocris.iaea.org/MVD/ (accessed on 6 September 2019).

Fletcher, A., (2003). *Gene Identified to Regulate Milk Content and Yield.* FoodProduction-Daily.com. Available at http://foodproductiondaily.com/news/news-NG.asp?id=29318 (accessed on 6 September 2019).

Fox, P. F., (1993). *Cheese Chemistry Physics and Microbiology* (Vol. 1, pp. 303–340). Chapman and Hall London.

Fox, P. F., McSweeney, P. L. H., & Lynch, C. M., (1998). Significance of non-starter LAB in cheddar cheese. *Australian Journal of Dairy Technology, 53*, 83.

Furet, J. P., Quenee, P., & Tailliez, P., (2004). Molecular quantification of LAB in fermented milk products using. *Journal of Food Microbiology, 103*, 131–142.

Galvez, A., Lucas, R. L., Abriouel, H., Valdivia, E., & Omar, N. B., (2008). Application of bacteriocins in the control of foodborne pathogenic and spoilage bacteria. *Critical Reviews in Biotechnology, 28*, 25–152.

Hynes, E., Ogier, J. C., & Delacroix-Buchet, A., (2000). Protocol for the manufacture of miniature washed curd cheeses in controlled microbiological conditions. *International Dairy Journal, 10*(10), 733–737.

Hynes, E., Ogier, J. C., & Delacroix-Buchet, A., (2001). Proteolysis during ripening of miniature washed curd cheeses manufactured with different strains of starter bacteria and a *Lactobacillus plantarum* adjunct culture. *International Dairy Journal, 11*(8), 587–597.

Ismail, B., & Nielsen, S. S., (2010). Invited review: Plasmin protease in milk: Current knowledge and relevance to dairy industry. *Journal of Dairy Science, 93*, 4999–5009.

Klaenhammer, T. R., Barrangou, R., Buck, B. L., Azcarate-Peril, M. A., & Alterman, E., (2005). Genomic features of LAB affecting bioprocessing and health. *FEMS Microbiology Reviews, 29*, 391–409.

Kondo, J. K., & McKay, L. L., (1984). Plasmid transformation of *Streptococcus lactis* protoplasts: Optimization and use in molecular cloning. *Applied and Environmental Microbiology, 48*, 252–259.

Kongo, J. M., Ho, A. J., Malcata, F. X., & Wiedmann, M., (2007). Characterization of dominant LAB isolated from Sao Jorge cheese, using biochemical and ribotyping methods. *Journal of Applied Microbiology, 103*, 1838–1844.

Kranenburg, R., Kleerebezem, M., Vlieg, J. H., Ursing, B., Boekhorst, J., Smit, B. A., Ayad, E. H. E., Smit, G., & Siezen, R. J., (2002). Flavor formation from amino acids by LAB: Predictions from genome sequence analysis. *International Dairy Journal, 12*, 111–121.

Lane, C. N., Fox, P. F., Walsh, E. M., Folkertsma, B., & McSweeney, P. L. H., (1997). Effect of compositional and environmental factors on the growth of indigenous nonstarter lab in cheddar cheese. *Lait., 77*, 561–573.

Law, B. A., (2001). Controlled and accelerated cheese ripening: The research base for new technology. *International Dairy Journal, 11*, 383–398.

Leroy, F., & De Vuyst, L., (2004). LAB as functional starter cultures for the food fermentation industry. *Food Science and Technology, 15*, 67–78.

Liu, G., Wang, H., Griffiths, M. W., & Li, P., (2011). Heterologous extracellular production of enterocin P in *Lactococcus lactis* by a food-grade expression system. *European Food Research & Technology, 233*, 123–129.

Luchansky, J. B., Muriana, P. M., & Klaenhammer, T. R., (1988). Application of electroporation for transfer of plasmid DNA to *Lactobacillus, Lactococcus, Leuconostoc, Listeria, Pediococcus, Bacillus, Staphylococcus, Enterococcus, Propionibacterium. Molecular Microbiology, 2*, 637–646.

Martley, F. G., & Crow, V. L., (1993). Interaction between non-starter microorganisms during cheese manufacture and ripening. *International Dairy Journal, 3*, 461–483.

McHughen, A., (2000). *Pandora's Picnic Basket: The Potential and Hazards of Genetically Modified Foods I* (pp. 7–61). Oxford University Press, New York

McSweeney, P. L. H., & Sousa, M. J., (2000). Biochemical pathways for the production of Flavor compounds in cheeses during ripening: A review. *Lait., 80*, 293–324.

McSweeney, P. L. H., Fox, P. F., Lucey, J. A., Kordan, K. N., & Cogan, T. M., (1993). Contribution of the indigenous microflora to the maturation of cheddar cheese. *International Dairy Journal, 3*(7), 613–624.

Mooney, J. S., Fox, P. F., Healy, A., & Leaver, J., (1998). Identification of the principal water-soluble peptides in cheddar cheese. *International Dairy Journal, 8*, 813–818.

Mullan, W. M. A., (2016). *Microbiologyogy of Starter Cultures*. Available from: https://www.dairyscience.info/index.php/cheese-starters/49-cheese-starters.html (accessed on 6 September 2019). Revised 2004, 2005, 2006, 2007, 2008, 2011., (2013). Last revision, February 2015.

Naylor, J., & Sharpe, M. E., (1958). *Lactobacilli* in cheddar cheese. III. The source of *lactobacilli* in cheese. *Journal of Dairy Research, 25*, 431.

O'Sullivan, L., O'Connor, E. B., Ross, R. P., & Hill, C., (2006). Evaluation of live-culture-producing lacticin 3147 as a treatment for the control of *Listeria monocytogenes* on the surface of smear-ripened cheese. *Journal of Applied Microbiology, 100*, 135–143.

Parente, E., & Cogan, T. M., (2004). Starter cultures: General aspects. In: Fox, P. F., McSweeney, P. J. H., Cogan, T. M., & Guninee, T. P., (eds.), *General Chemistry, Physics and Microbiologyogy* (Vol. I, pp. 123–148). Amsterdam Elsevier.

Patnaik, R. S., Louie, S., Gavrilovic, V., Perry, K., Stemmer, W. P. C., Ryan, C. M., & Del Cardayre, S., (2002). Genome shuffling of *Lactobacillus* for improved acid tolerance. *Nature Biotechnology, 20*, 707–712.

Prabhakar, V., Kocaoglu-Vurma, N., Harper, J., & Rodriguez-Saona, L., (2011). Classification of swiss cheese starter and adjunct cultures using Fourier transform infrared microspectroscopy. *Journal of Dairy Science, 94*, 4374–4382.

Rank, T. C., Grappin, R., & Olson, N. F., (1985). Secondary proteolysis of cheese during ripening: A review. *Journal of Dairy Science, 68*, 801–805.

Ray, B., (1992). The need for food bio-preservation. In: Ray, B., & Daeschel, M., (eds.), *Food Bio-Preservatives of Microbial Origin Boca Raton* (pp. 1–23). Florida: CRC Press.

Sandine, W. E., (1996). Commercial production of dairy starter cultures. In: Cogan, T. M., & Accolas, P., (eds.), *Dairy Starter Cultures* (pp. 233–248). Wiley-VCH, New York.

Smit, G, Smit, B. A., & Engels, W. M., (2005). Flavor formation by LAB and biochemical flavor profiling of cheeses products. *FEMS Microbiology Reviews, 29*, 591–610.

Songisepp, E., Kullisaar, T., Hutt, P., Elias, P., Brilene, T., Zilmer, M., & Mikelsaar, M., (2004). A new probiotic cheese with antioxidative and antimicrobial activity. *Journal of Dairy Science, 87*, 2017–2023.

Steenson, L. R., & Klaenhammer, T. R., (1987). Conjugal transfer of plasmid DNA between *Streptococci* immobilized in calcium alginate gel beads. *Applied and Environmental Microbiology, 53*, 898–900.

Swanson, E. B., Couman, M. P., Brown, G. L., Patel, J. D., & Beversdorf, W. D., (1988). The characterization of herbicide-tolerant plants in *Brassica napus* L. after in vitro selection of microspores and protoplasts. *Plant Cell Reports, 2*, 83–87.

Thomas, T., (1987). Cannibalism among Bacteria Found in Cheese. *New Zealand Journal of Dairy Science & Technology, 22*(3), 215–220.

Turner, K. W., Lawrence, R. C., & Lelievre, J., (1986). A microbiological specification for milk for aseptic cheese-making. *N.Z. Journal of Dairy Science and Technology, 21*(3), 249–254.

Verachia, W., (2005). *Application of Pediococcus sp. as Adjunct Cultures in Gouda Cheese.* (pp. 5–90), PhD dissertation Thesis, University of Pretoria, South Africa.

Weerkam, A. H., Klijn, N., Neeter, R., & Smit, G., (1996). Properties of mesophilic LAB from raw milk and naturally fermented raw products. *Netherland Milk Dairy Journal, 50*, 319–322.

Wood, B. J. B., (1997). *Microbiology of Fermented Foods* (pp. 217–262). Blackie Academic and Professional, London.

Wouters, J. T. M., Ayad, E. H. E., Hugenholtz, J., & Smit, G., (2002). Microbes from raw milk for fermented dairy products. *International Dairy Journal, 12*, 91–109.

Yadav, J. S., Grover, S., & Batish, V. K. A., (1993). *Comprehensive Dairy Microbiology* (pp. 463–524). Metropolitan Publisher, New Delhi, India.

Yvon, M., & Rijnen, L., (2001). Cheese flavor formation by amino acid catabolism. *International Dairy Journal, 11*, 185–201.

CHAPTER 3

Advances in Designing Starter Cultures for the Dairy and Cheese-Making Industry and Protecting Them Against Bacteriophages

VASILICA BARBU,[1] CĂTĂLIN IANCU,[2] DANIELA BORDA,[1] and ANCA IOANA NICOLAU[1*]

[1]*Faculty of Food Science and Engineering, Dunarea de Jos University of Galati, 47, Domneasca Street, 800008 Galati, Romania*

[2]*Micreos Food Safety BV, Nieuwe Kanaal 7P, 6709 PA Wageningen*

[*]*Corresponding author. E-mail: anca.nicolau@ugal.ro*

3.1 INTRODUCTION

Starter cultures or just starters are cultures of microorganisms that are used in the production of foods in order to speed-up the fermentative processes and to provide specific features in a certain environment with a predictable pathway in comparison with spontaneous fermentation. ISO 27205:2010 (IDF 149: 2010) defines starters as "*prepared cultures that contain one or several strains of microorganisms at high counts, being added to bring about a desirable enzymatic reaction (e.g., fermentation of lactose resulting in acid production, degradation of lactic acid to propionic acid or other metabolic activities directly related to specific product properties*)," while Deutsche Forschungsgemeinschaft Senate Commission (i.e., German Research Foundation) on Food Safety defines them as "*preparations of live microorganisms or their resting forms, whose metabolic activity has desired effects in the fermentation substrate, the food*" (DFG, 2010).

Starter cultures applied in the milk and dairy industry can be grouped from different points of view. Taking into consideration the number of pure cultures participating in the starter, we can talk about monocultures or single cultures, the ones containing just one strain of microorganism, mixed or compound cultures, the ones containing several strains, usually two or three that are naturally found together, and multiple cultures, the ones containing several strains, which are not found associated in nature, but are put to work together by food microbiologists.

Considering the category of microorganisms involved, we can differentiate them in lactic acid bacteria (LAB) and non-LABs. LABs include sixteen genera, but just five of them are important for dairy: *Lactobacillus, Streptococcus, Lactococcus, Leuconostoc, and Enterococcus.* The LABs can be differentiated in homofermentative, the ones producing lactic acid as a primary product when fermenting glucose as *Lactococcus spp.*, yogurt strains (*Streptococcus salivarius subsp. thermophilus, Lb. acidophilus, Lactobacillus delbruckii subspecies bulgaricus*), and *Lb. helveticus*, and heterofermentative, the ones producing lactic acid from glucose, but also ethanol, acetic acid, and carbon dioxide. Gram-positive cocci such as heterofermentative LAB consist of *Leuconostoc spp.*, and Gram-positive bacilli such as *Lactobacillus fermentum, Lb. brevis, Lb. buchneri* and *Lb. reuteri.* Certain strains of *Lactobacillus* (*Lb. casei, Lb. plantarum* and *Lb. curvatus)* are facultative heterofermentative species, able to produce CO_2 and other secondary metabolites in specific conditions. Non-LABs are bacteria from the genera *Propionibacterium, Bifidobacterium, Brevibacterium,* or molds from the *Penicillium* genera. A particular starter type is represented by probiotic cultures, certain cultures of either LABs or non-LABs that are bringing benefits for health when introduced in organism. The probiotics are represented by *Bifidobacterium bifidum, B. lactis, B. infantis, B. adolescentis, B. breve, B. longum, Enterococcus faecium, Lactobacillus acidophilus, Lb. rhamnosus, Lb. casei/paracasei, Lb. fermentum, Pediococcus acidilactici.*

From the point of the optimum temperature for producing lactic acid, starters can be mesophilic, which are acting better at 30–37°C, as *Lactococcus lactis* subsp. *lactis, Lactococcus lactis* subsp. *lactis* biovar *diacetylactis, Lactococcus lactis* subsp. *cremoris, Leuconostoc mesenteroides* subsp. *cremoris, Leconostoc lactis* or thermotropic, acting better at 45–50°C, as *Streptococcus salivarius* subsp. *thermophilus (S. thermophilus), Lactobacillus delbrueckii* subsp. *bulgaricus, Lb. delbrueckii* subsp. *lactis, Lb. casei, Lb. helveticus, Lb. plantarum, Lb. fermentum.* In the dairy industry and dairy literature, these starters are named thermophilic, although they are not because they are not able to optimum develop at 70°C.

Based on how well we know the strains participating in a starter culture, we differentiate them in defined cultures (DSS – defined strain starter), the ones where each strain is known, and undefined or artisanal cultures, the ones deriving from concentrates of spontaneously fermented products. The defined starters currently applied in cheese making technology as adjunct cultures are strains of *Lactococcus lactis, Streptococcus thermophilus, Lactobacillus, Leuconostoc* with significant roles in cheese flavor formation during ripening (Spus et al., 2015; Mahony et al., 2016). Artisanal cultures are still in use, although over time, they may change their composition due to strains disappearance or mutation or their properties after phage attacks (DFG, 2010). Despite their unpredictable performance, artisanal cultures are attractive for both industry and scientific community and are still used extensively in the manufacture of some cheeses (e.g., Mozzarella, Gouda). When maintained in laboratory as bulk starters (usually frozen) their variability is greatly reduced over time, and this method of starter use does not guarantee constant dairy products composition. Although relatively poor researched, these natural starters are mixtures of thermophilic as well as mesophilic mixed cultures known to contain lactobacilli, leuconstocs, lactococci, and frequently streptococci (Mullan, 2014) and are the starting point to create DSSs. Artisanal cultures include natural milk starters obtained by milk incubation in conditions that favor development of thermotropic LABs, which are then used as starters, back-slopped starters, when a part of a previously obtained product is used in a new process as starter culture, and whey and buttermilk cultures, which are obtained by incubating cheese whey respectively buttermilk under favorable conditions for the growth of active (typically thermotropic) LABs (Powell et al., 2011).

Cheese starters can also be differentiated in primary and secondary starters. Primary starters are responsible for lactic acid production in cheeses but can also contribute at flavor development to some extent, while secondary starters, also known as adjunct cultures, are not contributing significantly at acid formation, but influence the product flavor and the CO_2 release or other characteristics that cheeses develop over time. Secondary starters can be added simultaneously with primary starters (into curd) or added later (spiked or pulverized onto the surface).

Having in view the nature of LABs, they can be differentiated in non-genetically and genetically modified starters depicted in Table 3.1 (Geisen and Holzapfel, 1996; De Moreno de LeBlanc et al., 2015). Although paradoxically named non-genetically modified starters, this category includes not only natural LAB cultures, but also starters that have been genetically altered either uncontrolled or controlled. Natural LABs include the bacteria that are belonging to milk microflora and originate from plants and the human or animal intestinal tract. Such strains often suffer mutations caused by natural events. Examples of such events are sequence elements insertion (Visser et al., 2004), incorrect replication or transcription of DNA, radiation, and other factors (Sybesma, 2006). This explains why it is possible to select strains with improved fermentation characteristics, when screening of natural isolates of LABs. Mutant LABs belong to the category of uncontrolled genetically altered LABs and are obtained by artificially increasing the frequency of mutations through the LABs exposure to mutagenic conditions (*N*-methyl-*N*'-nitro-*N*-nitrosoguanidine (NNG), ethyl methyl sulfonate or UV light (Sybesma, 2006). Due to the difficulty in identifying the sites responsible for mutation, the applications of this technique decreased drastically since the apparition of genetic engineering, but seems to revive due to the progresses made by whole genome sequencing (WGS) (Bose, 2016), which makes easier the identification process of a LAB with desired traits for application in dairy industry. The uncontrolled genetic modified LABs are also obtained, making usage of natural gene transfer methods such as transduction, conjugation, and transformation. Controlled genetically modified LABs are obtained through genetic engineering techniques and are very controversial. Only those obtained by non-self cloning methods are considered to be GM LABs (de Moreno de LeBlanc et al., 2015).

Although at the moment a large variety of lactic starter cultures are commercially available, food microbiologists and biotechnologists continue to develop LAB strains with improved properties. The main reasons for doing this are the vulnerability of LABs against phages attack, the growing interest for functional foods with LABs as part of the daily diet in order to promote safe and healthy dietary habits and the constant changes in consumer preferences. The LAB properties that are considered relevant for to be upgraded: improved bacteriophage resistance, improved texture or flavor formation, increased stress tolerance, production of biologically active peptides as nutraceuticals and functional food ingredients.

TABLE 3.1 Categories of LAB Differentiated Based on Genetic Modifications

Type of LAB	LAB Category	Method Used for DNA Modification	Approved for Limited Use 90/219/EC	Approved for Deliberately Release 2001/18/EC
Non-genetically modified	Natural	Free mutations	Yes	Yes
	Uncontrolled genetically modified (random mutagenesis)	Induced mutations	Yes	Yes
		Insertional mutations	Yes	Yes
		Transduction	Yes	Yes
		Conjugation	Yes	Yes
	Controlled genetically modified	Self-cloning	Yes	No
Genetically modified	Controlled genetically modified	Non-self cloning	No	No

The aim of the chapter entitled "Advances in designing starter cultures for dairy and cheese-making industry and protecting them against bacteriophages" is on the one hand to present a method known as genome editing to obtain starter cultures with improved properties and on the other hand to explain how is possible to protect dairy starter cultures without using intervention at genome level. The genome-editing method, which makes use of the CRISPR-Cas system, generates strains considered to be non-genetically modified organisms, for which consumer acceptance and regulatory approval are easier obtained.

3.2 LAB DEVELOPMENT BASED ON CRISPR-CAS SYSTEM

Clustered regularly interspaced short palindromic repeats (CRISPR; acronym proposed by Francisco Mojica, in 2001) and *CRISPR-associated proteins* (Cas) represent a flexible immune frame in archaea and bacteria that provide protection versus foreign DNA originated from plasmids, phages or other mobile genetic elements (MGE). The capacity to accomplish sharpness genome engineering (using CRISP and Cas proteins) in LAB strains important for medicine or food industry will allow to increase the genetic and physiological potential of strains without compromising safety (van Pijkeren and Britton, 2014). Combined with other arising edge technologies, CRISPR-Cas technologies will be able to design new bacterial strains with improved phenotypes, safely by avoiding undesirable or unintended mutations. Accordingly, CRISPR-based practices are ready to bring major changes in many areas across food science, from farm to fork.

CRISPRs are interesting frames, which conceal complex biological mechanisms and could enlighten certain aspects of evolution. The CRISPR display consists of a leader sequence (range 300 to 500 base pairs) rich in AT, continued by short repeated sequences that are separated by unique sequences known as spacers (Jansen et al., 2002; Hille and Charpentier, 2016). These immune systems are estimated to occur in less than 50% of bacteria and in over 80% of archaea (Briner, 2015; Makarova et al., 2015). CRISPR repeats usually contain between 28 and 37 base pairs (bps), but may sometimes be less than 23 bps or higher than 55 bpb (Barrangou and Marraffini, 2014). Some of them are directed repeats, while others are inverted repeats (palindromes) with dyad symmetry, which cause the secondary configuration such as hairpin loop in the RNA. The unique spacers usually contain between 32 and 38 bps (but sometimes it can comprise between 21 and 72 bps). A CRISPER array often contains up to 50 repeat-spacer units (Barrangou and Marraffini, 2014). The spacers are originating from invaders MGEs like conjugative plasmids or phage genomes, being archives of previous infections. Foreign DNA sequences that match the spacers are known as protospacers. Next, to the CRISP cassette are located *cas* genes which codify the Cas proteins. Until now, 93 *cas* genes were characterized (Makarova et al., 2015). All CRISPR-Cas arrays hold the universal *cas1* gene. The newest classification of the CRISP-Cas systems, based on "signature gene," brings together 19 subtypes in 6 types and two classes (Westra et al., 2016). Besides the I-III systems, already known, the types IV–VI has been discovered lately (Wright et al., 2016). Class 1 system (I, III, and IV types) hold multi-subunit effector protein complexes which are involved in interference, while the type II, V, and VI systems (Class 2) possess a single-subunit effector (single large class protein), as can see in Table 3.2.

TABLE 3.2 General Features of the CRISPR-Cas Systems

CRISP-Cas Systems Type	Signature Gene	Protein Complexes Involved in Interference (Effectors)	Genes Involved in crRNA Biogenesis	Genes Involved in Acquisition/ Adaptation
I	Cas3	Cas3, Cas8, Csa5/Cse2	Cas6	Cas1, Cas2, Cas4
II	Cas9	Cas9	tracrRNA	Cas1, Cas2, Cas4/Csn2
III	Cas10	Cas10, Csm2/Cmr5, Csm3/Cmr4, Csm4/ Cmr3 Csm5/Cmr1	Cas6	Cas1, Cas2
IV	Csf1	Csf1, Csf2, Csf3	-	-
V	Cpf1, C2c1, C2c3	Cpf1	-	Cas1, Cas2, Cas4
VI	C2c2	C2c2	-	Cas1, Cas2,

Cas proteins are divided into four functional categories:

1. Proteins with nuclease and recombinase activities for acquisition and integration of new spacer sequences derived from invader DNA, such as Cas1 and Cas2;
2. Ribonucleases for processing CRISPR RNAs (crRNAs);
3. Proteins that assemble into crRNP complexes that include cascade (*CRISPR-associated complex for antiviral defense*), Cas9, Cmr, and Csm; and
4. Nucleases for degradation of foreign MGEs include Cas3 and Cas9 (Barrangou and Marraffini, 2014; Westra et al., 2016).

Bringing together simultaneously several criteria, CRISPR-Cas systems are classified from functional point of view in three categories. Type I has a *cas*3 signature gene, but always contain *cas*1, *cas*2 and *cas*3 genes, possess a palindromic repeat that form hairpin loops in crRNA, employ a large multi-protein complex (Cascade) and content seven subtypes (I-A, I-B, I-C, I-D, I-E, I-F, and I-U).

Type II CRISPR-Cas contain several *cas* genes (*cas*9, *cas*1, *cas*2, *csn*2 and *cas*4) and perform foreign DNA targeting and cleavage through only one protein, Cas9 (it has *cas*9 signature gene), and a supplementary RNA known as trans-activating CRISPR RNA (tracrRNA). Type II systems (II-A to II-C) are the least common (represented in ~ 5% of bacterial genomes) and are absent in archaea. Type II is prevalent among human pathogens and commensals like LAB and bifidobacteria (Chylinski et al., 2013) but also in seawater, soil, plant material, food or fermentation environments and even in harsh environments such as Antarctic ice, hot springs and deep-sea sediments (Fonfara et al., 2014).

Type III CRISPR-Cas systems are primarily found in archaea and, while phylogenetically different, are very similar in the structure and function. It has a *cas* 10 gene signature, but employs a *csm* in III-A or *cmr* in III-B, to target and cleave MGEs when guided by a crRNA. Type IIIB systems target RNA and not dsDNA (like all others systems). Diverse subtypes of Cas systems like Csa, Csh, Csm, and Cst are frequent in archaea, but also are found in bacteria. Csa is an exception because it is only found in the Archaea. Cas gene systems are disseminated by horizontal transfer; therefore, tight related species may have different Cas gene combinations, and highly divergent species can have very similar Cas systems (Terns and Terns, 2011).

The immune response has three stages:

1. *Acquisition* of novel spacer sequences originated from foreign DNA;
2. *Expression* and biogenesis of small CRISPR RNAs (crRNAs) that lead Cas proteins in the next step; and
3. *Interference* stage to destroy invaders.

The acquisition of a novel unique spacer in the CRISP array is carried out in a directional manner, preferential between the leader sequence and the first repeat sequence (Deveau et al., 2008; Erdmann and Garrett, 2012; Swarts et al., 2012). The Cas1 and Cas2 proteins are always liable for this step and any mutation in their encoding genes, stops new spacer acquisition (Nuñez et al., 2014). Team of Nuñez studies the Cas1–Cas2 complex (I-E subtype) from *E. coli* and reveals a 2:1 stoichiometry in which two Cas1 dimers binds a Cas2 dimer to form a crab-like design that indicates the site of integration of the new spacer at the leader end. These two proteins possess nuclease and recombinase activities, Cas2 play a non-enzymatic scaffolding role, binding double-stranded sequences of invading DNA, while Cas1 binds the single-stranded ends of the foreign DNA and catalyzes their integration (Nuñez et al., 2014; Wang et al., 2015). Cas1 acts as an integrase and causes a staggered cut across the leader-proximal repeat; the new double-stranded spacer DNA is ligated between the two single-stranded repeats (Briner, 2015). After DNA repair by the synthesis of the complementary strands, the new spacer is fully incorporated into the array between two repeats.

For the type I and II systems, the acquisition starts when the Cas proteins identify the protospacer-adjacent motif (PAM) that borders on one edge of a protospacer (PAMs are not important for the type III systems during acquisition). PAM contains approximately 2 to 4 nucleotides (Deveau et al., 2008). The last nucleotide from PAM, adjacent to the first nucleotide of the protospacer is strongly conserved and becomes the last base in the first direct repeat. A spacer is not adjacent to a PAM, therefore preventing the CRISPR-Cas systems of the cell from self-targeting and self-cutting their own host chromosome (Swarts et al., 2012). In Type I CRISPR-Cas systems, the PAM sequence is usually placed on the 5' end of the protospacer, while the PAM in Type II systems is located on the 3' flank (Deveau et al., 2008).

3.2.1 EXPRESSION/BIOGENESIS

The whole CRISPR cassette is co-transcribed into a precursor (a single long RNA transcript known as pre-crRNA). The mechanisms that transform the precursor into crRNAs differ among CRISPR-Cas systems. In Type I system, Cas6e/Cas6f proteins recognize stem-loops array, cleave the repeat-hairpins at the junction of ssRNA and dsRNA forming small interfering crRNAs with an entire spacer sequence bordered by eight nucleotides of the preceding repeat on the 5' end and followed by the remaining sequence of the downstream repeat on the 3'end (Li, 2015). Only in case of I-C subsystem, the responsible protein is Cas5d, but the mechanism is similar. The Cas5d protein can perform multiple roles in CRISPR-mediated immunity: possesses a specific endoribonuclease activity for CRISPR RNAs and nonspecific double-stranded DNA binding affinity (Koo et al., 2013). Type III systems use a Cas6 homolog, but it is possible that the Csm/Cmr complex to be involved. Functional type II systems encode an additional extra small RNA, known as a trans-activating crRNA (tracrRNA). This RNA is complementary to the repeat sequences from the primary CRISPR transcript and form together with a double-stranded RNA (Carte et al., 2014). This complementary region forms three structural modules that are conserved in tracrRNA: crRNA duplexes and are important for molecular stability and binding ability with Cas9: the upper stem, bulge, and lower stem. The first hairpin design in the tracrRNA is named the nexus. Lactobacilli usually contain a double-stemmed hairpin with a symmetrical, round bulge in the middle of the hairpin (Briner, 2015). This conformation allows RNase III (assisted by the Cas9 protein) to cleave the complex. The mature crRNAs do not possess the entire spacer, which is instead cut short at one end (only 19–22 nucleotides of a CRISPR spacer are kept in a mature crRNA) (Karvelis, 2013; Carte et al., 2014; Gasiunas et al., 2012).

3.2.2 INTERFERENCE

In the interference step, a complex of effector proteins uses the crRNA molecule to identify, target, and destroy any phage or plasmid which holds complementary sequence to the spacer sequence of the crRNA. Any MGE that penetrates the cell becomes the target recognized as non-self by the crRNP (effector complex). Many researchers have clarified the functionality and structures of couple complexes of CRISPR Cas systems. The target binding by means of crRNA is common to all three types and the mechanisms of target cleavage are very different. In type I system the crRNA leaded by the Cascade complex joins the DNA target sequence (protospacer) and then uses Cas3 to degrade the target. Cascade complex is composed of the crRNA, bound at both ends by Cas5 and Cas6. Along the crRNA are attached numerous Cas7 subunits and, moreover, a large subunit: Cas8, Cas10, Cse1 or Csy1, and sometimes small subunits: Cse2 or Csa5. In *E. coli* are five proteins: Cse1:Cse2:Cas5:Cas6:Cas7 in stoichiometric ratio 1:2:1:1:6 (Wright et al., 2016). The PAM sequence from foreign DNA is recognized by the large subunit of the Cascade complex and begin unwinding of the target DNA and annealing to the crRNA. The Cas3 protein cuts and then translocates throughout the displaced strand and gradually degrades it (Westra et al., 2016). The type III system has a similar multi-protein subunits effector based on Cmr/Csm complexes. The both ends of crRNA are bound by Cmr1/Csm5 and Cmr3/Csm4. The pillar of the complex is Csm3/Cmr4, the large subunit is Cas10, and the small subunit are Csm2/Cmr5. Cas10 catalyzes the split of DNA targets, while the backbone subunit provides for cleavage of the RNA targets at every sixth base, which is unpaired with the crRNA (Wright et al., 2016).

When foreign DNA enters a cell with a Type II CRISPR-Cas system, Cas9 first binds any potential PAM, looking for complementarity between the spacer and protospacer (Sternberg et al., 2014). If the PAM is not present, Cas9 will not bind the target thus preventing cleavage (Chen and Huang, 2014). Cas9 protein has two subunits: one of them has the nuclease role and the other is α-helical or REC lobe (Anders et al., 2014; Nishimasu et al., 2014). The nuclease subunit holds three areas: the HNH domain, the RuvC-like nuclease domain and the PAM-interacting domain. The 3' hairpins end of the tracrRNA join the nuclease subunit, and the stem loop and the spacer attaches to the groove between the two subunits. Attachment to a suitable PAM induces the HNH domain to cleave the hanging strand (target), while the displaced strand (non-target) is load into the RuvC active site for splitting (Anders et al., 2014; Wright et al., 2016).

LAB is a very heterogenic group (31 genera) from genetically and ecologically point of view, but at the same time also represents the largest cluster of Gram-positive bacteria used in industry, medicine, and pharmacy (Oh and Van Pijkeren, 2014). Over recent decades it has been proved that many strains of LAB, mostly members of the genus *Lactobacillus*, have healthy properties, which makes them suitable vectors for the development of biotherapeutic or industrial starter strains. The applications of CRISPR-based technologies in food science include genome editing, handling of microbial consortia, immunization against phages.

The 16S rDNA sequencing is quick and inexpensive for bacterial gross identification (genus and species), but sometimes, can be fickle for applications requiring high resolution (subspecies level or strains). Genome mapping *via* CRISPR-Cas array represents the latest technology (fast and affordable) for efficient functional genomics, up to the level of bacterial strains. Till now, CRISPR-based genotyping has been applied to fermentation starters such as *S. thermophilus* (Deveau et al., 2008), to probiotic bacteria such as *Lactobacillus casei* (Broadbent et al., 2012) or *Bifidobacterium* (Briner et al., 2015) and microorganisms that induce damages in foods such as *Lactobacillus buchneri* (Briner, 2015), or to foodborne pathogens such as *Salmonella* (Shariat et al., 2013) and *Escherichia coli* (Yin et al., 2013).

The classic phylogenetic analyze of the LAB genomes, based on 16S rDNA sequencing, compare the ancestral gene sets content across genera and species. Under the environmental pressure, a series of events were combined, such as an extensive gene loss accompanied by selective gene acquisitions through duplication and horizontal gene transfer which have resulted in the evolutionary adaptation of different groups in their various environmental niches. Horvath et al. (2009) identified eight distinct LAB CRISPR families (Blon1, Efam1, Ldbu1, Lhel1, Lsal1, Sthe1, Sthe2, and Sthe3) which do not classify suitable to phylogenetic kindships. About 36 bps long CRISPR repeats are present with accuracy in five of these families: Blon1, Efam1, Lsal1, Sthe1, Sthe3, so can be considered to have a common origin. The families Lhel1 (32–37 bps repeats), Sthe2 (36–37 bps repeats), and Ldbu1 (28–30 bps repeats) are more divergent (Horvath et al., 2009). The researchers highlight very similar CRISPR loci in relatively distant taxa. For example, in the Ldbu1 family is grouped four relatively distant genera: *Atopobium*, *Bifidobacterium*, *Lactobacillus*, and *Symbiobacterium*. Lhel1 family also contains members of the *Bifidobacterium*, *Lactobacillus*, *Streptococcus*, *Symbiobacterium* genera (Horvath et al., 2009). Contrariwise, *L. acidophilus* which belongs to Ldbu1 family and *L. helveticus* which belong to Lhel1 family though, based on 16S ribosomal DNA, are very similar species from phylogenetic standpoint (Ennahar et al., 2003). Their CRISPR arrays are very polymorphic, and the only Lhel1 contains a Cas gene set (Horvath et al., 2009). *Bifidobacterium* species are grouped in three CRISPR families: Blon1, Ldbu1, Lhel1, and *Lactobacillus* CRISPR loci were clustered in Ldbu1, Lhel1, Lsal1 CRISPR families (Horvath et al., 2009). This polymorphism of CRISPR pattern in LAB probably reflects their origin in lateral genomic clusters and their fast evolution due to adaptation at the environmental pressure, mainly correlated to MGEs (Barrangou and Marraffini, 2014).

CRISPR technology, by efficient editing, helps the timely improvement of microbial strains and will allow a deeper understanding of how health-promoting bacteria induce their effects. In the food industry, predatory bacteriophages (with a high rate of mutation and a higher capability for adjustment to the host cell) represent a significant menace to starters preservation and continue to be a major reason of loss or changeable quality in dairy. The producers seek and apply expensive and time-consuming control strategies such as multi-strain starter ratio, starters rotation, steam sterilization of manufacturing equipment or growth in the presence of chelating agents, etc. (Selle and Barrangou, 2015). In order to combat constant phage-attack, LAB has developed many phage-defense strategies such as: inhibiting absorption and DNA injection, restriction-modification systems (R-M systems), abortive infection and CRISPR-mediated phage defenses, but the last one is the only *adaptive* viral defense system that has the ability to capture resistance to specific, novel viral sequences (Barrangou and Horvath, 2012; Van Houte et al., 2016). Many fermentative industries have been able to exploit the natural immune function of the CRISPR system to confer resistance against undesirable DNA sequences including phages or plasmids carrying antibiotic resistance genes (Deveau et al., 2008; Terns and Terns, 2011; Gasiunas et al., 2012). In this way, by using CRISPR-Cas immune system, dairy, and biofermentation processors can relatively easy produce phage-resistant starter cultures and reduce loss caused by lytic phage infection (Briner, 2015). By mutagenesis experiments Barrangou demonstrated for the first time, in 2007, the existence of a direct correlation between the spacers nucleotide sequences of CRISP array and the phages genotype for which *Streptococcus* strains were resistant (correlated to its widespread use as a starter culture in yogurt and cheese fermentations) (Barrangou et al., 2007). The research undertaken by the laboratory headed by Mr. Barrangou has made important contributions regarding phylogenetic significance of CRISPR-Cas genetic locus. Because the newest spacers are always located at the leader end of the CRISPR pattern and the ancestral spacers are anchored at the distal end means that identical pattern is a sign of common ancestry between strains and the rare spacer sequences closer to the leader end of the array are a sign of strain divergence and evolution (Barrangou et al., 2007; Deveau et al., 2008). Currently, comparative genetic studies using CRISPR tool can clarify many controversial phylogenetic issues, but also epidemiological or ecological.

Lactobacillus buchneri is a relevant strain of acid- and halo-tolerant LAB with importance for ensilage, the fermentation of grains into animal fodder, for ethanol production and for fermentation of vegetables (as a common contaminant). Briner and Barrangou (2014) have studied the CRISPR diversity of *Lactobacillus buchneri,* and they identified a new PAM (5'-AAAA-3') which is implicated in new spacers acquirement and Cas9-mediated sequence-specific splitting of target. *Lactobacillus buchneri* genome contains a Type II-A CRISPR array with a 36 nt repeats (5'- GTTTTAGAAGGATGTTAAAT-CAATAAGGTTAAACCC-3') which interleaves with 9–29 variable spacers. The phylogenetic analyses of CRISPR-Cas systems in twenty-six *L. buchneri* isolates and closely related species *(Lactobacillus brevis* subsp. *gravensis* ATCC 27305, *Lactobacillus pentosus* KCA1, *Lactobacillus salivarius* UCC118 and *Pediococcus acidilactici* DSM 20284) showed that a cluster of widely conserved spacers at the ancestral end (positions 1 through 8) proves the common origin, and the leader-end polymorphism reflects a recent divergence (Briner and Barrangou, 2014). This system encodes the universal genes: Type II signature *cas9*, together with *cas1* and *cas2*, as well as *csn2*, which is own specific to Type II-A subtypes (Makarova et al., 2015).

S. thermophilus is endowed with two dynamic Type II-A CRISPR-Cas systems to protect versus the invader, foreign DNA: CRISPR1 (ubiquitous) and CRISPR3 (less frequent). Comparative genome analyses of certain S. *thermophilus* strains pointed out high similarity between those two systems (Sun et al., 2011, Selle, and Barrangou, 2015) but the CRISPR1 locus has 32 spacers compared to 12 spacers in the CRISPR3 array, which ensures a better adaptation for the CRISPR1-Cas system (Deveau et al., 2008; Horvath et al., 2009; Magadán et al., 2012). The acquisition of novel spacers is even required for acquiring immunological memory and continuously updating the spacers library by interference of MGEs is crucial for increasing the adaptability of strains. It appears that *S. thermophilus* genome is an evolution result by a combination of gene wane as well as horizontal gene transfer acquisition that concurred to its adaptation to the dairy environment, which is often contaminated with various consortia of virulent phages (Sun et al., 2011). For the CRISPR-Cas Type II-A system from *S. thermophilus* DGCC7710 PAMs sequences are 5'-NNAGAAW-3' for CRISPR1 and 5'-NGGNG-3' for CRISPR3 (Deveau et al., 2008; Magadán et al., 2012).

The Type I-E CRISPR-Cas system (present in multiple *Lactobacillus brevis* genomes) is recognized through an extremely conserved 28 nucleotides CRISPR repeat sequence and includes the universal genes: *cas1* and *cas2*, *cas3* signature gene, and Cascade and *cas6*, with a changeable number of spacers (Makarova et al., 2015).

The recent comparative genomic researches show that the genus *Bifidobacterium* includes 47 taxa (38 species and 9 subspecies) well-known thru their potential to positively affect human host health. The *Bifidobacterium* genome size varies between 1.73 Mb (*Bifidobacterium indicum*) to 3.25 Mb (*Bifidobacterium biavatii*), equivalent to 1,352 and 2,557 protein-encoding open reading frames (ORF), respectively (Milani et al., 2014). Adapting to a specific ecological niche (gastrointestinal tract) led the evolution of the *Bifidobacterium* genus towards genome decay by loosing of nonessential genes from common core ancestors and by acquiring new genes from other genres commensal in intestinal tract such as *Actinobacteria* class (28.5%), *Bacillus* (11.7%), *Gammaproteobacteria* (8.7%), *Clostridium* (8.7%), and *Alphaproteobacteria* (5.9%) (Milani et al., 2014). The same study shows that 35 from those 47 taxa analyzed (75%) possess CRISPR-Cas systems type I, II, and III. Briner and her team analyzed 48 *Bifidobacterium* genomes and identified 77% frequency of CRISPR-Cas systems: I-C, I-E, I-U, II-A, and II-C (type III systems seem to be absent). Certain strains can combine several distinct CRISPR-Cas systems, like *Bifidobacterium dentium* with three separate systems (Briner et al., 2015). The number of repeats between 4 and 172 (with 24–36nt each) denotes a huge diversity of CRISPR-Cas pattern. Some Type I-E systems encoding up to 171 unique spacers, type I-C systems may have up to 155 spacers and type I-U systems may hold a range of 6 to 52 spacers. Each unique spacer is gained from invaders MGEs like phages and plasmids, which means that Type I-E systems are the most efficient in genetic immunization events. Ten CRISPR spacers from *B. angulatum* (I-E) fit with *B. antenulatum* prophages, *B. dentium* (II-C) has six spacers that matches with *B. moukalabense* prophages, and *B. minimum* (I-E) has nine spacer sequences that target *B. mongoliense*. The explanation could be the co-evolution of these bacteria which inhabiting same ecological niches with a similar collection of phages (Briner et al., 2015).

Lactobacillus casei strains show a remarkable ecological adaptability and can be spread in diverse habitats: raw and fermented dairy (especially in cheese, as a "nonstarter" species to intensify flavor during ripening), fermented plant materials (pickles, silage, wine, and kimchi), or in gut and oral cavity of animals and humans (Broadbent et al., 2012). Comparative genome hybridization techniques suggested that their metabolic flexibility is especially based on gene decay and acquisition through horizontal gene transfer. Two distinct types of highly conserved (96–97%) CRISPR loci were identified (with CRISPRFinder web service tool) in the *L. casei* genomes: Type II-A with 5'-GTCTCAGGTAGATGTC-GAATCAATCAGTTCAAGAGC-3' repeats (in Lsal1 group) and Type I-E with 5'-GTTTTCCCCGCACATGCGGGGGTGATCCC-3' repeats (in Ldbu1 group) which have been recently identified in many species of lactobacilli (*Lb. casei, Lb. salivarius* and *Lb. rhamnosus* for Lsal1, and *Lb. acidophilus*, *Lb. brevis*, *Lb. delbrueckii* and *Lb. fermentum* for Ldbu1) (Horvath et al., 2009). Lsal1 family hold between 4–44 spacers, Ldbu1 group possess between 21 and 60 spacer sequences and many spacers are homologous to *Lactobacillus* bacteriophages (A2, Lc-Nu, Lrm1, and J1) or plasmids (pREN, pLgLA39 and pYIT356) (Broadbent et al., 2012). The same authors mention the presence of TGAAA proto-spacer associated motifs immediately after of the Lsal1 protospacers and AAY immediately upstream of the Ldbu1 protospacers.

The uptake and dissemination of antibiotic resistance genes in bacteria is an ongoing challenge for the medical community in treating pathogenic multi-resistance strains or in selection of probiotic strains and starter cultures. Antibiotic resistance genes can be genomically encoded or by MGEs like bacteriophages, plasmids, and transposable elements, all of which can be targeted by CRISPR-Cas systems. Thus, it is a natural way to vaccinate food-grade bacteria against transmissible antibiotic resistance genes, which can be achieved through incorporating a corresponding spacer sequence (Selle et al., 2015).

Besides the active role in immune function, CRISPR-Cas systems seem to have a regulatory role of prokaryotic cell metabolism in response to stress signals, aspect which has also been explored but is not yet fully elucidated and requires further study (Westra et al., 2016). Recent studies reported on the lethal autoimmune effect of CRISPR-Cas self-targeting in bacteria, which is harmful, can disturb the genome homeostasis by causing DNA damage, and may have repercussions for bacterial evolution and for their virulence or pathogenicity. This controlled genomic change can be designed for the targeted deletion of large regions of bacterial nucleoid, including the rearrangement or deletion of entire pathogenicity ORFs (Vercoe et al., 2013; Selle and Barrangou, 2015). Using this tool, Selle, and coworkers have achieved *S. thermophilus* transformants with 102 kbp deleted segment (approximately 5.5% of the entire nucleoid). The lost island encodes housekeeping genes such as: lactose catabolism genes, ABC transporters, regulatory genes, bacteriocin genes, or phage related proteins and several silent DNA regions with unknown function (Selle et al., 2015). The *lacZ* deficient *S. thermophilus* strains lost β-galactosidase activity. The same researcher obtained by CRISPR-Cas self-targeting technology a *Lb. acidophilus* transformant with lipoteichoic acid (LTA) deficiency, an important immunostimulant among Gram-positive bacteria. The lack of teichoic acid from the cell envelope of lactobacilli has pleiotropic effects such as an elongated cellular morphology and produce high inhibition in the presence of increased manganese concentrations because with deletion of teichoic acid encoding genes are simultaneously removed and the operon that encoding heavy metal resistance and several peptidoglycan hydrolases (Lightfoot et al., 2015).

3.3 NEW APPROACHES THAT MIGHT BE EMPLOYED TO PREVENT PHAGE INFECTION OF LAB

One of the main concerns with respect to the use of these bacteria in the dairy industry is their susceptibility to bacteriophage that is naturally occurring in the milk processing plants. The constant phage pressure has already opened a new field of research, about 30 years ago, which is addressing specifically the phage-host interaction in order to acquire a better understanding of the mechanisms involved in defending bacterial cells against phage predation (Mahony et al., 2014). Finding solutions to protect the dairy cultures in the industrial processes is still a high priority since viral contamination can cause to fermentation inefficiency or complete fermentation fails (McDonnell et al., 2016).

In general, the natural bacterial defense against phage predation includes different antiviral mechanisms such as locking of phage receptors, biosynthesis of extracellular matrix or competitive inhibitors, preventing DNA entry in the cell, cutting of phage nucleic acids (i.e., restriction-modification system and CRISPR-Cas system) and abortive infection systems (Ogata et al., 2000; Labrie et al., 2010). Only a few of these were investigated with respect to their potential on protecting starter cultures against viral infections. Among these, systems that prevent phage DNA injection called superinfection exclusion (Sie) systems were identified on *Lactococcus lactis* and *Streptococcus thermophilus* (Garvey et al., 1996; Mahony et al., 2008). Interestingly one of the Sie-like systems identified in a prophage of *Streptococcus thermophilus* confers resistance to certain lactococcal phages when transformed into *L. lactis* (Sun et al., 2006). CRISPR-Cas system was identified in LAB starter strains as well, and it is seen as a promising tool than would be employed for creating phage-resistant starters. However, it has been shown that *S. thermophilus*-specific phages can avoid the resistance provided by this system (Deveau et al., 2008). This proves that continuous co-evolution of phage and bacteria will be always a hurdle in the efforts of developing starters that can permanently resist to phage predation.

In this section, we will focus on new approaches that might be employed to prevent phage infection of LAB. These strategies, which were recently proposed, are not based on bioengineering of bacterial strains using recombinant DNA methods. This is mainly due to the fact that the recombinant DNA is still making these approaches not suitable for industrial implementation as the current food regulations, labeling issues and consumer acceptance are still a burden (Samson and Moineau, 2013). We aim to offer a practical insight, rather than a fundamental one on newly proposed methods that have the potential to revolutionize the design and development of phage resistant starters and strategies to prevent phage infections in fermentation processes in the next years. Two of the most recently studied strategies for prevention of phage infection in dairy and cheese-making processes are addressed here.

3.3.1 PROTECTING STARTER CULTURES BASED ON PEPTIDES

Thus far, two types of peptides have shown potential applicability in protecting starter cultures from phage attack: antimicrobial peptides (bacteriocins) and peptides from phage proteins hydrolysates (Brown et al., 2017). An interesting method for generating phage resistant LAB was proposed recently where *L. lactis* resistant to phage c2 could be isolated after few growth passages in a media containing a cell wall active antimicrobial peptide Lactococcin 972 (Lcn972). The bacteriocin is inducing genetic changing within the bacterial chromosome, including the partial deletion of *the pip* gene encoding the phage infection protein, which acts as the phage receptor protein on the bacterial cell wall. This was in line with previous studies where it has been shown that mutated variant of this protein is correlated with resistance to c2 (Mooney et al., 2006). In addition, the same *L. lactis* was less efficiently infected by phage sk1 even though the mechanism of lower infectivity of this phage is still unknown (Roces et al., 2012).

A different approach to prevent phage infections during dairy fermentation and/or cheese ripening was proposed recently, and it is based on using phage coat proteins hydrolysates in order to compete with viable phages for the attachment to the cell wall. Briefly, hydrolysates containing peptides were prepared from structural proteins of *Lactococcus lactis ssp. lactis* c2 phage and used to inhibited c2 proliferation in *L. lactis ssp. lactis* C2 cell host. However, there was a transient inhibition although the growing time previous to lysis was increased with 54 minutes when the culture was infected with 1 x 10^7 PFU/mL of c2 phage (Hicks et al., 2004). In this study, it was also suggested that ficin might be the hydrolase of choice due its low price, commercial availability and the fact that its hydrolysates were very efficient in prolonging the growth of C2 lactic culture in L-M17 medium and milk when infected with c2 phage. Importantly, the C2 culture was protected from phage infection through the ripening periods when bulk C2 starter was prepared in a medium comprising the c2 phage peptides. This consequence was observed when the c2 phage proteins hydrolysate was used only, while the culture media (M-17 Broth by Sigma-Aldrich) hydrolysate was not effective on preventing the phage infection.

3.3.2 BIOCIDES AND FUGICIDAL PROCESS TO PROTECT STARTER CULTURES

The activity of peroxy agents against the virulent P001 bacteriophage infecting LAB was evaluated. While potassium monopersulfate (MPS), phagicidal concentrations varied from 0.006 to 0.012%, chlorine was effective against P001 at either 0.05 or 0.125% in the suspension test or at 0.12–0.5% using the surface test (Morin et al., 2015). Peracetic acid products (PAP) were also highly effective against the *L. lactis* specific phage, only 0.005% leading to complete eradication as reported in the aforementioned study. The effectiveness of 14 commercially available sanitizers on inactivation of 36 phages from 936 groups was also evaluated in a more recent and extensive work (Hayes et al., 2017). The results of this study are in agreement with a previous one, and both are recommending sodium hydroxide as a secure sanitizing agent for the dairy industry for phage infection prevention (Murphy et al., 2014). Apparently, phage has wide variability in resistance to biocidal activity, this issue resulting from evaluation of biocides impact on phage structure and integrity, proteome corroborated with phylogenetic analysis. Phage resistance to biocides which must be closely monitored in the dairy industry and the selection of biocides applied in industrial equipment needs to be conducted according to the specific phage strains present in each dairy or cheese making facility (Murphy et al., 2014).

As an alternative to biocides an interposed "phage filtration" process step was proposed prior to subsequent processing of cheese whey to whey products to reduce fermentation risks due to phage contamination. It has been shown that using a 0.1 μm microfiltration membrane significantly reduced the phage numbers (3.4 log units) in cheese whey while the total protein transmission was 56.2%. Notably, experiments carried within a feasibility study conducted with a pilot plant microfiltration system provided higher phage retention (4.1 log units) and a significantly increased transmission of major whey proteins (up to 84%) in comparison with the laboratory plant (Samtlebe et al., 2017). In light of this pilot trial results, it is fair to conclude that such strategies might be applicable in the short term with a purpose to mitigate the risk of phage contaminations during fermentative processes in cheese processing.

3.4 CONCLUSION

Strain improvement of industrial LAB is an art, either if natural methods of gene transfer, genetic modification or genome editing techniques are used. Although it is possible to create 'food-grade' genetically modified organisms (GMO), it is hard to obtain for the consumer acceptance and regulatory approval, that is why the CRISPR-Cas9 system, which allows for precise modification in a genome of interest, is seen as promising technique to obtain industrial LAB strains being indistinguishable from the ones produced by classical strain improvement. Meanwhile, instead of creating LAB strains resistant to phages' attack, specialists were able to find alternatives to protect starter cultures for dairy and cheese making by using peptides, biocides, and phagicidal processes.

3.5 SUMMARY

This chapter starts with a presentation of the starter cultures types that are used by the dairy and cheese making industry, underlying that some are considered genetically modified, while some are not, although both categories are altered at genome level. Then, the chapter presents CRISPR-Cas technique as genome editing tool, a technique which allows for precise modification of genomes and is seen as promising to obtain industrial LAB strains that are indistinguishable from the ones produced by classical strain improvement. Several examples on how LAB have been transformed using CRISPR-Cas system are given. The use of peptides, biocides, and phagicidal processes are presented as an alternative to protect starter cultures for dairy and cheese-making instead to the creation of LAB strains resistant to phages' attack.

KEYWORDS

- **adjunct cultures**
- **antibiotic resistance**
- **artisanal cultures**
- **bacteriocins**
- **cas proteins**
- **clustered regularly interspaced short palindromic repeats (CRISPR)**
- **mutagenesis**
- **stress signal**
- **thermophilic**
- **thermotrophic**
- **transformant**

REFERENCES

Anders, C., Niewoehner, O., Duerst, A., & Jinek, M., (2014). Structural basis of PAM-dependent target DNA recognition by the Cas9 endonuclease. *Nature, 513*, 569–573.

Barrangou, R., & Horvath, P., (2012). CRISPR: New horizons in phage resistance and strain identification. *Annual Review of Food Science and Technology, 3*, 143–162.

Barrangou, R., & Marraffini, L. A., (2014). CRISPR-Cas systems: Prokaryotes upgrade to adaptive immunity. *Molecular Cell, 54*(2), 234–244.

Barrangou, R., Fremaux, C., Deveau, H., Richards, M., Boyaval, P., Moineau, S., Romero, D. A., & Horvath, P., (2007). CRISPR provides acquired resistance against viruses in prokaryotes. *Science, 315*, 1709–1712.

Bose, J. L., (2016). Chemical and UV mutagenesis. *Methods in Molecular Biology, 1373*, 111–115.

Briner, A. E., & Barrangou, R., (2014). *Lactobacillus buchneri* genotyping on the basis of clustered regularly interspaced short palindromic repeat (CRISPR) locus diversity. *Applied and Environmental Microbiology, 80*(3), 994–1001.

Briner, A. E., (2015). *CRISPR-Cas Systems in Lactic Acid Bacteria* (pp. 1–119). MSc Deseartation, Department of Food Science, North Carolina State University, USA.

Briner, A. E., Lugli, G. A., Milani, C., Duranti, S., Turroni, F., Gueimonde, M., Margolles, A., Van Sinderen, D., Ventura, M., & Barrangou, R., (2015). Occurrence and diversity of CRISPR-Cas systems in the genus *Bifidobacterium*. *PLoS One, 10*(7), e0133661.

Broadbent, J. R., Neeno-Eckwall, E. C., Stahl, B., Tandee, K., Cai, H., Morovic, W., et al. (2012). Analysis of the *Lactobacillus casei* supragenome and its influence in species evolution and lifestyle adaptation. *BMC Genomics, 13*(533), 1–18. doi: 10.1186/1471–2164–13–533.

Brown, L., Pingitore, E. V., Mozzi, F., Saavedra, L., Villegas, J. M., & Hebert, E. M., (2017). Lactic acid bacteria as cell factories for the generation of bioactive peptides. *Protein & Peptide Letters, 24*(2), 146–155.

Carte, J., Christopher, R. T., Smith, J. T., Olson, S., Barrangou, R., Moineau, S., Glover, C. V., Graveley, B. R., Terns, R. M., & Terns, M. P., (2014). The three major types of CRISPR-Cas systems function independently in CRISPR RNA biogenesis in *Streptococcus thermophilus. Molecular Microbiology, 93*, 98–112.

Chen, B., & Huang, B., (2014). Imaging genomic elements in living cells using CRISPR/Cas9. *Methods in Enzymology, 546*, 337–354.

Chylinski, K., Makarova, K. S., Charpentier, E., & Koonin, E. V., (2014). Classification and evolution of type II CRISPR-Cas systems. *Nucleic Acids Research, 42*, 6091–6105.

De Moreno, De LeBlanc, A., Del Carmen, S., Chatel, J. M., Miyoshi, A., Azevedo, V., Langella, P., Bermúdez-Humarán, L. G., & LeBlanc, J. G., (2015). Current review of genetically modified lactic acid bacteria for the prevention and treatment of colitis using murine models. *Gastroenterology Research and Practice* (pp. 1–8). Article ID 146972. http://dx.doi.org/10.1155/2015/146972 (accessed on 6 September 2019).

Deveau, H., Barrangou, R., Garneau, J. E., Labonté, J., Fremaux, C., Boyaval, P., Romero, D. A., Horvath, P., & Moineau, S., (2008). Phage response to CRISPR-encoded resistance in *Streptococcus thermophilus. Journal of Bacteriology, 190*(4), 1390–1400.

DFG (Deutsche Forschungsgemeinschaft), (2010). Microbial food cultures. In: *Deutsche Forschungsgemeinschaft Senate Commission on Food Safety SKLM*. URL: http://www.dfg.de/download/pdf/dfg_im_profil/reden_stellungnahmen/2010/sklm_mikrobielle_kulturen_101115_en.pdf (accessed on 6 September 2019).

Ennahar, S., Cai, Y., & Fujita, Y., (2003). Phylogenetic diversity of lactic acid bacteria associated with paddy rice silage as determined by 16S ribosomal DNA analysis, *Applied and Environmental Microbiology, 69*(1), 444–451.

Erdmann, S., & Garrett, R. A., (2012). Selective and hyperactive uptake of foreign DNA by adaptive immune systems of an archaeon via two distinct mechanisms. *Molecular Microbiology, 85*(6), 1044–1056.

Fonfara, I., Le Rhun, A., Chylinski, K., Makarova, K. S., Lécrivain, A. L., Bzdrenga, J., Koonin, E. V., & Charpentier, E., (2014). Phylogeny of Cas9 determines functional exchangeability of dual-RNA and Cas9 among orthologous type II CRISPR-Cas systems. *Nucleic Acids Research, 42*, 2577–2590.

Garvey, P., Hill, C., & Fitzgerald, G. F., (1996). The lactococcal plasmid pNP40 encodes a third bacteriophage resistance mechanism, one which affects phage DNA penetration. *Applied and Environmental Microbiology, 62*(2), 676–679.

Gasiunas, G., Barrangou, R., Horvath, P., & Siksnys, V., (2012). Cas9-crRNA ribonucleoprotein complex mediates specific DNA cleavage for adaptive immunity in bacteria. *Proceedings of the National Academy of Sciences of the United States of America, 109*(39), E2579–E2586.

Geisen, R., & Holzapfel, W. H., (1996). Genetically modified starter and protective cultures. *International Journal of Food Microbiology, 30*(3), 315–324.

Hayes, S., Murphy, J., Mahony, J., Lugli, G. A., Ventura, M., Noben, J. P., Franz, C. M., Neve, H., Nauta, A., & Van Sinderen, D., (2017). Biocidal inactivation of *lactococcus lactis* bacteriophages: Efficacy and targets of commonly used sanitizers. *Frontiers in Microbiology, 8*(107), 1–14.

Hicks, C. L., Clark-Safko, P. A., Surjawan, I., & O'Leary, J., (2004). Use of bacteriophage-derived peptides to delay phage infections. *Food Research International, 37*(2), 115–122.

Hille, F., & Charpentier, E., (2016). CRISPR-Cas: Biology, mechanisms and relevance. *Philosophical Transactions of the Royal Society of London: Series, B, Biological Sciences, 371*(1707), 20150496. doi: 10.1098/rstb.2015.0496.

Horvath, P., Coûté-Monvoisin, A. C., Romero, D. A., Boyaval, P., Fremaux, C., & Barr-angou, R., (2009). Comparative analysis of CRISPR loci in lactic acid bacteria genomes, *International Journal of Food Microbiology, 131*(1), 62–70.

ISO (International Organization for Standardization), (2017). *Fermented Milk Products-Bacterial Starter Cultures-Standard of Identit.* ISO 27205:2010, International Organization for Standardization, ISO Central Secretariat, Geneva, Switzerland. URL: https://www.iso.org/obp/ui/#iso:std:iso:27205:ed-1:v1:en (accessed on 6 September 2019).

Jansen, R., Embden, J. D., Gaastra, W., & Schouls, L. M., (2002). Identification of genes that are associated with DNA repeats in prokaryotes. *Molecular Microbiology, 43*, 1565–1575.

Karvelis, T., Gasiunas, G., Miksys, A., Barrangou, R., Horvath, P., & Siksnys, V., (2013). crRNA and tracrRNA guide Cas9-mediated DNA interference in *Streptococcus thermophilus*. *RNA Biology, 10*, 841–851.

Koo, Y., Ka, D., Kim, E. J., Suh, N., & Bae, E., (2013). Conservation and variability in the structure and function of the Cas5d endoribonuclease in the CRISPR-mediated microbial immune system. *Journal of Molecular Biology, 425*, 3799–3810.

Labrie, S. J., Samson, J. E., & Moineau, S., (2010). Bacteriophage resistance mechanisms. *Nature Reviews Microbiology, 8*(5), 317–327.

Li, H., (2015). Structural principles of CRISPR RNA processing. *Structure, 23*, 13–20.

Lightfoot, Y. L., Selle, K., Yang, T., Goh, Y. J., Sahay, B., Zadeh, M., Owen, J. L., Colliou, N., Li, E., Johannssen, T., Lepenies, B., Klaenhammer, T. R., & Mohamadzadeh, M., (2015). SIGNR3-dependent immune regulation by *Lactobacillus acidophilus* surface layer protein A in colitis, *EMBO Journal, 34*(7), 881–895.

Magadán, A. H., Dupuis, M. È., Villion, M., & Moineau, S., (2012). Cleavage of phage DNA by the *Streptococcus thermophilus* CRISPR3-Cas system. *PLoS One, 7*(7), e40913. doi: 10.1371/journal.pone.0040913.

Mahony, J., Bottacini, F., Van Sinderen, D., & Fitzgerald, G. F., (2014). Progress in lactic acid bacterial phage research. *Microbial Cell Factories, 13*(1), S1. doi: 10.1186/1475–2859–13-S1-S1.

Mahony, J., McDonnell, B., Casey, E., & Van Sinderen, D., (2016). Phage-host interactions of cheese-making lactic acid bacteria. *Annual Review of Food Science and Technology, 7*, 267–285.

Mahony, J., McGrath, S., Fitzgerald, G. F., & Van Sinderen, D., (2008). Identification and characterization of lactococcal-prophage-carried superinfection exclusion genes. *Applied and Environmental Microbiology, 74*(20), 6206–6215.

Makarova, K. S., Wolf, Y. I., Alkhnbashi, O. S., Costa, F., Shah, S. A., Saunders, S. J., Barrangou, R., Brouns, S. J. J., Charpentier, E., & Haft, D. H., (2015). An updated evolutionary classification of CRISPR-Cas systems. *Nature Reviews Microbiology, 13*, 722–736.

McDonnell, B., Mahony, J., Neve, H., Hanemaaijer, L., Noben, J. P., Kouwen, T., & Van Sinderen, D., (2016). Identification and analysis of a novel group of bacteriophages infecting the lactic acid bacterium *Streptococcus thermophilus*. *Applied and Environmental Microbiology, 82*(17), 5153–5165.

Milani, C., Lugli, G. A., Duranti, S., Turroni, F., Bottacini, F., Mangifesta, M., et al. (2014). Genomic encyclopedia of type strains of the genus *Bifidobacterium*. *Applied and Environmental Microbiology, 80*(20), 6290–6302.

Mooney, D. T., Jann, M., & Geller, B. L., (2006). Subcellular location of phage infection protein (Pip) in *Lactococcus lactis*. *Canadian Journal of Microbiology, 52*(7), 664–672.

Morin, T., Martin, H., Soumet, C., Fresnel, R., Lamaudiere, S., Le Sauvage, A. L., Deleurme, K., & Maris, P., (2015). Comparison of the virucidal efficacy of peracetic acid, potassium monopersulfate and sodium hypochlorite on bacteriophages P001 and MS2. *Journal of Applied Microbiol*ogy, *119*(3), 655–665.

Mullan, W. M. A., (2014). Starter cultures | Importance of selected genera. In: Batt, C. A., (ed.), *Reference Module in Food Science Encyclopedia of Food Microbiology* (2nd edn., pp. 515–521). Elsevier, B. V., Amsterdam, The Netherlands.

Murphy, J., Mahony, J., Bonestroo, M., Nauta, A., & Van Sinderen, D., (2014). Impact of thermal and biocidal treatments on lactococcal 936-type phages. *International Dairy Journal, 34*(1), 56–61.

Nishimasu, H., Ran, F. A., Hsu, P. D., Konermann, S., Shehata, S. I., Dohmae, N., Ishitani, R., Zhang, F., & Nureki, O., (2014). Crystal structure of Cas9 in complex with guide RNA and target DNA. *Cell, 156*, 935–949.

Nuñez, J. K., Kranzusch, P. J., Noeske, J., Wright, A. V., Davies, C. W., & Doudna, J. A., (2014). Cas1-Cas2 complex formation mediates spacer acquisition during CRISPR-Cas adaptive immunity. *Nature Structural & Molecular Biology, 21*(6), 528–534.

Ogata, S., Eguchi, T., & Doi, K., (2000). Protection against bacteriophage contamination in industrial fermentation processes--investigation and applications of phage resistance mechanisms in bacteria. *Uirusu, 50*(1), 17–26.

Oh, J. H., & Van Pijkeren, J. P., (2014). CRISPR–Cas9-assisted recombineering *in Lactobacillus reuteri*. *Nucleic Acids Research, 42*(17), e131. doi: 10.1093/nar/gku623.

Powell, I. B., Broome, M. C., & Limsowtin, G. K. Y., (2011). Starter cultures: General aspects. In: Fuquay, J. W., (ed.), *Encyclopedia of Dairy Sciences* (2nd edn., pp. 552–558). Elsevier, B.V. Amsterdam, The Netherlands.

Roces, C., Courtin, P., Kulakauskas, S., Rodriguez, A., Chapot-Chartier, M. P., & Martinez, B., (2012). Isolation of *Lactococcuslactis mutants* simultaneously resistant to the cell wall-active bacteriocin Lcn972, lysozyme, nisin, and bacteriophage c2. *Applied and Environmental Microbiology, 78*(12), 4157–4163.

Samson, J. E., & Moineau, S., (2013). Bacteriophages in food fermentations: New frontiers in a continuous arms race. *Annual Review of Food Science and Technology, 4*, 347–368.

Samtlebe, M., Wagner, N., Brinks, E., Neve, H., Heller, K. J., Hinrichs, J., & Atamer, Z., (2017). Production of phage free cheese whey: Design of a tubular laboratory membrane

filtration system and assessment of a feasibility study. *International Dairy Journal, 21*, 17–23.
Selle, K., & Barrangou, R., (2015). CRISPR-based technologies and the future of food science. *Journal of Food Science, 80*(11), R2367–R2372. doi: 10.1111/1750–3841.13094.
Selle, K., Klaenhammer, T. R., & Barrangou, R., (2015). CRISPR-based screening of genomic island excision events in bacteria. *Proceedings of the National Academy of Sciences of the United States of America, 112*(26), 8076–8081.
Shariat, N., DiMarzio, M. J., Yin, S., Dettinger, L., Sandt, C. H., Lute, J. R., Barrangou, R., & Dudley, E. G., (2013). The combination of CRISPR-MVLST and PFGE provides increased discriminatory power for differentiating human clinical isolates of *Salmonella enterica* subsp. *enterica* serovar Enteritidis. *Food Microbiology, 34*, 164–173.
Spus, M., Li, M., Alexeeva, S., Wolkers-Rooijackers, J. C., Zwietering, M. H., Abee, T., & Smid, E. J., (2015). Strain diversity and phage resistance in complex dairy starter cultures. *Journal of Dairy Science, 98*(8), 5173–5182.
Sternberg, S. H., Redding, S., Jinek, M., Greene, E. C., & Doudna, J. A., (2014). DNA interrogation by the CRISPR RNA-guided endonuclease Cas9. *Nature, 507*, 62–67.
Sun, X., Gohler, A., Heller, K. J., & Neve, H., (2006). The LTP gene of temperate *Streptococcus thermophilus* phage TP-J34 confers superinfection exclusion to *Streptococcus thermophilus* and *Lactococcus lactis*. *Virology, 350*(1), 146–157.
Sun, Z., Chen, X., Wang, J., Zhao, W., Shao, Y., Wu, L., Zhou, Z., Sun, T., Wang, L., Meng, H., Zhang, H., & Chen, W., (2011). Complete genome sequence of *Streptococcus thermophilus* strain ND03. *Journal of Bacteriology, 193*, 793–794.
Swarts, D. C., Mosterd, C., Van Passel, M. W., & Brouns, S. J., (2012). CRISPR interference directs strand specific spacer acquisition. *PloS One, 7*(4), e35888. doi: 10.1371/journal.pone.0035888.
Sybesma, W., (2006). Safe use of genetically modified lactic acid bacteria in food. Bridging the gap between consumers, green groups, and industry. *Electronic Journal of Biotechnology, 9*(4), 424–448.
Terns, M. P., & Terns, R. M., (2011). CRISPR-based adaptive immune systems. *Current Opinion in Microbiology, 14*(3), 321–327.
Van Houte, S., Buckling, A., & Westra, E. R., (2016). Evolutionary ecology of prokaryotic immune mechanisms. *Microbiology and Molecular Biology Reviews, 80*, 745–763.
Van Pijkeren, J. P., & Britton, R. A., (2014). Precision genome engineering in lactic acid bacteria. *Microbial Cell Factories, 29*(13), 1–10. doi: 10.1186/1475–2859–13-S1-S10.
Vercoe, R. B., Chang, J. T., Dy, R. L., Taylor, C., Gristwood, T., Clulow, J. S., Richter, C., Przybilski, R., Pitman, A. R., & Fineran, P. C., (2013). Cytotoxic chromosomal targeting by CRISPR/Cas systems can reshape bacterial genomes and expel or remodel pathogenicity islands. *PLoS Genet, 9*(4), e1003454. doi: 10.1371/journal.pgen.1003454.
Visser, J., Arjan, G. M., Akkermans, A. D. L., Hoekstra, R. F., & De Vos, W. M., (2004). Insertion-sequence-mediated mutations isolated during adaptation to growth and starvation in *Lactococcus lactis*. *Genetics, 168*(3), 1145–1157.
Wang, J., Li, J., Zhao, H., Sheng, G., Wang, M., Yin, M., & Wang, Y., (2015). Structural and mechanistic basis of PAM-dependent spacer acquisition in CRISPR-Cas systems. *Cell. 163*(4), 840–853.

Westra, E. R., Dowling, A. J., Broniewski, J. M., & Van Houte, S., (2016). Evolution and ecology of CRISPR. *Annual Review of Ecology, Evolution, and Systematics, 47*(1), 307–331.

Wright, A. V., Nuñez, J. K., & Doudna, J. A., (2016). Biology and applications of CRISPR systems: Harnessing nature's toolbox for genome engineering. *Cell, 164*, 28–44.

Yin, S., Jensen, M. A., Bai, J., Debroy, C., Barrangou, R., & Dudley, E. G., (2013). The evolutionary divergence of shiga toxin-producing *Escherichia coli* is reflected in clustered regularly interspaced short palindromic repeat (crispr) spacer composition. *Applied and Environmental Microbiology, 79*, 5710–5720.

PART II

Prospective Application of Food-Grade Microorganisms for Food Preservation and Food Safety

CHAPTER 4

Comparative Study of the Development and Probiotic Protection in Food Matrices

FERNANDA SILVA FARINAZZO, PAULO TERUMITSU SAITO, MARIA THEREZA CARLOS FERNANDES, CAROLINA SAORI ISHII MAURO, MARSILVIO LIMA DE MORAES FILHO, MARLI BUSANELLO, KARLA BIGETTI GUERGOLETTO, and SANDRA GARCIA

Department of Food Science and Technology, Londrina State University, Londrina-PR, 86057-970, Brazil,

Corresponding author. E-mail: sgarcia@uel.br; gassandra15@gmail.com

4.1 INTRODUCTION

Due to the growing interest from consumers and industry, exotic or unusual fruits and foods have been studied and their functional properties described, such as the antioxidant capacity to protect against oxidative stress; anti-inflammatory; anti-carcinogenic; anti-microbial properties and others (Keppler and Humpf, 2005; Espírito et al., 2010; Borges et al., 2011; Krumreich et al., 2015). The search for a better quality of life enhanced the introduction of probiotics in daily consumption. However, the application of probiotic microorganisms in industrial processes demands some care to maintain its viability and beneficial effects on human health.

In addition to inhibit the development of undesirable microorganisms (Das et al., 2017), food matrices can provide nutrients, stimulate or protect probiotic microorganisms (Chávarri et al., 2010; Agil et al., 2013) during preparation, processing, and digestion of foods (Moghadamtousi et al., 2014; Das et al., 2017).

On the other hand, probiotic lactic acid bacteria (LAB), depending on the species and strains, present unique characteristics of utilization, metabolization, and degradation of substrates as well as xenobiotics (Siezen and Van Hylckama Vlieg, 2011), which makes them even more interesting for application in foods with functional aspects. In *L. plantarum* WCFS1, for example, genes responsible for the fermentation of sugars were identified by both, glycolysis, and phosphoketolase pathways (Claesson et al., 2007). This species had found within a range of nutrient-rich habitats, in the oral cavity and in the gastrointestinal tract of humans and animals. The central genome of *L. plantarum* contains genes responsible for crucial roles such as DNA replication and central metabolism, and is highly conserved with similar (44.5%) G + C content. One hundred and twenty genes were found, having no homology with the other LAB, and therefore, are unique to *L. plantarum* serving as markers for this species (Claesson et al., 2007; Siezen et al., 2010).

Bacteroides present a large repertoire of genes involved in the acquisition and metabolism of polysaccharides, such as glycoside hydrolases (GHs); polysaccharide hydrolases (PLs) and membrane proteins involved in the recognition and import of specific structures (Candela et al., 2010).

The application of probiotic LAB in different matrices can maximize the beneficial effects of these foods, in addition to the potential protection that these foods confer to microorganisms during processing and storage. Therefore, the use of food matrices and specific probiotic species that enhance these functionalities should be evaluated with aiming to both adaptation and survival of the probiotic.

The objective of this chapter is to show the development of probiotics in organic foods or not, containing antioxidants, bioactive compounds, and mucilages and to verify that after fermentation in food matrices of different origins weather the functionality of the food was increased by the presence of metabolites and biotransformation of bioactive produced, as well as to compare the effect of the different combinations between microorganisms and matrices.

4.2 PROBIOTIC MICROORGANISMS AND PROTECTION

The protection of probiotics by substances considered prebiotics or non-prebiotic components such as mucilages and other polysaccharides is an alternative to increase the viability of microorganisms in the gastrointestinal system (GIT). In addition, ingredients such as inulin and galactooligosaccharides may exert a protective effect, enhancing the survival and activity of probiotic bacteria during food storage, as well as during simulated GIT passage (Buriti et al., 2010; Hernandez-Hernandez et al., 2012; Krasaekoopt and Watcharapoka, 2014). Other ingredients of the food matrix, such as whey proteins, native rice starch, potato starch, glucose, raffinose, and fructooligosaccharides may increase the resistance of the microorganism to simulated gastrointestinal conditions (Perrin et al., 2000; Charalampopoulos et al., 2003; Donkor et al., 2007; Avila-Reyes et al., 2014; Pinto et al., 2015; Peredo et al., 2016).

In addition to these compounds, studies have shown that phenolics with antioxidant properties can beneficially influence the gastrointestinal tract with the increase and maintenance of beneficial bacteria (Cueva et al., 2013; Flores et al., 2015; Das et al., 2017).

The fruit of the juçara palm (*Euterpe edulis*, Figure 4.1A), is known for its similarity to the popular açaí (*Euterpe oleracea*), whose beneficial properties have been reported due to the high content of antioxidant compounds (Cardoso et al., 2015; Schulz et al., 2015) in addition to flavonoids such as quercetin, rutin, and anthocyanins (Guergoletto et al., 2016; Schulz et al., 2017).

Therefore, extract of juçara containing 5% solids ($w.v^{-1}$) was added to mucilages and exopolysaccharides (EPS) extracted from kefir in the proportion of 0.08% ($w.v^{-1}$) okra mucilage (*Abelmoschus esculentus* L.–Figure 4.1C); 1.38% ($w.v^{-1}$) aloe mucilage (AM *Aloe vera*); 9.98% ($w.v^{-1}$) caraguata mucilage (*Bromelia antiacantha*–Figure 4.1B) and 1.63% ($w.v^{-1}$) EPS of kefir, to test the influence of these mucilages on the development of probiotics. After inoculation of probiotic lactobacilli 1% ($v.v^{-1}$) in these extracts and incubation at 32°C, the counts were determined and presented in Table 4.1.

(A) (B) (C)

FIGURE 4.1 (See color insert.) (A) Juçara (*Euterpe edulis Martius*), (B) Caraguata (*Bromelia antiacantha*), and (C) Okra (*Abelmoschus esculentus L*).

Table 4.1 shows that there were differences in the viability of probiotic microorganisms in substrates with or without addition of mucilage and EPS of kefir. The extract of juçara added by caraguata mucilage, since it contained high solids content, allowed a greater development of *L. plantarum* than the other matrices, reaching a cellular concentration of 9.59 ± 0.2 log $CFU.mL^{-1}$ at initial storage time.

TABLE 4.1 Mean Counts of *L. Reuteri* and *L. Plantarum* in 30 Days of Fermented Extract Storage of Juçara with Different Mucilages

Food Matrix	Microorganism (log CFU.mL^{-1})	Storage (Days)		
		0	15	30
Juçara extract	*L. reuteri* LR92	8.92±0.05	8.80±0.10	8.71±0.05
	L. plantarum BG112	8.89±0.05	8.73±0.06	8.89±0.05
Juçara extract + okra mucilage	*L. reuteri* LR92	7.74±0.05	8.20±0.04	8.40±0.09
Juçara extract + aloe mucilage	*L. reuteri* LR92	8.07±0.09	7.93±0.16	7.82±0.10
Juçara extract + caraguata mucilage	*L. plantarum* BG112	9.59±0.2	9.53±0.03	9.3±0.08
Juçara extract + EPS	*L. plantarum* BG112	8.46±0.24	7.93±0.08	7.60±0.17

EPS: Exopolysaccharides from kefir.

In relation to the microorganisms studied, *L. plantarum* presents adaptive regions that encode several nutrient utilization systems and extracellular functions. Molenaar et al. (2005) demonstrated that in these regions, there are genes involved in the fermentation of sugars, plantaricin production, and EPS biosynthesis. This cluster of genes is called "lifestyle adaptation region," suggesting a pool of adaptive genes whose G + C content is anomalous and different from that of the species. This was likely to be acquired horizontally and could explain the adaptive flexibility of this species (Claesson et al., 2007; Siezen and Van Hylckama Vlieg, 2011). *L. reuteri,* on the other hand, are Gram-positive hetero-fermentative LAB, facultative anaerobic or aero-tolerant bacteria, capable of using different sources of carbon and energy for fermentation. This microorganism has a wide range of hosts, residing in the gastrointestinal, vaginal, and oral tracts of man and other warm-blooded animals (Hammes and Hertel, 2006). The probiotic action of *L. reuteri* is attributed to the combination of several mechanisms, including the production of lactic acid, hydrogen peroxide and the ability of some strains to produce reuterin (β-hydroxypropionaldehyde) during anaerobic glycerol metabolism (Langa et al., 2013). The production of reuterin by some strains of *L. reuteri* is a competitive advantage in ecological niches where it is present, such as the gastrointestinal tract (Morita et al., 2008). This compound inhibits the growth of many Gram-positive and Gram-negative species such as *E. coli*, *S. enterica*, *L. monocytogenes*, and *S. aureus* (Ortiz-Rivera et al., 2017) as well as fungi and protozoa (Talarico and Dobrogosz, 1989). Some strains such as *Lactobacillus reuteri* 121 are producers of EPSs (α-glucans and β-fructans), sucrose, and maltodextrins of interest for food applications (Gangoiti et al., 2017).

The *Bromeliaceae* family is known not only for its ornamental features, but also for its fruit species such as pineapple and banana from the bush popularly known as caraguata. Besides containing polyphenols, flavonoids, and tannins in its composition, it has medicinal properties (Krumreich et al., 2015). For caraguata extract, expectorant action and benefits in the treatment of asthma and bronchitis were reported. Other studies have demonstrated anticancer action in rat fibroplast cells (Santos et al., 2009; Krumreich et al., 2015).

Okra mucilage (Figure 4.1C), a polysaccharide composed of units of rhamnose, galacturonic acid, galactose, glucose, and guluronic acid, all in the pyranosidic form, has been applied as emulsifier, antifoam (Mishra and Pal, 2007) and as probiotic encapsulant (Laurenti and Garcia, 2013; Rodrigues, 2017). Silva (2006) showed the influence of okra mucilage on the growth of *Bifidobacterium lactis* Bb-12 and *Lactobacillus acidophilus* La-5.

Mucilages may favor the protection and growth of probiotic microorganisms (Smolová et al., 2017). Hirozawa (2012) determined the viability under GIT simulation conditions (Table 4.2), and observed increase resistance comparing free cells and those encapsulated in alginate. Whereas in the capsules with mucilages under 120 minutes of gastric simulation (pH 1.55) and an additional 150 minutes of enteric simulation (pH 7.43 and 0.6% bile salts), *L. plantarum* ATCC 14917^T cells, encapsulated in alginate and okra mucilage survived at final counts of 5.99 log UFC.g^{-1}.

According to the methodology used by Bedani et al. (2014), Rodrigues (2017) tested the survival of *L. casei* BGP 93 encapsulated in alginate under simulated gastrointestinal conditions. Author found that sodium alginate (2%) and linseed mucilage (0.3° Brix) microspheres, with addition of FOS (1.5% w.v^{-1}) after 6 h of gastric and enteral simulation, reduced only 3 log cycles, whereas those containing just alginate reduced 3.57 log (statistical difference at the 5% level of significance).

Bustamante et al. (2017) used mucilage and soluble protein extracted from chia (*Salvia hispanica L.*) and linseed (*Linum usitatissimum L.*) as encapsulating material combined with maltodextrin, for *Bifidobacterium infantis* and *Lactobacillus plantarum*, by spray drying. The probiotic bacteria encapsulated with the mixtures (maltodextrin: chia mucilage:

soluble protein-7.5: 0.6: 7.5%, w/w/w) and (maltodextrin: flax mucilage: soluble linseed protein-7.5: 0.2: 7.5%, w/w/w) incorporated into instant powdered juices showed high viability (> 9 log CFU.g^{-1}) after 45 days of refrigerated storage. In addition, they were resistant to gastrointestinal simulation *in vitro*, revealing that chia seed and flaxseed are excellent encapsulating agents for probiotics.

TABLE 4.2 Number of Free and Microencapsulated Viable *Lactobacillus Plantarum* ATCC 14917^{T} in Simulated Gastrointestinal System

	Number of Viable Cell (log CFU.g^{-1})				
L. plantarum	**0 min**	**30 min**	**60 min**	**90 min**	**120 min**
Free cells	$9.11^{a} \pm 0.00$	$4.21^{d} \pm 0.02$	$2.48^{c} \pm 0.02$	$0.00^{d} \pm 0.00$	$0.00^{b} \pm 0.00$
MQ	$8.90^{b} \pm 0.02$	$7.62^{a} \pm 0.03$	$7.18^{a} \pm 0.02$	$6.49^{a} \pm 0.02$	$5.99^{a} \pm 0.00$
ML	$9.12^{a} \pm 0.07$	$7.33^{b} \pm 0.11$	$7.00^{a} \pm 0.00$	$6.17^{b} \pm 0.04$	$0.00^{b} \pm 0.00$
AL	$9.05^{a,b} \pm 0.02$	$6.10^{c} \pm 0.04$	$4.85^{b} \pm 0.14$	$2.78^{c} \pm 0.10$	$0.00^{b} \pm 0.00$

Source: Hirozawa (2012).

Mean ± deviation in the same column, followed by the same letter, did not differ from each other at the 5% level of significance ($p \leq 0.05$).

MQ: cells encapsulated in alginate and okra mucilage;

ML: cells encapsulated in alginate and flaxseed mucilage;

AL: cells encapsulated in alginate.

Guevara-Arauza et al. (2012) evaluated the prebiotic effect of Nopal cactus-derivative oligosaccharide mucilage (*Opuntia ficus-indica*) on healthy adult colon microbiota. Mucilage treatment provided a 23.8% increase in lactobacilli growth, while derivative-pectic oligosaccharides increased the concentration of bifidobacteria by 25%. The addition of mucilage reduced the population of enterococci, enterobacter, staphylococci, and clostridia by about 4%. The mixture of mucilage and oligosaccharides of Nopal derivative-pectin presented prebiotic action.

Guergoletto et al. (2016) observed that the fermentation of juçara pulp caused a significant increase in the population of *Bifidobacterium* spp. In *in vitro* intestinal system after 24 h of fermentation and after *in vitro* digestion, there was a remnant of 46% of the initial phenolic content.

Plant polysaccharides are degraded by a consortium of colonic microorganisms producing SCFA; mainly acetate, propionate, and butyrate. The first has an impact on lipid metabolism and cholesterol synthesis in the liver, and the latter as an energy source for enterocytes (Candela et al., 2010).

Das and Fams (2002) observed that long-chain polyunsaturated fatty acids (LCPUFAs) have similar effects on linoleic acid as well as beneficial actions such as the ability to restore normal and healthy intestinal microbiota, and to promote the adhesion of probiotics to the intestinal mucosa. Additionally, they have an anti-inflammatory action in relation to the alleviation of allergic inflammations (Kankaanpa et al., 2001). Espírito et al. (2010) demonstrated that açai pulp increased the levels of mono and polyunsaturated fatty acids in yogurt and increased the production of linolenic acid and conjugated linolenic acid, when skim milk was fermented by two strains of *B. animalis* sp. *lactis*.

4.3 PROBIOTIC CULTURES VIABILITY IN DIFFERENT MATRICES DURING STORAGE

Processing conditions during food production containing probiotic microorganisms can lead to significant losses in cell viability due to heating, mechanical, and osmotic stresses, causing cellular injuries. These may be aggravated during storage, especially when carried out under inadequate conditions (Fu and Chen, 2011). However, several studies have shown the influence of food matrix and the addition of some compounds as prebiotics in the maintenance of cellular viability during processing and storage (Saarela et al., 2006; Nualkaekul et al., 2012).

Tables 4.3 and 4.4 show the stability of probiotic cellular concentration during the storage of different dairy products and soy derivatives, respectively.

Among products containing probiotics, functional dairy products are the most common in the world market, and the introduction of bioactive ingredients adds value and increases the demand for these already traditional products. In this sense, the addition of compounds with antioxidant, prebiotic, or bioactive properties may interfere with the growth of microorganisms.

Moreover, it is important to note that banana flour is rich in resistant starch, vitamins, minerals, phenolic compounds (Sarawong et al., 2014). Some fibers and compounds present in fruits may favor the growth of LAB (Espírito et al., 2012). In Table 4.3, the addition of GBF to milk did not change the counts of LAB. Likewise, the addition of curcuma in fermented *viili* or cheese milk maintained population levels, with the benefits of introducing an ingredient with antioxidant, coloring, anti-inflammatory, and immunomodulatory properties. At the end of the 30-day period for both matrices, there was a drop in counts, which may be due to the antimicrobial power of curcuma (Moghadamtousi et al., 2014).

TABLE 4.3 Average Survival of Probiotic Microorganisms During Storage Under Refrigeration or Freezing in Different Dairy Matrices

Food Matrix	Microorganism (log CFU.mL–1)	Storage (Days)		
		0	15	30
Milk	*L. plantarum*	9.67±0.16	9.03±0.01	8.95±0.04
	L. helveticus	9.16±0.01	9.09±0.01	9.06±0.02
Viili with curcuma*	*Lactococcus lactis*	9.30±0.02	9.18±0.01	8.72±0.01
Milk with GBF	*L. plantarum*	9.72±0.03	9.18±0.01	9.03±0.02
	L. helveticus	9.57±0.01	8.94±0.04	8.14±0.08
Cheese	*L. helveticus*	7.81±0.15	8.73±0.06	8.60±0.05
	E. faecium	7.30±0.23	7.34±0.08	7.30±0.11
Cheese with curcuma**	*L. helveticus*	7.64±0.05	8.43±0.04	7.80±0.03
	E. faecium	7.00±0.07	7.59±0.26	6.85±0.05
Ice cream with inulin	*L. kefiri*	8.45±0.05	8.45±0.01	8.49±0.03

*curcuma-0.6% ($w.v^{-1}$);

**curcuma-0.005 $g.100g^{-1}$ dry weight;

GBF: green banana flour 1.08% ($w.v^{-1}$);

TABLE 4.4 Mean Survival During Storage Under Refrigeration of Probiotic Fermented Soybean in Different Soybean-Derived Food Matrices

Food Matrix	Microorganism (log $CFU.mL^{-1}$)	Storage (Days)		
		0	15	30
Soybean extract	*L. reuteri* LR92	7.71±0.36	7.69±0.25	7.88±0.21
Soybean extract with soybean germ	*L. reuteri* LR92	8.44±0.05	8.45±0.06	8.30±0.04
Petit suisse* from black soybean	*L. acidophilus*	7.95±0.01	7.81±0.05	7.59±0.11
	B. animalis subsp. *lactis*	9.96±0.02	9.67±0.04	9.31±0.05

*Added mulberry pulp (*Morus nigra*) 20% ($w.v^{-1}$) and sugar 12% ($w.v^{-1}$)

According to Saito (2016), LAB showed growth in ice cream formulation and maintenance of numbers after freezing. Ice cream with inulin did not present a decrease in viable cell counts (Table 4.3). The addition of inulin to dairy products may aid the survival and stability of probiotic microorganisms applied. This is due to their cryoprotective properties, helping to reduce ice crystals formation during possible temperature fluctuations that may occur during product storage (Akalin and Erisir, 2008). Frighetto (2012) studied the viability of *L. paracasei* in symbiotic ice creams added with inulin and polydextrose, and observed that the storage at –18°C during 90 days had no negative effect on the viability of this bacterium, maintaining counts of 8 log $CFU.g^{-1}$. Miguel (2009) studied the viability of *L. acidophilus* in symbiotic ice creams with yacon *(Smallanthus sonchifolius)* extract, which contains inulin, and observed 2.2 x 10^9 CFU mL^{-1} in 1st day and 8.2 x 10^8 mL^{-1} after 180 days of storage. The storage conditions for yogurt ice cream are not considered ideal for the survival of LAB, and the process of freezing can cause a reduction of ½ to 1 logarithmic cycle in the viable cells counts, according to Davidson et al. (2000).

Other food matrices such as soybean derivatives have been gaining market share as a non-dairy vehicle for the consumption of probiotic cultures. These foods were developed to meet the demand of part of the population intolerant to lactose, vegans or soy products consumers (Moraes et al., 2014; Baú et al., 2015; Fernandes et al., 2017).

Soy and derivatives are substrates that support the growth of LAB due to their nutritional properties and the presence of sugars such as sucrose and oligosaccharides called α galactosides, such as raffinose and stachyose. LAB contains α-galactosidases responsible for the hydrolysis of these sugars (Donkor et al., 2007; Alazzeh et al., 2009; Baú et al., 2015). In the case of this matrix, the addition of other compounds to the base formulation alters the growth. In comparison with Table 4.4, the counts in soybean extract with and without the addition of soybean germ (3% $m.v^{-1}$), showed an increase in the counts of *L. reuteri.*

Soybean germ contains increased levels of isoflavones besides di, tri, and tetrasaccharides already mentioned. This can cause an increased maintenance of the viability during storage. For the petit Suisse derivative, the counts of *B. animalis* subsp. *lactis* were high (Table 4.4), demonstrating the suitability of this matrix to the growth of microorganisms. Likewise, the conversion of β-glucosides into aglycones is widely reported in literature (Wong et al., 2012; Moraes et al., 2016; Fernandes et al., 2017). Soybean germ concentrates isoflavones, which play an important role in the prevention of cardiovascular diseases, osteoporosis, and hormone-dependent cancers (Song et al., 1999; Steinberg et al., 2011). The aglycone fractions of isoflavones (daidzein, glycitein, and genistein) are rapidly absorbed in the body, have antioxidant activity and help prevent diseases triggered by the deleterious action of free radicals in tissues (Wong et al., 2012).

4.4 VIABILITY OF PROBIOTIC CULTURES IN PLANT MATRICES DURING FERMENTATION

Historically, the fermentative process has been used for thousands of years to improve properties and conservation of foods. This is the result of many biochemical changes that occur during fermentation; leading to changes in product components, affecting bioactivity, digestibility, and sensory aspects (Dordevic et al., 2010).

Although probiotic microorganisms are commonly applied in milk and dairy products with good growth and survival rates (Soccol et al., 2010), application in plant matrices has been shown to be promising for fermentation and bacterial maintenance. Moreover, fermentation of probiotic microorganisms in non-dairy matrices may lead to an increase in food functionality, combining both the benefits of probiotics and of fruits, vegetables, and cereals (Peres et al., 2012).

Figure 4.2 shows the growth curve of probiotic microorganisms in different food matrices: juçara, organic apple, conventional apple, black soybean extract and black soybean extract with pectin.

According to Figure 4.2, the black soybean extract (pH 6.4) allowed a higher growth of *L. plantarum* BG 112 (1% $w.v^{-1}$) than in juçara extract. This is justified by the fact that soybean extract contains nutrients with greater nutritional value in its composition (approximate composition: 2.3% protein, 0.9% lipids, 1.0% carbohydrates, 0.25% ash and 95.55% moisture); whereas juçara extract allowed a much lower content of protein (0.09% $w.v^{-1}$) but higher sugar content (2.8% $w.v^{-1}$) and acids (pH 4.9). In spite of this, the juçara extract presented similar behavior to other substrates at the end of 48 hours of fermentation. Both juçara and black soybean extracts contain anthocyanins, 966 and 388 mg gallic acid equivalent.$100g^{-1}$, respectively, cyanidin-3-rutinoside being the most abundant in juçara. According to Hidalgo et al. (2012), anthocyanins may exert a positive effect on modulation of the intestinal population and all anthocyanins tested significantly increased the growth of *Bifidobacterium* spp. and *Lactobacillus-Enterococcus* spp.

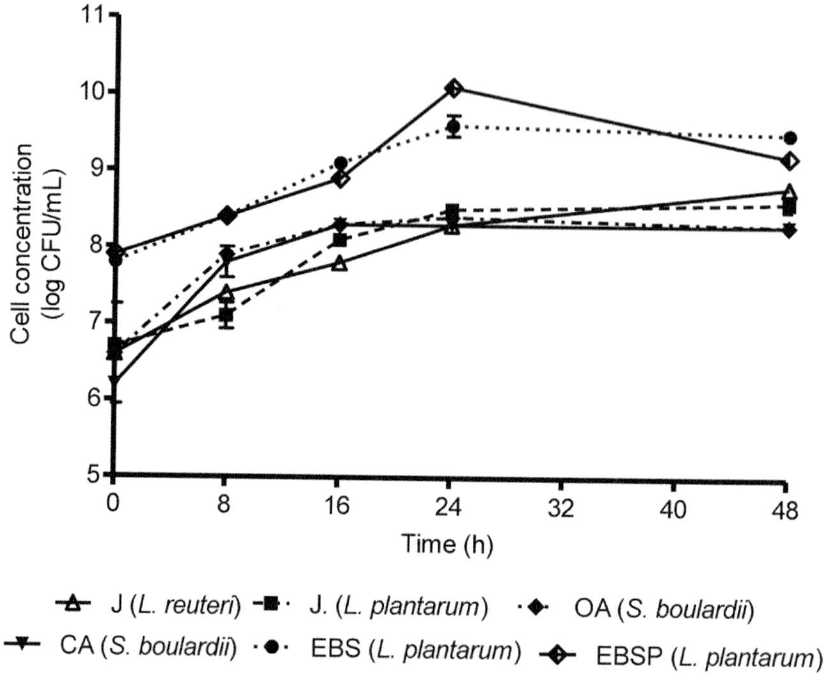

FIGURE 4.2 Cellular concentration of probiotic microorganisms during 48 hours of fermentation, in different food matrices.

[J: juçara; OA: organic apple; CA: conventional apple; EBS: extract from black soybean; EBSP: extract from black soybean with pectin].

There is evidence to support the concept that oral administration of probiotics may aid gastrointestinal tract disorders. According to Hidalgo et al. (2012), anthocyanins are rapidly fermented between 4 and 5 h, and the metabolites in the large intestine modulate the composition of the microbiota, especially *Lactobacillus* and *Bifidobacterium* spp. This is due to the fact that major groups of colonic microbiota have β-glucosidase activity. Since a small part of the anthocyanins is absorbed, the other part reaches the colon. According to Keppler and Humpf (2005), cyanidin-3-rutinoside is first transformed into a glycoside by the colonic microbiota, and then transformed into phenolic acids, that can be further metabolized by the intestinal microbiota.

4.5 ANTIOXIDANT ACTIVITY OF PROBIOTIC PRODUCTS IN DIFFERENT FOOD MATRICES

The antioxidant activity of fermented products can be modified throughout the fermentation process, mainly due to the action of enzymes, promoting bioconversion of bioactive components present in the food matrices, or due to the oxidation of some phenolic compounds (Wang et al., 2010). Each case should be analyzed separately, since the fermentation conditions are important to determine the final antioxidant activity.

In the human body, antioxidant effects of phenolic compounds help to protect cells against oxidative damage caused by free radicals, stabilizing or deactivating these radicals before the attack to the cells (Borges et al., 2013). Recently, phenolic compounds have attracted attention as potential agents for the prevention of many diseases related to oxidative stress (Schulz et al., 2017).

The use of plant phenolic compounds as antioxidants in food and pharmaceutical industry requires evidence of their antioxidant effect *in vivo,* as well as their bioavailability and toxicological properties (Borges et al., 2013).

Marazza et al. (2009) found that there was a significant increase in the antioxidant capacity of the *L. rhamnosus* fermented extract when compared to the unfermented soybean extract.

In Table 4.5, the concentration of phenolic compounds and antioxidant activity (DPPH – 2,2-diphenyl-1-picrylhydrazyl) during fermentation along with kinetics of four different treatments are reported; two of them being juice extracts from fruits of the juçara palm (*Euterpe edulis Mart.*) fermented with *L. reuteri* and *L. plantarum*, and the other two, 'Fuji' apple musts (*Malus domestica*), one organic and the other conventional, both fermented with *S. boulardii.*

The content of phenolic compounds (mg GAE.mL^{-1}) in organic apple pulp (112.8 ± 0.02 (0 h), 135.9 ± 0.06 (16 h)) is higher than in conventional apple (94.20 ± 0.31 (0 h), 100.5 ± 1.03 (16 h)) mainly in the first 16 hours of fermentation, as well as in antioxidant activity (DPPH) (Table 4.5). The differences found between the two apples substrates are probably due to the treatment used during the fruits cultivation. In organic apples, fertilizers, and pesticides are not apply, which artificially defend the plant. This provides greater production of phenolic compounds, functioning as a natural defense of organic plants (Soleas et al., 1997).

TABLE 4.5 Antioxidant Activity and Total Phenolic Compounds After Probiotic Fermentation in Different Food Matrices

Food Matrix	Antioxidant Activity	Kinetic (h)*			
		0	16	24	32
Juçara extract (*L. reuteri*)	Total phenolics (mg GAE.mL^{-1})	473.7±0.67	489.3±5.35	482.7±16.06	537.14±5.35
	DPPH (μmol trolox.mL^{-1})	138.3±5.43	160.2±0.88	165±10.61	160.8±3.68
Juçara extract (*L. plantarum*)	Total phenolics (mg GAE.mL^{-1})	447.3±2.00	454.6±2.71	514±7.36	536.7±4.01
	DPPH (μmol trolox.mL^{-1})	136.4±10.61	133.6±5.74	156.4±7.98	133.3±2.56
Organic apple pulp (*S. boulardii*)	Total phenolics (mg GAE.mL^{-1})	112.8±0.02	135.9±0.06	121.9±1.14	110.7±0.31
	DPPH (μmol trolox.mL^{-1})	15.4± 1.17	15.1±1.32	13.6±0.59	11.5±3.48
Conventional apple pulp (*S. boulardii*)	Total phenolics (mg GAE.mL^{-1})	94.20±0.31	100.5±1.03	90.4±0.31	87.7±0.21
	DPPH (μmol trolox.mL^{-1})	12.73±0.44	11.27±1.61	11.41±2.05	10.0±4.94

*Mean values of three replicates

The content of phenolics for apple pulps (Table 4.5) presented a drop after 16 hours of fermentation. At the end of the 32 hours, the organic apple pulp showed phenolic compounds content of 110.7 ± 0.31 mg GAE.mL^{-1} and conventional apple of 87.7 ± 0.21 mg GAE.mL^{-1}. This drop is a consequence of the bioactive compounds oxidation in the presence of atmospheric air, since in order to improve the fermentation efficiency with *S. boulardii*, musts agitation was promoted. Phenolic compounds in apple are formed mainly by the following classes: flavan-3-ol, hydroxycinnamic acids, glycosidic florethane, quercetin, and cyanidin (Alonso-Salces et al., 2004). However, quercetin is considered as the most effective antioxidant present in apple. The beneficial effects of this flavonoid have been attributed to the ability to reduce oxidative stress levels, generating anti-inflammatory, anti-carcinogenic, antihypertensive, anti-amnesic, and anti-atherogenic action (Molina et al., 2003; Perez-Vizcaino et al., 2006; Rogerio et al., 2010; Choi et al., 2012).

At the beginning of the fermentation (time 0 h), juçara fruit extract wort (*L. reuteri* LR92) presented antioxidant activity 9 times and phenolic compound 4.2 times higher than organic apple (*S. boulardii*). This difference is markedly higher at the end of the 32 hours of fermentation, since the fermented juçara does not show a decrease in the concentration of bioactive compounds all over the fermentation process (Table 4.5). Unlike the aerobic fermentation with probiotic yeast, fermentation with *Lactobacillus* spp. does not require the presence of oxygen, which reduces the chance of phenolic compounds oxidation; in addition, fermentation induces structural rupture of plant cell walls, leading to the release or synthesis of other antioxidant compounds (Hur et al., 2014).

The fruits of the juçara *Euterpe edulis* palm, native of the Brazilian Atlantic Forest, have good nutritional value. Its high content of phenolic compounds has in recent years been of great interest due to its antioxidant properties, mainly caused by the presence of anthocyanins and other flavonoids (Borges et al., 2011; Schulz et al., 2017). The anthocyanin content of the juçara fruit (36.9 $g.kg^{-1}$) is much higher than açai (*Euterpe oleracea* Mart.) (3.3 $g.kg^{-1}$), which also has gained importance in the world scenario due to its great antioxidant potential (Inácio et al., 2013).

As described in Table 4.5, during the fermentation process, some modifications in the substrates can occur, affecting directly the concentration of phenolic compounds, and consequently the antioxidant properties of the final products. Other modifications that may alter the product bioactivity may also occur along the days of storage. Figure 4.3 shows the antioxidant activity and content of phenolic compounds during storage of fermented probiotic products in different food matrices.

With the addition of turmeric (*Curcuma longa L.*), both fermented milk with *L. lactis* and cheese with *L. helveticus* and *E. faecium* presented higher phenolic compound contents (Figure 4.3C), and consequently higher antioxidant activities DPPH and ABTS (Figure 4.3A and B) than when the milk or cheese was not supplemented with turmeric. This spice has functional power attributed mainly to curcumin, its main phenolic compound. In literature, the rhizome antioxidant power is widely known, as well as its antimicrobial, antifungal, anti-inflammatory, anticancer, and antiviral activities (Aggarwal et al., 2013; Moghadamtousi et al., 2014).

At zero storage time, the fermented milk with turmeric presented 1.5 times more phenolic compounds and 6 times more antioxidant activity (DPPH) than the fermented milk *viili*. The same occurred for cheese with turmeric, which presented 2 times more phenolic compounds and 3.5 times more antioxidant activity (DPPH) than cheese without the addition of spice. At the 30th day of storage, products supplemented with rhizome had an increase in the concentrations of phenolic compounds, although the antioxidant activity remained significantly constant throughout the period.

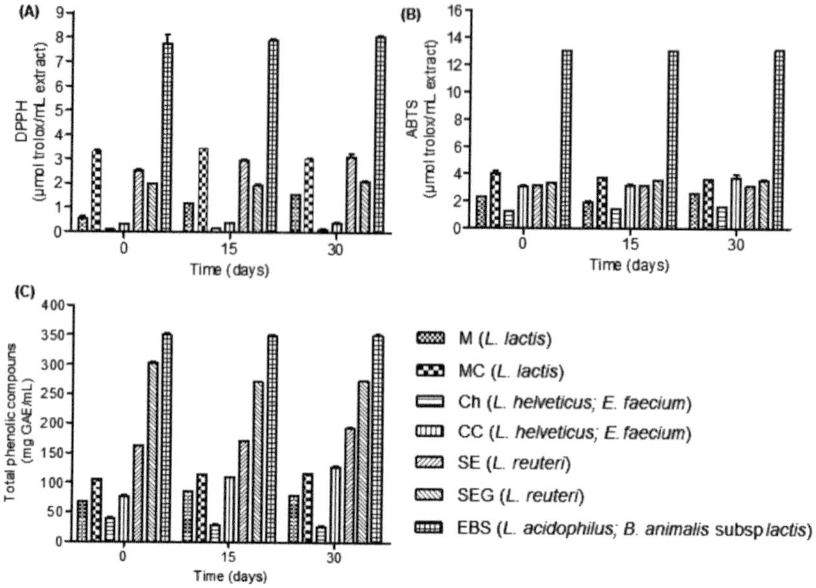

FIGURE 4.3 Antioxidant activity during storage of probiotic food matrices. (A) DPPH, (B) ABTS and (C) Concentration of phenolic compounds.

[M: milk; MC: milk and curcuma; Ch: cheese; CC: cheese and curcuma; SE: soybean extract; SEG: soybean extract and soybean germ; EBS: extract from black soybean].

The fermented (*L. reuteri*) yellow soybean extract (*Glycine max*) fortified with soybean germ has 1.9-fold phenolic compound concentrations (Figure 4.3C) and 1.1-fold antioxidant activity (ABTS) (Figure 4.3A and B) higher than fermented soybean extract (*L. reuteri*) not supplemented. The antioxidant power of soybeans is due to the presence of isoflavones, a phenolic compound, which have attracted attention due to the ability to reduce the risks of cardiovascular diseases, inhibit the growth of cancer cells, and prevent osteoporosis and menopausal symptoms (Esteves and Monteiro, 2001).

The four main forms of soy isoflavones are glycosylated (daidzin, genistin, and glycitin), acetylglycosylated (acetylidaidzin, acetylgenistin, and acetylglycitin), malonylglycosylated (malonyldaidzin, malonylgenistin, and malonylglycitin) and aglycones (daidzein, genistein, and glycitein). Aglycones, although they are the smallest portion, they are the forms that have greater biological activity, which consequently leads to a greater antioxidant potential (Kao and Chen, 2002).

The grain differs from the soybean germ, both by higher levels of isoflavones present in the latter, and by the different proportion in which the glycitein, genistein, and daidzein derivatives are present. In the germ, genistein, and glycitein derivatives occur in greater proportion in the grain, and genistein has a higher antioxidant capacity among the three-aglycone forms present in soybean (Barbosa et al., 2006). This should be taken into account, once the fermented supplement with germ was the one with the highest antioxidant activity when compared to the fermented soybean extract (Figure 4.3).

It is important to note that several studies have reported the capacity of *Lactobacillus* species (*L. acidophilus*, *L. casei*, *L. plantarum*, *L. rhamnosus*) and *Bifidobacterium* (*B. animalis* subsp. *lactis*) to increase the content of isoflavones aglycones in fermented soybean extract (Chien et al., 2006; Otieno et al., 2006; Wang et al., 2006; Pham and Shah, 2008; Marazza et al., 2009). This is only possible because these bacteria have high enzymatic activity, which makes the bioconversion of the glycosylated forms into aglycones, which have greater biological activity (Bao et al., 2012).

Moraes et al. (2016), found a positive linear correlation between the growths of *L. plantarum* BG 112 and *Lactobacillus acidophilus* LA3 and the conversion of isoflavones in fermented soybean extract, that is, an increase in the conversion of isoflavones as the multiplication of *Lactobacilli* spp. occurred. This caused an increase in final antioxidant activity of 72% (DPPH$^{\cdot}$) and 28% (ABTS$^{\cdot}$) compared to the initial activities in the *L. plantarum* fermented soybean extract. For the *L. acidophilus* fermented soybean extract, this increase was 54 and 23%, respectively, according to the radical DPPH and ABTS.

Baú et al. (2015) evaluated changes in the isoflavone aglycone content of soybean extract during fermentation with commercial kefir culture. The culture showed to be able to hydrolyze the β-glycosides of the soybean extract at 25°C for 30 h, which resulted in a 100% bioconversion of glycitin and daidzin and 89% of genistin in their corresponding aglycones forms.

Fernandes et al. (2017) also studied the fermentation of soybean extract with kefir culture. However, these authors evaluated the total isoflavone and phenolic contents during storage of the fermented soybean extract and after the *in vitro* gastrointestinal simulation. The storage process did not change the content of aglycone isoflavones and total phenolic content. However, the product fermented and stored for 1, 2 and 4 days at 4°C had the aglycone isoflavone content increased 2-fold, and total phenolic compounds increased 9-fold after *in vitro* gastrointestinal simulation. That is, soybean extract fermented with kefir is a good source of isoflavones, since the content of these compounds increased significantly after simulation of the product in the digestive system.

Fermentation of black soybean extracts (*Glycine max* (L.) Merrill) with *L. acidophilus* and *B. animalis* subsp. *lactis*, at time zero, presented an antioxidant activity (ABTS) of 4.2 times greater than fermented soybean with *L. reuteri* and 10.7 times greater than the cheese with *L. helveticus* + *E. faecium*, which has lower antioxidant power in relation to the products analyzed in Figure 4.3. In this study, the highest concentrations of phenolic compounds (351 mg GAE.mL^{-1}) and antioxidant activity, DPPH (8 μmol trolox.mL^{-1}) and ABTS (13 μmol trolox.mL^{-1}) were observed at day 30. However, substrates of yellow soybean and black soybean maintained the antioxidant activity and phenolic compounds during the 30 days of storage. As it occurs during the fermentation of yellow soybeans, the fermented black soybeans undergo biotransformation, which increases the content of aglycone isoflavones including daidzein, glycytein, and genistein (Moraes et al., 2014).

Black soybeans are a variety of soybeans, in which addition to the composition of isoflavones, also have a black tegument in the bark due to the presence of anthocyanins. This flavonoid provides black soybeans with a considerably higher degree of antioxidant activity than yellow soybeans. The glycosylated derivatives of cyanidin, petunidine, and delphinidin are usually the major constituents of the dark integument (anthocyanin), cyanidin-3-glycoside being the most abundant (Lee et al., 2009).

Black soybeans have aroused consumers' interest for the potential benefits of their consumption. Several researchers have demonstrated that it has anti-diabetes, anticancer, anti-inflammatory, hypolipidemic, and anti-mutagenic activities (Hung et al., 2007; Kim et al., 2008, Slavin et al., 2009; Kanamoto et al., 2011; Zou and Chang, 2011).

4.6 PROTECTION ABILITY OF *LACTOBACILLUS* FERMENTED PRODUCTS AGAINST PLASMID DNA OXIDATION

The action of microorganisms or proteolytic enzymes may give rise to peptides with bioactive properties. Important characteristics of peptides have been described such as antioxidant, anti-inflammatory, antimicrobial, and antihypertensive properties (Korhonen and Pihlanto, 2006; Phelan et al., 2009).

In Figure 4.4, the effects of probiotic fermentation during storage can be compared. The *in vitro* protective action of plasmid DNA in extract of milk added green banana flour against the action of an oxidizing agent (Fenton) was more effective for the case of *L. helveticus* than for *L. plantarum* (Figure 4.4 and B). For the case of soybean with addition of okara flour 3% (m.v^{-1}) *L. acidophilus* was less effective (Figure 4.4C) than *L. plantarum* (Figure 4.4D). Both matrices contain antioxidants such as green banana flour and isoflavones, but different microorganisms presented different effects of protection. Marazza et al. (2012) also studied the protection of plasmid DNA by fermented *L. rhamnosus* CRL 981 soybean extract, and there was inhibition of DNA oxidation induced by Fenton reagent.

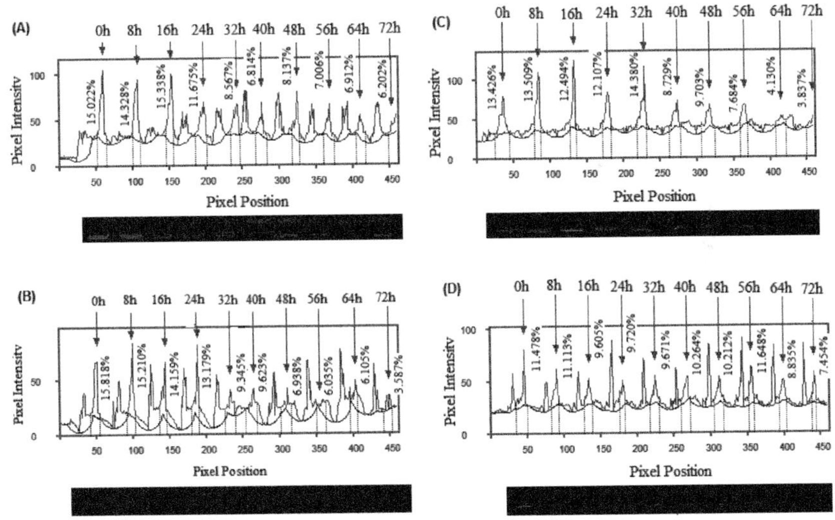

FIGURE 4.4 **(See color insert.)** Electrophoresis pattern on agarose gel (1%), DNA incubated with Fenton reagent at 0, 8, 16, 24, 32, 40, 48, 56, 64 and 72 h times. They are; (A) Fermented milk by *L. helveticus* (29°C), (B) By *L. plantarum* (27°C) with green banana flour (1.08% m.v^{-1}), (C) Fermented soybean extract by *L. acidophilus* (37°C), and (D) *L. plantarum* (37°C) with okara (3% m.v^{-1}).

Note: Image processing by João Paulo Andrade de Araujo. The quantification of plasmid DNA bands was done using LabImage 1D software version 3.3.2 (Kapeplan Bio-imaging Solutions, 2006), represented in Brazil by the company Loccus Biotecnologia.

L. helveticus is known for its high proteolytic action on casein, being used in commercial products (Phelan et al., 2009) for the production of bioactive peptides. In the study, it was not determined which compounds were responsible for DNA protection, but they exerted antioxidant action during 72h of fermentation.

Unlike many reports on the prebiotic activity of oligosaccharides, little has been described in literature on the prebiotic activity of peptides. Glycopeptides may exert this function of stimulating probiotics due to the glycosidic portion of the molecule, although studies describe that non-glycosylated peptides also stimulated the growth of bifidobacteria (Arihara, 2006).

4.7 CONCLUSION

With the growing concern about healthy eating, the demand and research on probiotics, bioactive, and functional ingredients have increased significantly. The combined application of several strategies can generate products with potential benefits to human health. However, the introduction of different ingredients and microorganisms does not promote the same effect in the food. Consequently, further case studies are needed in order to demonstrate the viability and bioavailability of these products in each food matrix.

4.8 SUMMARY

In recent years, there has been a search for food, quality of life and healthy habits. The use of exotic or unusual foods as a source of active components fulfills this demand and provides some additional benefits, being nourishing one of them. Fermented foods have linked healthy aspects to their image and safe use by mankind. The application of probiotics, despite technological challenges to maintain its viability, offers advantages in terms of value aggregation and desirable sensory, nutritional or functional characteristics. Foods, that when transformed by the growth of microorganisms have some properties changed to a pleasant and favorable effect, are presented in this chapter. Each array carries elements, as well as unique and specific components. Plant extracts, juices, soybean extract, or dairy comparisons are made in terms of microbiological and functional aspects.

ACKNOWLEDGMENT

The authors would like to thank the Londrina State University, that provided the infrastructure and facilities, and to the National Council of Technological and Scientific Development and Coordination-CNPq (479623/2012-0) and Coordination of Improvement of Higher Level Personnel-CAPES, for scholarships and financial support.

KEYWORDS

- ***Abelmoschus esculentus***
- ***aloe vera***
- **antioxidant activity**
- **black soybean extract**
- ***Bromelia antiacantha***
- **Curcuma**
- **functional ingredients**
- **mucilages**
- **okra**
- **quercetin**
- **soybean extract**

REFERENCES

Aggarwal, B. B., Yuan, W., Li, S., & Gupta, S. C., (2013). Review: Curcumin-free turmeric exhibits anti-inflammatory and anticancer activities: Identification of novel components of turmeric. *Molecular Nutrition & Food Research, 57*(9), 1529–1542.

Agil. R., Gaget, A., Gliwa, J., Avis, T. J., Willmore, W. G., & Hosseinian, F., (2013). Lentils enhance probiotic growth in yogurt and provide added benefit of antioxidant protection. *LWT - Food Science and Technology, 50*(1), 45–49.

Akalin, A. S., & Erisir, D., (2008). Effects of inulin and oligofrutose on the rheological characteristics and probiotic culture survival in low-fat probiotic ice cream. *Journal of Food Science, 73*(4), 184–188.

Alazzeh, A. Y., Ibrahim, S. A., Song, D., Shahbazi, A., & Ghazaleh, A. A. A., (2009). Carbohydrate and protein sources influence the induction of α- and β-galactosidases in *Lactobacillus reuteri*. *Food Chemistry, 117*(4), 654–659.

Alonso-Salces, R. M., Barranco, A., Abad, B., Berrueta, L. A., Gallo, B., & Vicente, F., (2004). Polyphenolic profiles of Basque cider apple cultivars and their technological properties. *Journal of Agricultural and Food Chemistry, 52*(10), 2938–2952.

Arihara, K., (2006). Strategies for designing novel functional meat products. *Meat Science, 74*(1), 219–229.

Avila-Reyes, A. V., Garcia-Suarez, F. J., Jiménez, M. T., Martín-Gonzalezc, M. F. S., & Bello-Perez, L. A., (2014). Protection of *L. rhamnosus* by spray-drying using two prebiotics colloids to enhance the viability. *Carbohydrate Polymers, 102*, 423–430.

Bao, Y., Zhang, Y., Li, H., Liu, Y., Wang, S., Dong, X., Su, F., Yao, G., Sun, T., & Zhang, H., (2012). *In vitro* screen of *Lactobacillus plantarum* as probiotic bacteria and their fermented characteristics in soymilk. *Annals of Microbiology, 62*(3), 1311–1320.

Barbosa, A. C. L., Hassimotto, N. M. A., Lajolo, F. M., & Genovese, M. I., (2006). Isoflavone content and profile and antioxidant activity of soy and soy products. *Food Science and Technology* (*Campinas*), *26*(4), 921–926.

Baú, T. B., Garcia, S., & Ida, E. I., (2015). Changes in soymilk during fermentation with kefir culture: Oligosaccharides hydrolysis and isoflavone aglycone production. *International Journal of Food Sciences and Nutrition, 66*(8), 845–850.

Bedani, R., Vieira, A. D. S., Rossi, E. A., & Saad, S. M. I., (2014). Tropical fruit pulps decreased probiotic survival to *in vitro* gastrointestinal stress in synbiotic soy yogurt with okara during storage. *LWT - Food Science and Technology, 55*(2), 436–443.

Borges, G. S. C., Gonzaga, L. V., Jardini, F. A., Marcini, F. J., Heller, M., Micke, G., Costa, A. C. O., & Fett, R., (2013). Protective effect of *Euterpe edulis* M. on Vero cell culture and antioxidant evaluation based on phenolic composition using HPLC−ESI-MS/MS. *Food Research International, 51*(1), 363–369.

Borges, G. S. C., Vieira, F. G. K., Copetti, C., Gonzaga, L. V., Zambiazi, R. C., Mancini, F. J., & Fett, R., (2011). Chemical characterization, bioactive compounds, and antioxidant capacity of Jussara (*Euterpe edulis*) fruit from the Atlantic forest in southern Brazil. *Food Research International, 44*(7), 2128–2133.

Buriti, F. C. A., Castro, I. A., & Saad, S. M. I., (2010). Viability of *Lactobacillus acidophilus* in synbiotic guava mousses and its survival under *in vitro* simulated gastrointestinal conditions. *International Journal of Food Microbiology, 137*(2/3), 121–129.

Bustamante, M., Oomah, B. D., Rubilar, M., & Shene, C., (2017). Effective *Lactobacillus plantarum* and *Bifidobacterium infantis* encapsulation with chia seed (*Salvia hispanica* L.) and flaxseed (*Linum usitatissimum* L.) mucilage and soluble protein by spray drying. *Food Chemistry, 216*, 97–105.

Candela, M., Maccaferri, S., Turroni, S., Carnevali, P., & Brigidi, P., (2010). Functional intestinal microbiome, new frontiers in prebiotic design. *International Journal of Food Microbiology, 140*(2/3), 93–101.

Cardoso, A. L., Di Pietro, P. F., Vieira, F. G. K., Boaventura, B. C. B., Liz, S., Borges, G. S. C., Fett, R., Andrade, D. F., & Silva, E. L., (2015). Acute consumption of Juçara juice (*Euterpe edulis*) and antioxidant activity in health individuals. *Journal of Functional Foods, 17*, 152–162.

Charalampopoulos, D., Pandiella, S. S., & Webb, C., (2003). Evaluation of the effect of malt, wheat and barley extracts on the viability of potentially probiotic lactic acid bacteria under acidic conditions. *International Journal of Food Microbiology, 82*(2), 133–142.

Chávarri, M., Marañón, I., & Ares, R., (2010). Microencapsulation of a probiotic and prebiotic in alginate-chitosan capsules improves survival in simulated gastrointestinal conditions. *International Journal of Food Microbiology, 142*(1/2), 185–189.

Chien, H. L., Huang, H. Y., & Chou, C. C., (2006). Transformation of isoflavone phytoestrogens during the fermentation of soymilk with lactic acid bacteria and bifidobacteria. *Food Microbiology, 23*(8), 772–778.

Choi, G. N., Kim, J. H., Kwak, J. H., Jeong, C. H., Jeong, H. R., Lee, U., & Heo, H. J., (2012). Effect of quercetin on learning and memory performance in ICR mice under neurotoxic trimethyltin exposure. *Food Chemistry, 132*(2), 1019–1024.

Claesson, M. J., Sinderen, D. V., & O'Toole, P. W., (2007). The genus *Lactobacillus* a genomic basis for understanding its diversity. *FEMS Microbiology Letters, 269*(1), 22–28.

Cueva, C., Sánchez-Patán, F., Monagas, M., Walton, G. E., Gibson, G. R., Martín-Álvarez, P. J., Bartolomé, B., & Moreno-Arribas, V., (2013). *In vitro* fermentation of grape seed flavan-3-ol fractions by human fecal microbiota: Changes in microbial groups and phenolic metabolites. *FEMS Microbiology Ecology, 83*(3), 791–805.

Das, Q., Islam, R. M. D., Marcone, M. F., Warriner, K., & Diarra, M. S., (2017). Potential of berry extracts to control foodborne pathogens. *Food Control, 73*(B), 650–662.

Das, U. N., & Fams, M. D., (2002). Essential fatty acid as possible enhancers of the beneficial actions of probiotics. *Nutrition, 18*(9), 786–789.

Davidson, R. H., Duncan, S. E., Hackney, C. R., Eigel, W. N., & Boling, J. W., (2000). Probiotic culture survival and implications in fermented frozen yogurt characteristics. *Journal of Dairy Science, 83*(4), 666–673.

Donkor, O. N., Henriksson, A., Vasiljevic, T., & Shah, N. P., (2007). α-Galactosidase and proteolytic activities of selected probiotic and dairy cultures in fermented soymilk. *Food Chemistry, 104*(1), 10–20.

Dordevic, T. M., Siler-Marinkovic, S., & Dimitrijevic-Brankovic, S. I., (2010). Effect of fermentation on antioxidant properties of some cerelas and pseudo cereals. *Food Chemistry, 119*(3), 957–963.

Espírito, S. A. P., Cartolano, N. S., Silva, T. F., Soares, F. A., Gioielli, L. A., Perego, P., Converti, A., & Oliveira, M. N., (2012). Fibers from fruit by-products enhance probiotic viability and fatty acid profile and increase CLA content in yoghurts. *International Journal of Food Microbiology, 154*(3), 135–144.

Espírito, S. A. P., Silva, R. C., Soares, F. A. S. M., Anjos, D., Gioielli, L. A., & Oliveira, M. N., (2010). Açaí pulp addition improves fatty acid profile and probiotic viability in yoghurt. *International Dairy Journal, 20*(6), 415–422.

Esteves, E. A., & Monteiro, J. B. R., (2001). Beneficial effects of soy isoflavones on chronic diseases. *Brazilian Journal of Nutrition, 14*(1), 43–52.

Fernandes, M., Lima, F. S., Rodrigues, D., Handa, C., Guelfi, M., Garcia, S., & Ida, E. I., (2017). Evaluation of the isoflavone and total phenolic contents of kefir-fermented soymilk storage and after the *in vitro* digestive system simulation. *Food Chemistry, 229*, 373–380.

Flores, G., Del, C. M. L., Costabile, A., Klee, A., Guergoletto, K. B., & Gibson, G. R., (2015). *In vitro* fermentation of anthocyanins encapsulated with cyclodextrins: Release, metabolism and influence on gut microbiota growth. *Journal of Functional Foods, 16*, 50–57.

Frighetto, J. M., (2012). *Manufacture of Symbiotic Ice Cream and Survival of Lactobacillus Paracasei Under Simulated Gastrointestinal Conditions* (pp. 15–22). MSc (food science and technology) dissertation, area of concentration in food science and technology, Federal University of Santa Maria, Santa Maria, Brazil.

Fu, N., & Chen, X. D., (2011). Towards a maximal cell survival in convective thermal drying processes. *Food Research International, 44*(5), 1127–1149.

Gangoiti, J., Lammerts, V. B., A., & Dijkhuizen, L., (2017). Draft genome sequence of *lactobacillus reuteri* 121, a source of α-glucan and β-fructan exopolysaccharides. *Genome Announcements, 5*(10), e01691–16.

Guergoletto, K. B., Costabile, A., Flores, G., Garcia, S., & Gibson, G. R., (2016). *In vitro* fermentation of juçara pulp (*Euterpe edulis*) by human colonic microbiota. *Food Chemistry, 196*, 251–258.

Guevara-Arauza, J. C., Ornelas-Paz, J. J., Pimentel-González, D. J., Mendonza, S. R., Guerra, R. E. S., & Maldonado, L. M. T. P., (2012). Prebiotic effect of mucilage and pectic-derived oligosaccharides from Nopal (*Opuntia ficus-indica*). *Food Science and Biotechnology, 21*(4), 997–1003.

Hammes, W., & Hertel, C., (2006). The genera *Lactobacillus* and *Carnobacterium. Prokaryotes, 4*, 320–403.

Hernandez-Hernandez, O., Muthaiyan, A., Moreno, F. J., Montilla, A., Sanz, M. L., & Ricke, S. C., (2012). Effect of prebiotic carbohydrates on the growth and tolerance of *Lactobacillus. Food Microbiology, 30*(2), 355–361.

Hidalgo, M., Oruna-Concha, M. J., Kolida, S., Walton, G. E., Kallithraka, S., Spencer, J. P., & De Pascual-Teresa, S., (2012). Metabolism of anthocyanins by human gut microflora and their influence on gut bacterial growth. *Journal of Agricultural and Food Chemistry, 60*(15), 3882–3890.

Hirozawa, S. S., (2012). *Study of Natural Polymers in the Protection of Lactobacillus Plantarum ATCC 14917*[T] *Under Simulated Gastrointestinal Conditions* (pp. 10–25). Undergraduate Dissertation in Biology, Department of Biology, State University of Londrina, Londrina, Brazil.

Hung, Y. H., Huang, H. Y., & Chou, C. C., (2007). Mutagenic and antimutagenic effects of methanol extracts of unfermented and fermented black soybeans. *International Journal of Food Microbiology, 118*(1), 62–68.

Hur, S. J., Lee, S. Y., Kim, Y. C., Choi, I., & Kim, G. B., (2014). Effect of fermentation on the antioxidant activity in plant-based foods. *Food Chemistry, 160*, 346–356.

Inácio, M. R. C., De Lima, K. M. G., Lopes, V. G., Pessoa, J. C. D., & De Almeida, T. G. H., (2013). Total anthocyanin content determination in intact açaí (*Euterpe oleracea* Mart.) and palmitero-juçara (*Euterpe edulis* Mart.) fruit using near infrared spectroscopy (NIR) and multivariate calibration. *Food Chemistry, 136*(3/4), 1160–1164.

Kanamoto, Y., Yamashita, Y., Nanba, F., Yoshida, T., Tsuda, T., Fukuda, I., Nakamura-Tsurata, S., & Ashida, H., (2011). A black soybean seed coat extract prevents obesity and glucose intolerance by up-regulating upcoupling proteins and down-regulating inflammatory cytokines in high-fat diet-fed mice. *Journal of Agricultural and Food Chemistry, 59*(16), 8985–8993.

Kankaanpa, A. P. E., Salminen, S. J., Isolauri, E., & Lee, Y. K., (2001). The influence of polyunsaturated fatty acid on probiotic growth and adhesion. FEMS *Microbiology Letters, 194*(2), 149–153.

Kao, T. H., & Chen, B. H., (2002). An improved method for determination of isoflavones in soybean powder by liquid chromatography. *Chromatographia, 56*, 423–430.

Keppler, K., & Humpf, H. U., (2005). Metabolism of anthocyanins and their phenolic degradation products by the intestinal microflora. *Bioorganic & Medicinal Chemistry, 13*(17), 5195–5205.

Kim, J. M., Kim, J. S., Yoo, H., Choung, M. G., & Sung, M. K., (2008). Effects of black soybean [Glycine max (L.) Merr.] seed coats and its anthocyanidins on colonic inflammation and cell proliferation in vitro and *in vivo*. *Journal of Agricultural and Food Chemistry, 56*(18), 8427–8433.

Korhonen, H., & Pihlanto, A., (2006). Bioactive peptides: Production and functionality. *International Dairy Journal, 16*(9), 945–960.

Krasaekoopt, W., & Watcharapoka, S., (2014). Effect of addition of inulin and galactooligosaccharide on the survival of microencapsulated probiotics in alginate beads coated with chitosan in simulated digestive system, yogurt and fruit juice. *LWT - Food Science and Technology, 57*(2), 761–766.

Krumreich, F. D., Corrêa, A. P. A., Da Silva, S. D. S., & Zambiazi, R. C., (2015). Physical and chemical composition and bioactive compounds in *Bromelia antiacantha* Bertol. *Brazilian Journal of Fruitculture, 37*(2), 450–456.

Langa, S., Landete, J. M., Martín-Cabrejas, I., Rodrígues, E., Arqués, J. L., & Medina, M., (2013). *In situ* reuterin production by *Lactobacillus reuteri* in dairy products. *Food Control, 33*(1), 200–206.

Laurenti, E. S., & Garcia, S., (2013). Efficiency of natural and commercial encapsulating materials on controlled release of encapsulated probiotic. *Brazilian Journal of Food Technology, 16*(2), 107–115.

Lee, J., Kang, N. S., Shin, S. O., Shin, S. H., Lim, S. G., Suh, D. Y., Baek, I. Y., Park, K. Y., & Ha, T. J., (2009). Characterization of anthocyanins in the black soybean (*Glycine max L.*) by HPLC-DADESI/ MS analysis. *Food Chemistry, 112*(1), 226–231.

Marazza, J. A., Garro, M. S., & Giori, G. S., (2009). Aglycone production by *Lactobacillus rhamnosus* CRL981 during soymilk fermentation. *Food Microbiology, 26*(3), 333–339.

Marazza, J. A., Nazarenob, M. A., Gioria, G. S., & Garro, M. S., (2012). Enhancement of the antioxidant capacity of soymilk by fermentation with *Lactobacillus rhamnosus*. *Journal of Functional Foods, 4*(3), 594–601.

Miguel, M. G. C. P., (2009). *Identification of Microorganisms Isolated from Kefir Grains in Milk and Water of Different Localities I* (pp. 1–71). MSc (Agricultural Microbiology) Dissertation, Department of Agriculture, Federal University of Lavras, Lavras, Brazil.

Mishra, S. P., & Pal, S., (2007). Polyacrylonitrile-grafted okra mucilage: A renewable reservoir to polymeric materials. *Carbohydrate Polymers, 68*(1), 95–100.

Moghadamtousi, S. Z., Kadir, H. A., Hassandarvish, P., Tajik, H., Abubakar, S., & Zandi, K. A., (2014). Review on antibacterial, antiviral, and antifungal activity of curcumin. *BioMed. Research International, 1*, 1–12.

Molenaar, D., Bringel, F., Schuren, F. H., De Vos, W. M., Siezen, R. J., & Kleerebezem, M., (2005). Exploring *Lactobacillus plantarum* genome diversity by using microarrays. *Journal of Bacteriology, 187*(17), 6119–6127.

Molina, M. F., Sanchez-Reus, I., Iglesias, I., & Benedi, J., (2003). Quercetin a flavonoid antioxidant, prevents and protects against ethanol-induced oxidative stress in mouse liver. *Biological and Pharmaceutical Bulleti, 26*(10), 1398–1402.

Moraes, F. M. L., Busanello, B., & Garcia, S., (2016). Optimization of the fermentation parameters for the growth of *Lactobacillus* in soymilk with okara flour. *LWT - Food Science and Technology, 74*, 456–464.

Moraes, F. M. L., Hirozawa, S. S., Prudencio, S. H., Ida, E. I., & Garcia, S., (2014). Petit Suisse from black soybean: Bioactive compounds and antioxidant properties during development process. *International Journal of Food Sciences and Nutrition, 65*(4), 470–475.

Morita, H., Toh, H., Fukuda, S., Horikawa, H., Oshima, K., Suziki, T., et al. (2008). Comparative genome analysis of *Lactobacillus reuteri* and *Lactobacillus fermentum* reveal a genomic island for reuterin and cobalamin production. *DNA Research, 15*(3), 151–161.

Nualkaekul, S., Deepika, G., & Charalampopoulos, D., (2012). Survival of freeze dried *Lactobacillus plantarum* in instant fruit powders and reconstituted fruit juices. *Food Research International, 48*(2), 627–633.

Ortiz-Rivera, Y., Sánchez-Vega, R., Gutiérrez-Méndez, N., León-Félix, J., Acosta-Moñiz, C., & Sepulveda, D. R., (2017). Production of reuterin in a fermented milk product by *Lactobacillus reuteri* : Inhibition of pathogens, spoilage microorganisms, and lactic acid bacteria. *Journal of Dairy Science, 100*(6), 4258–4268.

Otieno, D. O., Ashton, J. F., & Shah, N. P., (2006). Evaluation of enzymic potential for biotransformation of isoflavones phytoestrogen in soymilk by *Bifidobacterium animalis, Lactobacillus acidophilus* and *Lactobacillus casei. Food Research International, 39*(4), 394–407.

Peredo, A. G., Beristain, C. I., Pascual, L. A., Azuara, E., & Jimenez, M., (2016). The effect of prebiotics on the viability of encapsulated probiotic bacteria. *LWT - Food Science and Technology, 73*, 191–196.

Peres, C. M., Peres, C., Hernandez-Mendoza, A., & Malcata, F. X., (2012). Review on fermented plant materials as carriers and sources of potentially probiotic lactic acid bacteria with an emphasis on table olives. *Trends in Food Science & Technology, 26*(1), 31–42.

Perez-Vizcaino, F., Bishop-Bailley, D., Lodi, F., Duarte, J., Cogolludo, A., Moreno, L., Bosca, L., Mitchell, J. A., & Warner, T. D., (2006). The flavonoid quercetin induces apoptosis and inhibits JNK activation in intimal vascular smooth muscle cells. *Biochemical and Biophysical Research Communications, 346*(3), 919–925.

Perrin, S., Grill, J. P., & Scheneider, F., (2000). Effects of fructooligosaccharides and their monomeric components on bile salt resistance in three species of bifidobacteria. *Journal of Applied Microbiology, 88*(6), 968–974.

Pham, T. T., & Shah, N. P., (2008). Effect of lactulose on biotransformation of isoflavone glycosides to aglycones in soymilk by Lactobacilli. *Journal of Food Science, 73*(3), M158–M165.

Phelan, M., Aherne, A., Fitzgerald, R. J. B., & O'Brien, N. M., (2009). Casein-derived bioactive peptides: Biological effects, industrial uses, safety aspects and regulatory status. *International Dairy Journal, 19*(11), 643–654.

Pinto, S. S., Carlise, B., Fritzen-Freire, B., Benedetti, S., Murakami, F. S., Petrus, J. C., Prudêncio, E. S., & Amboni, R. D. M. C., (2015). Potential use of whey concentrate and prebiotics as carrier agents to protect *Bifidobacterium*-BB-12 microencapsulated by spray drying. *Food Research International, 67*, 400–408.

Rodrigues, F. J., (2017). *Natural Polymers in Microencapsulation by Extrusion and in the Development of Edible Toppings of Probiotic Culture for Yacon Minimally Processed.* MSc (Food Science) Dissertation, Department of Food Science and Technology, State University of Londrina, Londrina, Brazil.

Rogerio, A. P., Dora, C. L., Andrade, E. L., Chaves, J. S., Silva, K. F. C., Lemos-Senna, E., & Calixto, J. B., (2010). Anti-inflammatory effect of quercetin-loaded microemulsion in the airways allergic inflammatory model in mice. *Pharmacological Research, 61*(4), 288–297.

Saarela, M., Virkajarvi, I., Nohynek, L., Vaari, A., & Matto, J., (2006). Fibre as carriers for *Lactobacillus rhamnosus* during freeze-drying and stotage in apple juice and chocolate-coated breakfast cereals. *International Journal of Food Microbiology, 112*(2), 171–178.

Saito, P. T., (2016). *Elaboration of Functional Ice Cream Based on Whey Protein Isolate Fermented with Kefir.* MSc (Food Science) Dissertation, Department of Food Science and Technology, State University of Londrina, Londrina, Brazil.

Santos, V. N. C., Freitas, R. A., Deschamps, F. C., & Biavatti, M. W., (2009). Ripe fruits of *Bromelia antiacantha*: Investigations on the chemical and bioactivity profile. *Brazilian Journal of Pharmacognosy, 19*(2A), 358–365.

Sarawong, C., Schoenlechner, R., Sekiguchi, K., Berghofer, E., & Ng, P. K. W., (2014). Effect of extrusion cooking on the physicochemical properties, resistant starch, phenolic content and antioxidant capacities of green banana flour. *Food Chemistry, 143,* 33–39.

Schulz, M., Biluca, F. C., Gonzaga, L. V., Borges, G. D., Vitali, L., Micke, G. A., De Gois, J. S., De Almeida, T. S., Borges, D. L., Miller, P. R., Costa, A. C., & Fett, R., (2017). Bioaccessibility of bioactive compounds and antioxidant potential of juçara fruits (*Euterpe edulis* Martius) subjected to *in vitro* gastrointestinal digestion. *Food Chemistry, 228,* 447–454.

Schulz, M., Borges, G. S. C., Gonzaga, L. V., Seraglio, S. K. T., Olivo, I. S., Azevedo, M. S., et al. (2015). Chemical composition, bioactive compounds and antioxidant capacity of juçara fruit (*Euterpe edulis* Martius) during ripening. *Food Research International, 77*(2), 125–131.

Siezen, R. J., Tzeneva, V. A., Castioni, A., Wels, M., Phan, H. T., Rademaker, J. L., Starrenburg, M. J., Kleerebezem, M., Molenaar, D., Van Hylckama., & Vlieg, J. E., (2010). Phenotypic and genomic diversity of *Lactobacillus plantarum* strains isolated from various environmental niches. *Environmental Microbiology, 12*(3), 758–773.

Siezen, R. J., Van Hylckama., & Vlieg, J. E., (2011). Genomic diversity and versatility of *Lactobacillus plantarum*, a natural metabolic engineer. *Microbial Cell Factories, 10*(1), S3.

Silva, V. S. N., (2006). *Study of the Nutritionals Effects of the Pulp Flour and Mucilage Extracted from Okara (Hibiscus esculentus L).* PhD Thesis, Department of Food and Nutrition, Campinas State University, Campinas, Brazil.

Slavin, M., Kenwirthy, W., & Yu, L. L., (2009). Antioxidant properties, phytochemical composition, and antiproliferative activity of Maryland-grown soybeans with colored seed coats. *Journal of Agricultural and Food Chemistry, 57*(23), 11174–11185.

Smolová, J., Němečková, I., Klimešová, M., & Kyselka, J., (2017). Flaxseed varieties: Composition and influence on the growth of probiotic microorganisms in milk. *Czech Journal of Animal Science, 35*(1), 18–23.

Soccol, C. R., Vandenberghe, L. P. S., Spier, M. R., Medeiros, A. B. P., Yamaguishi, C. T., Lindner, J. D., Pandey, A., & Soccol, V. T., (2010). The potential of probiotics. *Food Technology and Biotechnology, 48*(4), 413–434.

Soleas, G. J., Diamandis, E. P., & Goldberg, D. M., (1997). Resveratrol: A molecule whose time has come? and gone? *Clinical Biochemistry, 30*(2), 91–113.

Song, T. T., Hendrich, S.,&Murphy, P. A., (1999). Estrogenic activity of glycitein, a soy isoflavone. *Journal of Agricultural and Food Chemistry, 47*(4), 1607–1610.

Steinberg, F. M., Murray, M. J., Lewis, R. D., Cramer, M. A., Amato, P., Young, R. L., et al. (2011). Clinical outcomes of a 2-y soy isoflavones supplementation in menopausal woman. *American Journal* of *Clinical Nutrition, 93*(2), 356–367.

Talarico, T. L., & Dobrogosz, W. J., (1989). Chemical characterization of an antimicrobial substance produced by *Lactobacillus reuteri. Antimicrobial Agents and Chemotherapy, 33*(5), 674–679.

Wang, A., Zhang, F., Huang, L., Yin, X., Li, H., Wang, Q., Zeng, Z., & Xie, T., (2010). New progress in biocatalysis and biotransformation of flavonoids. *Journal* of *Medicinal Plants Research, 4*(10), 847–856.

Wang, Y. C., Yu, R. C., & Chou, C. C., (2006). Antioxidative activities of soymilk fermented with lactic acid bacteria and bifidobacteria. *Food Microbiology, 23*(2), 128–135.

Wong, W. W., Taylor, A. A., Smith, E. O., Barnes, S., & Hachey, D. L., (2012). Effect of soy isoflavone supplementation on nitric oxide metabolism and blood pressure in menopausal women. *American Journal of Clinical Nutrition, 95*(6), 1487–1494.

Zou, Y., & Chang, S. K. C., (2011). Effect of black soybean extract on the suppression of the proliferation of human AGS gastric cancer cells via the induction of apoptosis. *Journal of Agricultural and Food Chemistry, 59*(9), 4597–4605.

CHAPTER 5

Indigenous Food and Food Products of West Africa: Employed Microorganisms and Their Antimicrobial and Antifungal Activities

ESSODOLOM TAALE,[1*] BOURAÏMA DJERY,[1] HAZIZ SINA,[2] ESSOZIMNA KOGNO,[1] SIMPLICE D. KAROU,[1] ALFRED S. TRAORE,[3] LAMINE BABA-MOUSSA,[2] ALY SAVADOGO,[3] and YAOVI AMEYAPOH[1]

[1]*Laboratory of Microbiology and Quality Control of Foodstuffs (LAMICODA), School of Biological and Food Techniques (ESTBA), University of Lomé, Lomé, Togo*

[2]*Laboratory of Molecular Biology and Typing in Microbiology, Department of Biochemistry and Cellular Biology, Faculty of Science and Technology (FAST), University of Abomey-Calavi; 05BP 1604 Cotonou (Benin)*

[3]*Laboratory of Biotechnology and Microbiology (LaBM), Center for Research in Biological, Food and Nutritional Sciences (CRSBAN), Department of Biochemistry-Microbiology (DBM), Training and Research Unit in Life and Earth Sciences (UFR)-SVT, Ouaga University I Pr Joseph KI-ZERBO, 03BP7131 Ouagadougou 03, Ouagadougou (Burkina Faso)*

**Corresponding author. E-mail: taaleernest12@hotmail.com*

5.1 INTRODUCTION

In the western sub-region of Africa (Benin, Burkina Faso, Cape Verde, Côte d'Ivoire, Gambia, Ghana, Guinea, Guinea-Bissau, Liberia, Mali, Mauritanie, Niger, Nigeria, Sénégal, Sierra Leone and Togo) *ogi, kunun-zaki, gowê, koko or akasa, kenkey, ben-saalga, doklu, agbélima, attiéké, wara, wagashie, dolo, tchoukoutou, tchakpalo, ablo, akluii*, etc., are some of indigenous fermented dairy and food products that plays an essential role in the diet. Those different fermented foods are derived from raw material such as maize, sorghum, millet, cassava, and milk. Therefore, West Africa countries have a rich culture and tradition in food technology by preparing those foods because the skill is transmitted from generation to generation and some are specific to ethnic group. Even the household level of these traditional foods and their informal marketing, they play key roles in the culture, economy, and food security of West Africa countries (Adeyeye, 2017). The beneficial effects of these traditionally fermented foods are their preservation and the increase in their self-life and organoleptic characteristics because of the production of lactic acid and other metabolites like bacteriocins, Bacteriocin-like inhibitory Substances (BLIS), nonribosomal peptide synthetase (NRPS), synthesized by a complex and significant microbial biodiversity such as lactic acid bacteria (LAB), *Bacillus* spp., yeasts, and molds (Kalui et al., 2009; Savadogo et al., 2011; Adimpong et al., 2012; Achi and Ukwuru, 2015; Taale et al., 2015, 2016a, b). For Achi and Ukwuru (2015), the relation between microbial diversity and products characteristics are linked between the food microbiota and health benefits.

The predominant microbial species in African indigenous fermented food such those coming from West Africa Countries are LAB and are used for different applications in the food and biotechnology industries nowadays. Numerous studies have described antimicrobial susceptibility and therapeutics profiles of LAB and are accepted as probiotics from different parts of the world (Adimpong et al., 2012; Mokoena et al., 2016).

The aim of spontaneous fermentation is to improve flavor, texture, and increased shelf-life, bioavailability of micronutrients, and reduction or absence of antinutritional and toxic compounds in the indigenous fermented food. As reported by Mokoena et al. (2016), fermented foods are preferred by consumers because of their characteristic taste, texture, and color in spite of their unfermented counterparts.

According to Panda and Ray (2016), food, and food products made with cereals and tropical roots crops are produced by allying human skills and microbes ability to convert or use cereals and roots crops nutrients. Fermented foods could be defined as the action of enzymes (amylases, proteases, and lipases) synthesized by microorganisms such as bacteria, yeasts, and molds in order to produce for human a non-toxic food products with attractive and pleasant aromas, flavors, and texture. As previously reported by Steinkraus (2002) and cited by Panda and Ray (2016), fermentation plays roles in food processing such as (i) enrichment of the human diet through development of a wide diversity of flavors, aromas, and textures in foods; (ii) preservation of foods through inhibitory metabolites such as organic acids (lactic, acetic, formic, and propionic acids), ethanol, and bacteriocins; (iii) enrichment of food substrates biologically with vitamins, proteins, essential amino acids and essential fatty acids; (iv) detoxification and inhibition of pathogens during food fermentation processing; and (v) decrease in cooking times and fuel requirements.

In West Africa, cereal-based fermented foods (*ogi, akassa, koko, kekkey, gowê, tchoukoutou, dolo, tchapalo,* etc.), roots crop food (*attiété, gari,* etc.) and dairy products (*wagashie* and *wara*) have been fully gained researchers attention. But the data information availability is not the same among countries. Thus, in Nigeria, there is enough literature on local traditional fermented foods (*pito, ogi, wara, nunu, gari*) but there is less information's in other countries. This paper aimed at highlighting some popular indigenous foods widely consumed by West Africa citizen focusing to (i) understand their process technology; (ii) know the microorganisms involved in their production, and (iii) show antimicrobial and antifungal activities of those microorganisms.

5.2 DIFFERENT TYPE OF WEST AFRICAN INDIGENOUS FOOD AND FOOD PRODUCTS

5.2.1 CEREAL BASE FOOD

5.2.1.1 AKAMU

Akamu is popular uncontrolled cereal fermented gruel or porridge and infant weaning food among the Igbo-speaking people of Nigeria. Maize, millet, and sorghum are the cereals used to prepare *akamu*. As traditionally gruel, its quality may depend on the skills of the producers (Chelule et al., 2010). *Akamu* contributes substantially to the daily diet of both rural and urban communities especially for nursing mothers and weaning ration for infants between the ages of 1–2 years (Adams, 1998) and food for invalids because of its lightness and easily digests (Afolayan et al., 2010). *Akamu* is produced by spontaneous fermentation, the reason of presence of several *akumu* with variable quality. Soak clean maize grains in water for 2 to3 days, followed by washing and grounding into paste, sieve of paste into slurry are main steps essential in the production of *akamu*. To have *akamu* porridge, the hot water is added to the slurry (Nwokoro and Chukwu, 2012). *Akamu* could be fortified with sugar, legumes, milk, or chocolate to improve taste or soothe the sour taste (Osungbaro, 2009a; Adelekan and Oyewole, 2010). This supplementation is also done to improve the low quality of maize protein and to replenish the substantial loss of nutrient at different stages of production.

5.2.1.2 DOKLU OR KOM

Doklu or *Kom* is a fermented corn dump cooked in dry corn leaves. It is widely consumed by the populations of southern regions of Côte d'Ivoire, Ghana, and Togo but also in Burkina Faso and Benin republic. It is a traditional food much appreciated for its sour taste linked to the spontaneous fermentation. For its preparation, the corn kernels are soaked for 2 to 3 days and then ground. The flour obtained is mixed with water and then allowed to ferment at room temperature for 24 to 72 hours. The fermented dough often mixed with *agbélima* (fermented cassava paste) is pre-cooked and pellets are then formed followed by packing in dry maize leaves. The pellets are then cooked for about 2 hours (Ola, 2010; Assohoun et al., 2013). The populations consume it mostly accompanied by chili paste + onion + fish. The consumption of *doklu* with protein-rich sauces (fish, turkeys, etc.) and vegetables is recommended to meet the nutritional requirements of the body. The high organic acid content of *doklu* could be an undeniable benefit for the safety of the product by inhibiting the growth of microorganisms, particularly those that do not support high acid levels (Assohounet al., 2013).

5.2.1.3 FURA

Fura is one way of transforming willet into a good, attractive, and pleasant food product for human, especially those living in Burkina Faso, Ghana, Nigeria, and northern Togo (Jideani et al., 2001; Obodai et al., 2014). In those countries *fura* is mostly consumed by Muslim peoples that are why *fura* could easily found in Zongo communities. Kordylasi (1990), Jideani et al. (2001), and Owusu-Kwarteng et al. (2010) have described *fura* production process as sowed by Figure 5.1. Steep millet grains in water, washing, dehulling, milling, dough fermentation, initial cooking, pounding, and second molding are main steps involved in the *fura* bioprocessing. Aromatic ingredients are added in the milling stage. Yogurt, fresh milk, or water could be mixed with *fura* (a small prepared millet ball) to make porridge (Kordylasi, 1990). "*Fura de nunu*" is a product obtained after mixture of milk and *fura*. Sugar is used as taste quencher. *Fura* is consumed by adults and use as weaning food for babies and children. The specificity of *fura* is its pounding and molding occurred after fermentation.

5.2.1.4 GOWÉ

Gowé, the non-alcoholic cereal-based beverage is a fermented paste traditionally made from pre-germinated corn very popular in the south and central Benin. It's consumed by all ages and the social strata. *Gowé* is generally consumed as beverage after homogenizing the paste with water, sugar, milk, and ice (Tchekessi et al., 2013; Adinsi et al., 2014). Sorghum, millet or maize can be used alone or mixture to produce *gowé* (Adinsi et al., 2014). The fermented dough from germinated cereals called *gowé* in Fon, a local language can only be obtained from maize, especially white sprouted. It is, moreover, this white color of maize from which this product is generally prepared which gives it this gouged name. On the other hand, the fermented pulp from germinated sorghum is called *abotin* (in Fon) and not *gowé*. It should be noted that most of the studies carried out on the *gowé* produced in Benin were found to be done on *gowé* prepared from sorghum (Tchekessi et al., 2013). *Abotin* is a reddish beverage with a pleasant aroma, a slightly sweet and acid flavor; while the beverage obtained from *gowé* is whitish in color.

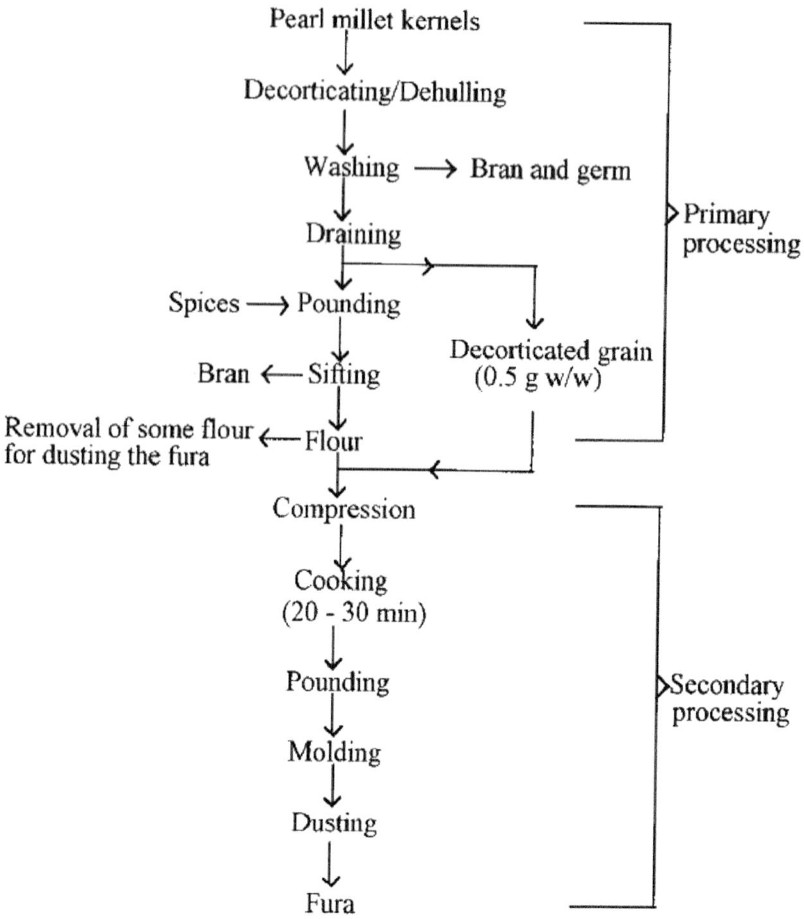

FIGURE 5.1 Schematic diagram for process and product technology of *Fura* production (Jideani et al., 2001).

[Reprint from: Jideani, V. A., Nkama, I., Agbo, E. B., & Jideani, I. A., (2001). Fura production in some northern states of Nigeria-A survey. *Plant Foods for Human Nutrition, 56*, 23–36. With permission].

The *gowé* production sector plays a major role in the Beninese economy. It generates jobs for thousands of people especially women. In addition to its socio-economic importance, *gowé* has nutritional interest not only by its richness of iron, B vitamins and other minerals (Tchekessi et al., 2013), but also by the many advantages of fermentation cereals.

In Benin, it is a favorite food of all classes of age (children, pregnant women, the sick and the elderly) socio-cultural groups, and educational levels (Adinsi et al., 2014). Production technology of *gowé* shown in Figure 5.2 proposed by Vieira-Dalodé et al. (2007) and Tchekessi et al. (2013).

A study conducted by Adinsi et al. (2013) on behalf of the African Food Tradition Revisited by Research, a project of CIRAD (Center for International Cooperation in Economic Research for Development) in 2013 had make available biochemical and nutritional quality of the kind of *Gowé* (maize, sorghum, and mixed sorghum and maize). The results showed that the amount of lactic acid is 3.8%, 3.3%, and 3.1%, respectively for maize *gowé*, sorghum *gowé* and mixed sorghum and maize *gowé*. The protein content for sorghum *gowé,* maize *gowé,* steam cooked *gowé,* and mixed sorghum and maize *gowé* are respectively 11.1%, 10%, 9.9%, and 9.5%. On dry basis (db), total minerals content varies from 0.7% to 2.0% respectively for maize *gowé* and sorghum one. The maize *gowé* contains more total fiber content which are 1.9 (% db). Maize *gowé* has the highest crude fat content (1.5% db). They also identified the presence of sugars like maltose, glucose, sucrose, and fructose in large amount. Methionine and lysine are the most limiting amino-acid because their amount is less than the Recommended Daily Intake (RDI).

5.2.1.5 KENKEY

Kenkey, a fermented maize dough product not only eaten by the people of Ghana (Nyanzi and Jooste, 2012), but in other West Africa countries like Benin, Togo, Burkina Faso, Nigeria, and Côte d'Ivoire. According to Halm et al. (2004) and Nout (2009), *Kenkey* is fermented food produced traditionally by steeping whole or dehulled maize grains in water for about two days. Then steeped grains are milled and kneaded into a dough which is allowed to ferment spontaneously for about two to four days to obtain a sourdough. After half of the sourdough is cooked into aflata and mixed with other uncooked dough to obtain a characteristic sticky texture. To have kenkey, the obtained mixture is molded into balls as tennis ball size, wrapped in several leaves and cooked for at least 3 hours.

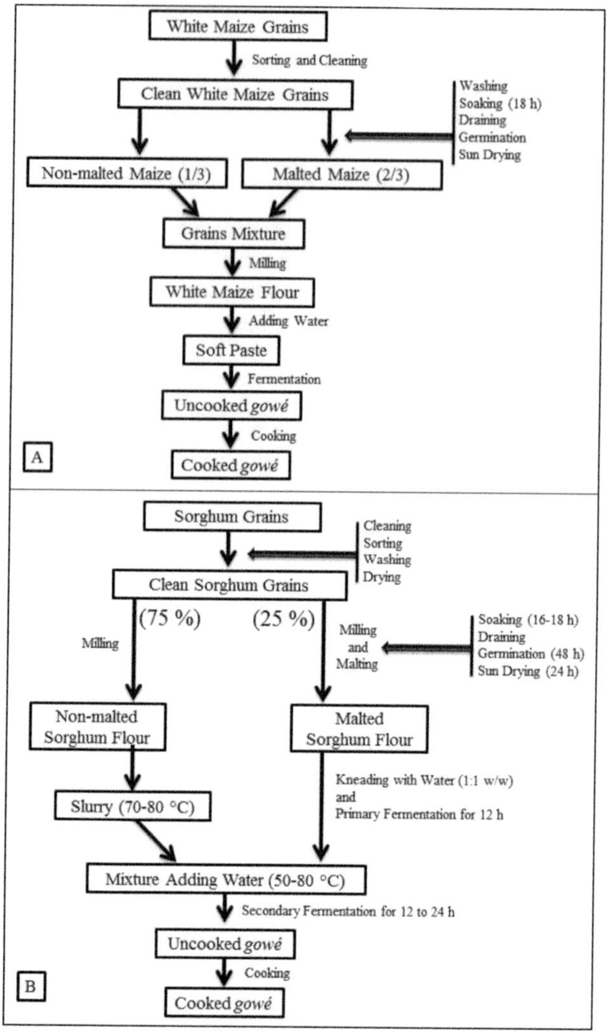

FIGURE 5.2 Schematic diagram for process and product technology of *Gowé* production. A) Production technology from white maize grains B) Production technology from sorghum grains (A: Reprinted with permission from Tchekessi, et al., 2013.; B: Reprinted with modification with permission from Vieira-Dalodé, et al., 2007. © 2007 John Wiley and Sons.)

Ga-kenkey or *Komi, Fanti-kenkey* or *Dokono*, and *Nsiho* or *Akporhi* are mainly types of *kenkey* (Table 5.1). *Nsiho* or *Akporhi* produced from dehulled or polished maize and widely consumed in Eastern, Western,

Central, and Volta regions of Ghana is another type of *kenkey* (Annan, 2013). Ga *kenkey*, *Fanti kenkey*, and *Nsiho* differ in their organoleptic quality and the processing procedures as previously described by Halm et al. (2004). They are both produced from whole maize grains. *Fanti-kenkey* and *Ga-kenkey* are respectively fermented for at least 3 days and 2 days. Only *Ga-kenkey* contains salt. *Fanti-kenkey* and *Ga-kenkey* are respectively wrapped in maize husks and in several plantains or banana leaves (Halm et al., 2004). *Nsiho* is kneaded with water into dough and left to ferment spontaneously for 24 hours, and the dough obtained is precooked, molded into balls, wrapped in maize husks and steamed for about two hours (Annan, 2013). *Kenkey* consumed as a ready-to-eat meal together with other foods has a shelf life of several days (Nout, 2009; Franz et al., 2014). The physicochemical characteristics of the three type of *kenkey* are listed in Table 5.1.

TABLE 5.1 Different Types of *Kenkey* and Their Physico-Chemical Characteristics

Physico-Chemical Characteristics	Ga-Kenkey	Fanti-Kenkey	White Kenkey or Nsiho
Moisture content	64.5%	-	-
Crude ash content (g/100g)	0.92	0.71	0.87
Protein content (g/100g)	8.9–9.8	5.1	2.4
Fiber content (g/100g)	-	1.22	0.13
Fat content (g/100g)	1.3–3.2	1.02	0.16
Total carbohydrates (%, db)	74.3–87.1	-	-
Glucose (%, db)	2.4–2.5	0.7%	2.4–2.5
Calcium (mg/100g)	10.6–78.6	-	-
Phosphorus (mg/100g)	202.4–213.8	-	-
Zinc (mg/100g)	0.68	0.78	0.18
Iron (mg/100g)	6.5–12.6	2.5	2.2
Vitamin B_1 (μg/100g, db)	200	113	17
Vitamin B_3 (μg/100g, db)	643	807	138
Vitamin B_6 (μg/100g, db)	241	235	30
Vitamin B_8 (μg/100g, db)	10	5	6
Vitamin E (μg/100g, db)	694	1193	405

Source: Annan-Prah and Agyeman, (1997); Obiri-Danso et al. (1997); Halm et al. (2004); Adinsi et al. (2013).

According to the project of CIRAD in 2013 conducted by the team of Adinsi (Adinsiet al., 2013); *Nsiho* has the lowest mineral content (Table 5.1).

Nsiho has also the lower lysine content, and methionine and lysine are the most limiting amino-acid if compared to RDI. Table 5.1 shows that *Nsiho* is very poor in vitamins compared to the other Kenkey types. Those values obtained for *Nsiho* could probably due to the dehulling and degerming step during its production, which eliminate a large part of the vitamins and mineral which are mainly located in the germ and aleurone layer.

5.2.1.6 KOKO

Koko is one of the millet porridge produced and consumed daily by many people (children, teenagers, adults, and elderly) in Ghana and other West Africa countries like Burkina Faso, Togo, Benin, Nigeria, etc., as lunch or an in-between meal. According to Lei and Jakobsen (2004), *koko* production starts with overnight pearl millet steeping, followed by a grains wet-milling with spices (ginger, chili pepper, black pepper, and cloves), and obtaining of a thick slurry after adding water. Afterwards, the sieved slurry is let for fermentation and sedimentation about three (3) hours. The top-layer liquid is boiled for two (2) hours. Finally, the bottom thicker slurry is added to the boiling top-layer liquid until the desired consistency of *koko* is achieved. Figure 5.3 shows *koko* production process as previously reported by Lei and Jakobsen (2004) with few modifications.

5.2.1.7 KUNUN DRINK/KUNUN-ZAKI

Known as a nutritional drink, *kunun* drink has widely used by Nigerian as food complement, taste quencher and also used as refresher. According to Mbachu et al. (2014) and Umaru et al. (2014), *kunun* drink, non-alcoholic beverage, milky cream, and appetizer can be found in several public places such as markets, offices, schools, motor parks and used as drinks during festivities, weddings, and naming ceremonies. It's also used as a substitute of imported milk for poor people especially in rural areas. The investigations of Adelekan et al. (2013) and Umaru et al. (2014) showed that sorghum (*Sorghum bicolor*), millet (*Penisetum typhoides*), maize (*Zea mays*), rice (*Oryza sativa*), wheat (*Triticum aesstivum*), and acha (*Digitalis exilis*) are some plants used for the preparation of *Kunun* drink. The preparation of *Kunun* drink involved those steps: steeping the whole grains for 6–24 hour, wet milling with spices such as ginger (*Zingiber officinales*), aligator pepper (*Afromonium melegueta*), red pepper (*Capsicum* species), black pepper (*Piper guinense*), and sweet potato, gelatinization of one part in which the ungelled part is added; afterwards the mixture is let for overnight fermentation. Sugar is commonly added (Oranusi et al., 2003; Ogbonna et al., 2011; Umaru et al., 2014; Orutugu et al., 2015). Those ingredients used to prepare *kunun* drink (for all kinds of non-alcoholic beverages that are cereal-based) had let to distinguish several type namely *Kunun zaki, Kunun gyada, Kunun akamu, Kunun tsamiya, Kunun baule, Kunun jiko and Kunun gayamba*. But amount them *kunun zaki* is the most widely produced and consumed (Amusa and Ashaye, 2009; Amusa and Odunbaku, 2009; Ayo et al., 2010; Nahemiah et al., 2014). For example the millet-based *Kunu zaki* technology involved steeping of millet grains, wet milling with spices (ginger, cloves, and pepper), partial gelatinization of the slurry, cooling (34°C), fermenting at 34°C for 12hours, sieving, sweetening with sucrose and bottling (Ayo et al., 2011). As previously reported by Amusa and Ashaye (2009) *kunun* drink may contain 2.31–3.63% of protein, 3.55–3.63% of fat, 1.16–1.21% of ash, and 82.92–83.55% of carbohydrate. The main problem encountered by *kunun* drink is their self-life because it must be consumed within few hours after their production (Adeleke and Abiodun, 2010).

5.2.1.8 MAWÈ OR ÉBLIMA

Mawè or *éblima* is an uncooked dehulled fermented maize dough largely consumed in West Africa like Benin, Ghana, Togo which preparation still in house scale as the result of spontaneous fermentation processes (Greppiet al., 2013a). In Benin it's used as important ingredient in the preparation of cooked beverages, steamed cooked bread (*ablo*), stiff gels (*akassa, agidi, eko*), pre-gelatinized yoghurt-like product as *akpan*, fritter as *massa* and *paté*, couscous as *yeke-yeke*, and porridge (*koko, aklui, akluiyonou*) (Nout, 2009; Nyanzi and Jooste, 2012; Greppi et al., 2013a; Franz et al., 2014). We distinguish *éblima* and *époma* respectively *mawè* prepare with maize and millet. In Togo, *mawè* is used to prepare the following meals: *akassa, ablo, koko, aklui, akpan*, and *massa*. For *ablo* and *massa* the producers often used rice or both maize and rice. The *mawè* production involved maize grains washing, wet extraction of the endosperm, and knead to dough, leave for about three days spontaneously fermentation (Nout, 2009;

Nyanzi and Jooste, 2012; Kohajdová, 2014, 2017). *Mawè* manufacturing processes have been described by Hounhouigan et al. (1993a, b).

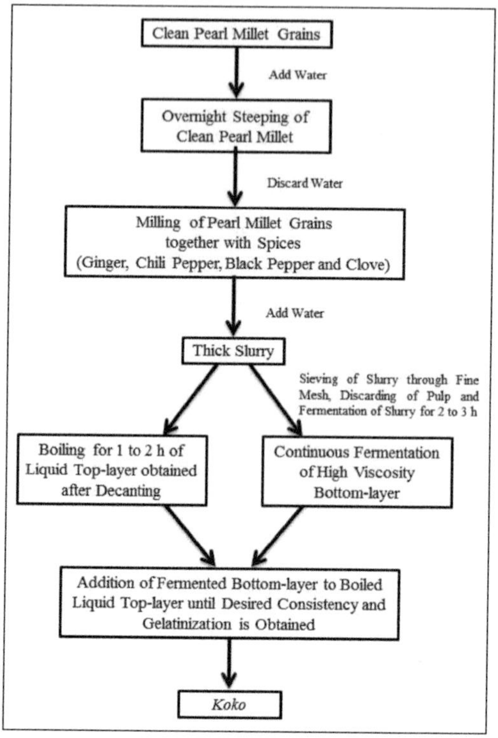

FIGURE 5.3 Schematic diagram for process and product technology of *koko* production from *Pennisetum glaucum* (Lei and Jakobsen, 2004).

5.2.1.9 OBIOLOR

Obiolor is a Nigerian Igala tribe sweet non-alcoholic beverage produced from fermented sorghum and millet malts (Ajiboye et al., 2016). The technology of *obiolor* had been described by Achi (1990) and cited by Ajiboye et al. (2016) and involved the following steps: after steeping in water overnight, sorghum, and millet grains were wrapped in fresh banana leaves and allowed to germinate for 3 days. Then the germinated grains were wet milled and prepared into slurry and add boiled water (ratio 1:4 v/v). The obtained mash was cooled, filtered (residue discarded), and the filtrate was concentrated by boiling for 30 min with continuous stirring. Finally, the gruel obtained was cooled rapidly and let for 24 hours spontaneously fermentation, and kept frozen till used. The acidity, taste, and aroma of *obiolor* is due to the activity of *Lactobacillus plantarum* and *Lactococcus lactis* (Achi, 1990). *Obiolor*, less known as others Nigerian traditionally fermented beverage is a beverage with available carbohydrate (8.90%), crude fat (0.39%), crude protein (7.80%), crude fiber (0.30%) and high energy content (459.30 kJ/g) (Ajiboyeet al., 2014).

5.2.1.10 OGI

Ogi which has a sour flavor and a characteristic aroma are Nigerian and northern of the republic of Benin acid fermented cereal gruel or porridge prepared either with maize, or sorghum, or millet (Nyanzi and Jooste, 2012). In West Africa, according to the cereal used, *ogi* has various names such as *ogi, ogi-baba, and ogi-ogero,* respectively, when white maize, sorghum, and millet was used to prepare this gruel. For Onyekwere et al. (1989) cited by Nyanzi and Jooste (2012), *ogi* from maize is known as '*akamu*' or '*ekogbona*' in Northern Nigeria, while it is known as '*koko*' in the Republics of Togo, Benin, and Ghana. Teniola and Odunfa (2001) had reported that *ogi* is the major traditional weaning food commonly served to babies and eaten as a breakfast meal in some West Africa countries. Steeping clean maize or sorghum or millet grains in water at room temperature, after 48–72 hours the steep water is decanted and the fermented grain is washed with clean water before been wet-milled. Finally the bran is removed by wet sieving and the filtrate is let to settle to ferment for 24 to 48 hours to obtain sour *ogi*: those are the main steeps the producer have to follow to produce *ogi* (Teniola and Odunfa, 2001; Taiwo, 2009; Omemu, 2011; Banwo et al., 2012; Oyedeji et al., 2013). Cream *ogi* is prepared with maize, while light brown *ogi* and greenish to grey *ogi* are respectively prepared with sorghum and millet. *Ogi* is one of Nigerian cereal-based fermented food most studied on the microbial diversity, starters used and *ogi* fortification (Teniola and Odunfa, 2002; Teniola et al., 2005; Adebayo-Tayo and Onilude, 2008; Oguntoyinbo et al., 2011; Adebolu et al., 2012; Banwo et al., 2012; Oguntoyinbo and Narbad, 2012; Esther et al., 2013; Greppi et al., 2013a, b; Oluseyi et al., 2013; Oyedeji et al., 2013; Onwuakor et al., 2014a; Ijeoma et al., 2015; Okeke et al., 2015).

5.2.2 CASSAVA DERIVED FOODS

5.2.2.1 AGBÉLIMA

Agbélima is a popular fermented product in Côte d'Ivoire, Ghana, and Togo. *Banku*, *akplé*, and *kenkey* are some of several meals prepare with *agbélima*. *Agbélima* can easily be produced in larger quantities at a relatively low cost (Ellis et al., 1997; Panda et al., 2016). The main characteristic of *agbélima* is the use of an inoculum locally called *kudeme*. For Sefa-Dadeh (1989), the main purpose for using *kudeme* is for souring and texture degrading, which helps to improve the texture, color, and flavor. *Agbélima* production process steps are: after peeling and steeping in water (initial fermentation), the cassava roots are mashed into a paste. The paste put in jute sacks sustained for two days fermentation. After fermentation, the sacks are pressed by tightening and pressing by putting on the sacks heavy stones. Finally, when the liquid stop flowing the cassava paste is removed and crumbled. *Agbélima* is widely produced in Togo, especially in the forest area like the area going from Kpalimé to Badou. It's also found in big cities like Lomé, Notsè, Kpalimé, Badou, Atakpamé, etc. People usually used it to prepare a paste locally called *émakumé* or *agbélima kumé*. It can be used only or with fermented maize flour or with non-fermented maize flour. *Emakumé* or *agbélimakumé* is a gelatinized paste and can be consumed with any kind of sauce but they prefer fresh tomatoes sauce with fried fish.

5.2.2.2 ATTIÉKÉ

Attiéké is a steamed granular obtained after processing cassava (*Manihot esculenta* Crantz). It's a meal, couscous-like product, with slightly sour taste and whitish color which can be consumed all day with meat, fish or vegetables. If nowadays *attiéké* is consumed in several West African countries such as Benin, Burkina Faso, Togo; originally this food was prepared and consumed exclusively by the ethnic groups like Adjoukrou, Alladjan, Ebrié, Avikam, and Ahizi, living in the Laguna area in the south of Côte d'Ivoire (Assanvo et al., 2006; Djeni et al., 2011). *Attiéké* production always begins by preparing a traditional starter from cassava which takes about 2 to 3 days. The starter constitutes the main source of microorganisms which are active in the later fermentation. The choice of traditional starter to use is the basis of the different organoleptic qualities of the various types of *attiéke* sold on the market (Westby, 1991; Assanvo et al., 2006). That is why there exist several types of *Attiéké* in Cote d'Ivoire linked to the ethnic groups who produce them and amount them we can distinguish some such as *attiéké* Adjoukrou, *attiéké* Ebrié, and *attiéké* Alladjan because of their highly appreciation (Assanvoet al., 2006). Assanvo et al. (2006) suggested that having a standardized starter could be beneficial for producers and consumers because quality and attractive *attiéké* would be available. This will help producers to rise the *attiéké* exported amount especially increase their turnover or their benefice. They continue their suggestion by saying that those are reasons which make *attiéké* quality variable even from the same producer and the same cassava variety. The cassava tuber is harvested, peeled, and washed, mash the tuber, dewatering, and fermentation of the mash, followed by sieving, granulating, sundrying, and steaming of the granular product are the main steps involved in the *attiéké* production (Djeni et al., 2011). Ready to eat food either hot or cold, his cheapness, and lower bulk make the popularity of *attiéké* (Assanvo et al., 2006; Djeni et al., 2011).

The bitterness of the cassava is eliminated during the long fermentation process of *attiéké* production and also reduced or removed the cyanide content (Fortin et al., 1998; Assanvo et al., 2006). During the fermentation of ground fresh cassava, changes in physical properties take place, allowing the formation of small granules (similar to couscous in the final stages) of the process (Fortinet al., 1998).

5.2.2.3 GARI

One of the known cassava derived food is *gari*. *Gari* is widely produced and consumed in several West Africa countries, especially in Nigeria (the biggest producer), Ghana, Togo, and Benin. In West Africa, 70% of produced cassava is processed into *gari*. That is why Felber et al. (2016) and Sanni et al. (2016) said that *gari* is the most produce and consume food product from cassava. *Gari* is a ready-to-eat food obtained by a solid fermentation of cassava roots. *Gari* could be eaten by soaking it in cold water with sugar, and or coconut, roasted groundnuts. *Gari* could also be used with dry fish, or boiled cowpea as complements or as a paste made with hot water and eaten with vegetable sauce (Oguntoyinbo and Dodd, 2010; Evans et al., 2013). According to Oguntoyinbo and Dodd (2010); Evans et al. (2013) and Ojo et al. (2016) the production of *gari* is by spontaneous fermentation following peeling and grating of cassava. Then the grated pulp is filled into cloth bags and tied securely and heavy stones are placed on the sack in order to press out cassava juice. After three to five days of fermentation, the pulp is afterwards roasted on a shallow metal pot, and the final product *gari* is a dry, farinaceous cream-colored powder. *Gari* produce is packing in polythene bags and store in a cool and dry place. Irtwange and Achimba (2009) reported that in Nigeria and in many African countries, *gari* is consumed in various forms either as a snack or a full meal. In Togo, a paste called "*pinon*" is obtained by pouring *gari* in boiling water and continuous stirring up to obtain a gelatinize paste. According to producers or consumer taste, tomatoes, piper oil (preferring red palm oil) are added, making it as a ready-to-eat food. If prepare simply, *pinon* have to be accompanied with a sauce of choice. Evans et al. (2013) reported three types of *gari* which are Rough-sour *gari,* medium *gari* (usually used to prepare "*pinon*" which are prepare by mixing *gari* with a boiling water with continuous stirring. Pinon is consumed by Togolese people) and smooth *gari.* Felber et al. (2016) cassava variety, climate, soil conditions, harvesting time, and *gari* processing are different parameters influencing *gari* quality.

5.2.2.4 LAFUN

Known as cassava fermented flour, *lafun* is popular in the Southwestern States of Nigeria and Republic of Benin. As *gari, lafun* could be conserved for a long time at ambient temperature. Its cheapness had made him as popular meal especially in the rural areas. Compare to *gari, lafun* could be prepare easily and in reduce time (Latunde-Dada, 1997). *Lafun* is generally used to prepare a stiff porridge called *Oka* usually consumed with vegetable soups or tomatoes. Cassava root peelings, washing, retting, crumbling, and drying are the most important operations involved in *lafun* production. We distinguish ordinary *lafun* and *Chigan lafun* (Figure 5.4) (Hounhouigan et al., 2003; Padonou et al., 2009a). In the production of *chiganlafun*, the washing performed after retting contribute to reduce cassava strong aroma and also the toxic substrates of cassava such as cyanohydric acid (Cereda and Mattos, 1996). Differences of processing amount producers, season (changes in temperature of fermentation), and cassava age and cassava variety are factors which may influence *lafun* quality (Akingbala et al., 1991; Blansherd et al., 1994; Idowu and Akindele, 1994).

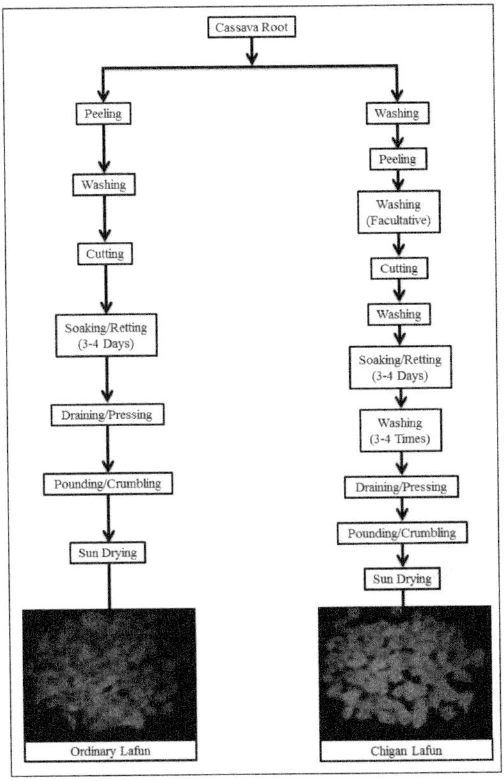

FIGURE 5.4 (See color insert.) Schematic diagram for process and product technology of ordinary *lafun* and *chigan lafun* production. (Reprinted from Padonou, Hounhouigan, & Nago, 2009a. Open access.)

5.2.3 INDIGENOUS BEERS

Sorghum beer is also produced in several West African countries and known as *pito* and *burukutu* (Nigeria and Ghana), *otika* in Nigeria, *tchoukoutou*and *tchapalo*" in Benin, Burkina Faso, Togo, and Côte d'Ivoire (N'tcha et al., 2016). If *dolo* is found in Burkina Faso, its equivalent are available in Ghana and Nigeria known as *pito*. On the other hand, *tchoukoutou*, and *tchapalo* are widely found in Western African subregion, but the quality varies from one country to another and from producers in the same country.

5.2.3.1 PITO

For Evans et al. (2013), the steps involved in the production of *pito* includes grains soaking, malting, mash filtration, boiling, and fermentation. Indeed, those steps are involved in the production of pito: maize grains or sorghum grains or combination of maize and sorghum are soaked in water for two days, then the water is removed, and the grains are put in moistened banana leaves and let about 5 days for germination (malting). After, the ground grains are mixed with water and boiled for one hour at least. The cooled mash is filtered through a fine-mesh basket. Finally, the filtrate is allowed to stand overnight until it assumes a slightly sour flavor, and boiled again up to concentrate. The *pito* is obtained after adding a starter culture coming from the previous brew. After overnight fermentation, the alcoholic beverage *pito* obtained is a dark brown liquid. *Pito* taste varies from sweat (no fermented *pito*) to bitter. Lactic acid, sugars, amino acids, and ethanol 3% are the main content of *pito* (Evans et al., 2013).

5.2.3.2 TCHAPALO AND TCHOUKOUTOU

Tchoukoutou and *tchapalo* are traditional alcoholic beverages produced in Benin. Sorghum malt is mainly used (red and brown varieties), but millet and corn can be used as partial or total substitutes (Kayodé et al., 2005; Osseyi et al., 2011). *Tchoukoutou* and *tchapalo*, found also in Togo, Burkina Faso, and Côte d'Ivoire; are distinguished by their appearance and taste. Indeed, *tchoukoutou* is an opaque and acid beer (see Figure 5.5) while *tchapalo* is a light and slightly sweet fluid beer. There are two types of *tchoukoutou*: light *tchoukoutou* which undergoes filtration after the second fermentation and heavy *tchoukoutou*. The *tchoukoutou* is produced mainly by women with various unitary operations. In general, as in the case of conventional Lager beers, production takes place in three phases: malting, brewing, and fermentation (Kayodéet al., 2007b). The process of traditional *tchoukoutou* brewing has been fully described. In Togo, *tchoukoutou* is widely appreciated and most produce by kabyè ethnic group women. *Tchapalo* is most appreciated by the people from the northern region of Togo like Dapaong and Mango. Nowadays, *tchoukoutou* and *tchapalo* are widely consumed by all type of Togo ethnic groups, by people with low income and people with high revenue. *Tchoukoutou* and *tchapalo* contribute in Togolese economy because it generates many jobs especially for women.

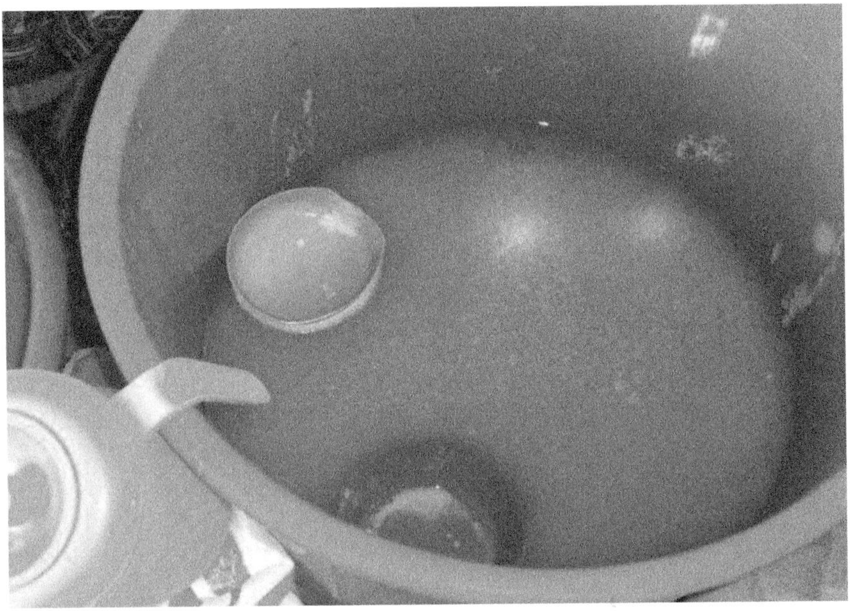

FIGURE 5.5 (See color insert.) Traditional *tchoukoutou* beer aspect. (Reprinted from N'tcha, 2016. With permission.)

5.2.4 DAIRY-BASED FOOD PRODUCTS

5.2.4.1 DÉGUÉ

Dégue is a traditional fermented millet-based food which is consumed in West Africa countries like Benin, Burkina Faso, Togo, Cote d'Ivoire, etc. *Dégue* can be consumed by children, adults, and old persons and accompanied by sugar, milk, honey, tamarind juice, and citron juice (Hama et al., 2009). According to Abriouel et al. (2006); Hama et al. (2009) and Todorov and Holzapfel (2015), *dégue*, a ready-to-eat food which preparation involves dehulling and grinding of the millet grains, modeling with water and steam cooking in the production of gelatinous balls, then the balls are stored to allow a further 24 hours spontaneous and uncontrolled fermentation. Nowadays, couscous is used. In Togo, *dégué* is sell in schools, markets, shops, and other attractive places. In Lomé, there are several restaurants that specialize in *dégué*.

5.2.4.2 FERMENTED MILK

Traditionally fermented milk is prepared domestically, and fermentation occurs spontaneously at ambient temperatures. In rural areas in West Africa, fermented milk products are consumed as food and beverages and also serve as a source of income. Traditional fermented milk products could be found in Africa, Asia, the Middle East, and Europe (Savadogo et al., 2004b). In West Africa countries, the milk is processed into various products such as yogurt, curd, and *wagashie*. In West Africa, most milk is produced by smallholders who sell it in informal markets.

5.2.4.3 NUNU

Nunu is a spontaneously unpasteurized fermented yogurt-like milk product consumed as a staple food commodity in some parts of the Saharan West Africa, especially in Ghana and Nigeria (Akabanda et al., 2010, 2013). As previously reported by Akabanda et al. (2010), in northern Ghana, *nunu* processing technology is: fresh cow milk is collected in the morning in calabashes, sieved, and left to spontaneously fermentation without adding starter cultures at ambient temperature for a minimum of one day (in hot season) or a maximum of two days (in cold season from October to February). The particularity of *nunu* production is that no pre-fermentation heating of the milk is needed. The fermented milk is then churned using a wooden ladle, and the accumulating fat is removed. Finally, excess whey is drained off in order to obtain a final product, namely *nunu*, with a thick consistency. *Nunu* can be consumed alone or with sugar and *fura*. *Nono* and *wagashie* from Nigeria, Benin, and Togo, and the Fulani fermented milk in Burkina Faso are similar fermented milk products such as *nunu* find in other parts of West African countries (Savadogo et al., 2004b; Akabanda et al., 2013).

5.2.4.4 CHEESE

The production of cheese from milk had been started about 8000 years ago in the fertile crescent between Tigris and Euphrates rivers (Hayologlu et al., 2002). Milk, rennet, microorganisms, and salt are main ingredients used together to produce cheese and includes, gel formation, acid production, whey expulsion, salt addition and ripening (Beresford et al., 2001). The rennet coagulates the milk and breakdown protein in the cheese, thus contributing to the ripening process. The cheese found in West Africa is *wara* or *warakanshi* in Nigeria (Adetunji and Babalobi, 2011; Ojo et al., 2016), *wagashie* in Togo, Bénin and Ghana.

Wagashie or wara is a traditional cheese from an artisanal process developed by the Fulani ethnic group in Benin, Togo, and other West Africa countries. As previously reported by Aïssi et al. (2009) and Adetunji and Arigbede (2011), *wagashie* is the most popular and most consumed milk derivate products especially in Benin, Togo, and Nigeria where it's called *wara* or *warakanshi. wagashie* or *wara* is known as West Africa soft cheese (WASC). According to Adetunji and Arigbede (2011), *wara* is an unrippened soft and moist cheese prepared by coagulating fresh cow milk with the leaf extract of the Sodom apple (*Calotropis pocera*) or pawpaw (*Carica papaya*). The preferred coagulant namely calotropin -an enzyme that curdles milk proteins- comes from *C. procera* because the cheese made with calotropin has a sweeter flavor versus the cheese made with the other coagulant (Adetunjiaet al., 2008). For Kèkè et al. (2008), *wagashie* is an important source of animal protein, fats, and minerals such as calcium, iron, and phosphorus, vitamins, and essential amino acids. Thus, *wagashie* could efficaciously contribute to solving proteins deficiency in the diets in West Africa countries because of his cheapness (Sangoyomiet al., 2010). *Wagashie* has self-life like other milk products in West Africa due to the lack of a cold chain. Several studies have been conducted by researchers especially from Benin (Kèkè et al., 2008; Sessou et al., 2013a, b); Nigeria (Belewu et al., 2005; Adetunji and Adegoke, 2007; Adesokan et al., 2008), and Ghana (Tohibu, 2009) in order to contribute to its long preservation. Producers of *wagashie*, traditionally, preserve them by boiling in water, or salt solution, or by frying, or by smoking or by drying (Tohibu, 2009). Let notice that *wagashie* is largely produced and consumed in Togo, but we have lack of information's about this sector in this country. We encourage several researchers to focus their research topic on that dairy product widely consumed by many Togolese. *Wagashie* was firstly described in Northern Togo early 1899, especially in Sokodé and Bassar region (Rühe, 1938). The production technology of *wagashie* is reported in Figure 5.6. *Wagashie* is white (Figure 5.6A) but could be colored in red with *Sorghum vulgaris* (Figure

5.6B) and in light red with teak leaves (Figure 5.6C) in order to make them more attractive and also to enhance their conservation.

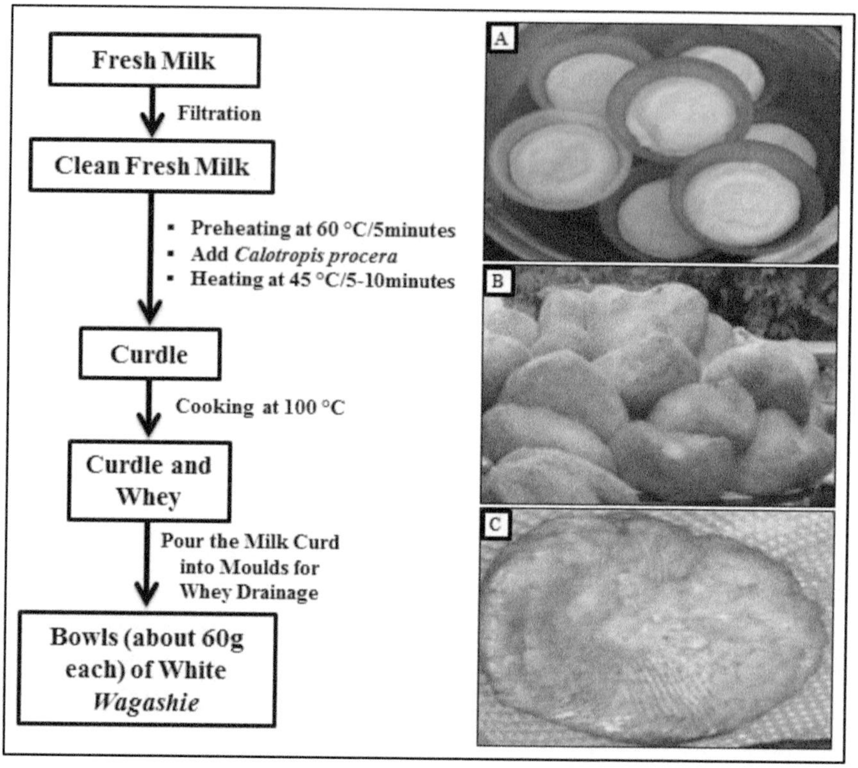

FIGURE 5.6 (See color insert.) Schematic diagram for process and product technology of white *wagashie* production (NWFASUAC, 2006). A) White *wagashie*, B) Red *wagashie*, and C) Light red *wagashie*. (Reprinted from NWFASUAC, 2006.)

5.3 MICROORGANISMS INVOLVED IN BIOPROCESSING AND TECHNOLOGY OF WEST AFRICAN INDIGENOUS FOOD AND FOOD PRODUCTS

Several works on African fermented food products showed a significant complex microbial biodiversity responsible for their known characteristics (Adimpong et al., 2012). As reported by Shetty and Jespersen (2006) LAB and yeasts are the predominant microorganisms population in these spontaneous fermented food products and also used as starters cultures by agribusiness industries.

5.3.1 AGBÉLIMA

Lactobacillus spp. (*Lb. brevis*, *Lb. plantarum*) and *Leuconostoc mesenteroides*, as well as yeasts such as *C. krusei*, *C. tropicalis* and *Zygosaccharomyces bailii* are the microorganisms involved in the *agbélima* production (Amoa-Awuaet al., 1996). A study conducted by Kogno showed the presence of amylolitic LAB in *agbélima* (Kogno, 2016).

5.3.2 AKAMU

In a study conducted by Nwachukwu et al. (2010); Obinna-Echem et al. (2014a, b); the LAB population of a selected Nigerian traditional fermented maize food called *akamu* was found to be dominated by strains of *Lactobacillus plantarum*, *Lb. fermentum*, *Lb. celibiosus*, *Lb. delbrueckii* subsp. *bulgaricus*, *Lb. acidophilus*, *Lb. pentosus*, *Lb. lactis*, *Lb. casei*, *Lb. helveticus*, *Pediococcus pentosaceus and Lactococcus mesenteroides* while the identified yeasts included *Candida tropicalis*, *C. albicans*, *Clavispora lusitaniae* and *Saccharomyces paradoxus*. *Akamu* initial microflora consisted of a heterogeneous mixture of microorganisms, namely *Lb. delbrueckii*, *Lb. plantarum*, *Lb. fermentum*, *Lb. amylovorus*, *Pseudomonas aeruginosa*, *Pseudo. alkaligenes*, *Bacillus cereus*, *B. licheniformis*, *B. subtilis*, *Candida utilis*, *C. tropicalis*, *Saccharomyces cerevisiae*, *Aspergillus oryzae*, *A. niger*, *Penicillium citrinum*, *Rhizopus microsporus*, and *R. oligosporus* (Nwokoro and Chukwu, 2012).

5.3.3 ATTIÉKÉ

Leuconostoc mesenteroides subsp. *mesenteroides*, *Lactobacillus salivarius*, *Lb. delbrueckii* subsp. *delbrueckii*, *Lb. plantarum*, *Lb. fermentum*, *Lb. confuses*, *Weissella cibaria*, *W. confuse*, *Bacillus* spp. (*B. circulans*, *B. lentus*), Enterococci, yeasts, and molds are the microbes found in *attiéké*. The presence of coliforms like *Enterobacter sakazakii*, *Ent. cloacae* and *Klebsiella pneumoniae* subsp. *pneumoniae* may indicate contamination from the environment and producers during production (Assanvo et al., 2006; Coulin et al., 2006; Assamoi et al., 2016; Krabi et al., 2016).

5.3.4 CHEESE

The genera *Lactobacillus, Leuconostoc, Streptococcus, Pediococcus, Escherichia coli, Klebsiella, Enterobacter,* and *Staphylococci* were isolated from *wara* (Sangoyomi et al., 2010; Ojo et al., 2016).

5.3.5 DÉGUÉ

Enterococcus spp., *Lactococcus curvatus, Lc. lactis* subsp.*lactis, Lactobacillus acidophilus, Lb. brevis, Lb. buchneri, Lb. casei, Lb. cellobiosus, Lb. crispatus, Lb. curvatus, Lb. delbrueckii* subsp.*delbrueckii, Lb. fermentum, Lb. gasseri, Lb. paracasei subsp paracasei, Lb. pentosus, Lb. plantarum, Leuconostoc lactis, Ln. mesenteroides* subsp. *mesenteroides, Pediococcus acidilactici, Ped. pentosaceus, Ped. damnosus,* and *Tetragenococcus halophilus* are the microorganisms isolated from *dégué* (Abriouel et al., 2006; Ouattara et al., 2015; Todorov and Holzapfel, 2015).

5.3.6 DOKLU OR KOM

Lactobacillus plantarum, Lb. fermentum, Pediococcus acidilactici, Ped. pentosaceus and *Weissella cibaria* are the species of LAB identified by molecular methods (PCR-TTGE-polymerase chain reaction–temporal temperature gradient gel electrophoresis) in fermented maize pulp according to Assohounet al. (2013). The 16S rRNA gene of *Pediococcus acidilacti* and *Ped. pentosaceus* isolated from *doklu* consumed in Togo were partially sequenced (Kogno et al., 2017).

5.3.7 FERMENTED MILK

For Schutte (2013), LAB are responsible for milk fermentation during which they convert the milk carbohydrates to lactic acid, carbon dioxide, alcohol, and other organic metabolites. *Microorganisms involved in Fulani traditional fermented milk in Burkina Faso are Lactobacillus confuses, Lb. delbrueckii subsp. lactis, Lb. plantarum, Lactococcus lactis spp. lactis, Lc. Lactis subsp. lactis biovar diacetylactis, and Leuconostoc lactis* (Savadogo et al., 2004b).

5.3.8 FURA

Owusu-Kwarteng et al. (2010) reported that *fura* process was dominated by LAB and yeasts. The identified LAB belong to *Lb. plantarum, Lb. brevis, Lb. delbrukii* subsp. *delbrukii*, and *Lb. fermentum*. Among yeast isolates we distinguish *Candida tropicalis, Galactomyces geotricum, Issatchenkia orientalis, Pichia anomala, S. pastorianus, S. cerevisiae*, and *Yarrowia lipolytica* (Pedersen et al., 2012). Another work conducted by Owusu-Kwarteng et al. (2012) identified from traditional *fura* production sites in northern Ghana, homo-and heterofermentative species of *P. acidilactici, Weisella confusa, Lb. fermentum, Lb. reuteri, Lb. salivarius, Lb. paraplantarum*, and *Lb. fermentum*.

5.3.9 GARI

Lactobacillus plantarum, Lb. fermentum, Lb. pentosus, Lb. acidophilus, Lb. casei, and *Leuconostoc mesenteroides*, are the microorganisms involved in the fermentation of cassava dough for *gari* production (Oguntoyinbo and Dodd, 2010). For Coulin et al. (2006), in the production of *gari*, the LAB species metabolize the starch in the cassava pulp leading to the production of organic acids, such as lactic acid, that lowers the pH.

5.3.10 GOWÉ

All reported works showed that the LAB, yeast, and molds are the microorganisms involved in the production of non-alcoholic cereal-based beverage *gowé*. Yeast strains involved in *gowé* bioprocess are: *Clavispora lusitaniae, C. tropicalis, C. krusei, and Kluyveromyces marxianus*, and *Pichia anomala. Lb. fermentum, Lb. mucosae, Weissella confusa, Ped. acidilactici, Ped. pentosaceus*, and *Weissella kimchii* are LAB responsible for *gowé* fermentation (Vieira-Dalodé et al., 2007, 2016; Greppi et al., 2013b; Tchekessi et al., 2013).

5.3.11 KENKEY

Lactobacillus fermentum, Lb. reuteri, Lb. plantarum, Lb. brevis, Latococcus mesenteroides, Pediococcus pentosaceus, and *Ped. acidilactici* as LAB and *C. krusei, S. cerevisiae,* and *G. candidum* as yeast strains are the microorganisms found in fermented *kenkey* (Nyanzi and Jooste, 2012; Franz et al., 2014; Kohajdová, 2014, 2017; Soni et al., 2014).

5.3.12 KOKO

Lei et al. (2006) reported that *Lb. fermentum* and *Weissella confuse* are the predominant microbes in *koko*.

5.3.13 KUNUN DRINK / KUNUN-ZAKI

Fungi species are*Aspergillus* spp., *A. fumigatus, A. niger, Candida albicans, Fusarium* spp., *Penicillium* spp., *Rhizopus nigrican, Saccharomyces cerevisiae.* LAB and *Bacillus* species found in *kunun zaki* drink are *Lactobacillus* spp., *Lb. plantarium, Streptococcus lactis,* and *Bacillus subtilis* respectively. *Citrobacter* spp., *Enterobacter aerogenes Escherichia coli, Micrococcus acidophilus, Klebsiella* spp., *Pseudomonas aerogenosa, Salomonella typhi, Shigella* spp., and *Staphylococcus aureus* are the pathogens bacteria found in *kunun zaki* beverage (Ogbonna et al., 2011; Abimbola et al., 2013; Ikpoh et al., 2013; Aboh and Oladosu, 2014; Mbachu et al., 2014; Umaru et al., 2014; Orutugu et al., 2015).

5.3.14 LAFUN

Omafuvbe et al. (2007) reported that the LAB consistently isolated from *lafun* were *Lb. plantarum, Lb. bulgaricus, Lb. fermentum, Lb. casei* and *Lactococcus sp.* Yeasts species isolated were *Candida* sp., *Debaryomyces* sp. and *Saccharomyces* sp. Other bacteria species isolated were *Bacillus subtilis*, *B. pumilus*, *B. cereus*, *B. macerans* and *B. circulans*, *Klebsiella* sp., *Corynebacterium* sp., *Propionibacterium*, and *Citrobacter freundii*. A recent study conducted by Padonou et al. (2009b) showed that the aerobic bacteria (*B. cereus*, *Klebsiella pneumoniae* and *Pantoea agglomerans*), LAB and yeasts are associated with the fermentation of *Lafun* under traditional conditions. The predominant yeast species associated with *Lafun* fermentation are *Saccharomyces cerevisiae*, *Candida tropicalis*, *Candida glabrata*, *Pichia scutulata*, *P. kudriavzevii*, *Kluyveromyces marxianus*, *Hanseniaspora guilliermondii*, *P. rhodanensis*, and *Trichosporon asahii*.

5.3.15 MAWÈ

During the fermentation process, microbes population dominated by LAB followed by yeasts such as *Clavispora lusitaniae*, *Candida krusei*, *C. kefyr*, *C. glabrata*, and *S. cerevisiae*. *Lactobacillus fermentum*, *Lb. brevis*, *Lb. buchneri*, *Lb. confusus*, *Lb. curvatus*, *Lb. reuteri*, *Lb. salivarius*, *Lactococcus lactis*, *Ped. acidilactici*, *Ped. pentosaceus*, and *Leuconostoc mesenteroides* are the LAB involved the production of *mawè* (Hounhouigan et al., 1993a, b; Nout, 2009; Nyanzi and Jooste, 2012; Franz et al., 2014; Kohajdová, 2014).

5.3.16 NUNU

Several microorganisms were isolated from *nunu* such as LAB, yeasts, and *enterobacteriaceae*. LAB isolated are *Enterococcus faecium*, *E. italicus*, *Lb. fermentum*, *Lb. helveticus*, *Lb. plantarum*, *Lactococcus* spp., *Leuconostoc* spp., *Ln. mesenteroides*, *Streptococcus* spp., and *Weissella confuse*. The involved yeasts are *Candida kefyr*, *C. parapsilosis*, *C. rugosa*, *C. stellata*, *C. tropicalis*, *Galactomyces geotrichum*, *Kluyveromyce maxianus*, *Pichia kudriavzevii*, *Saccharomyces cerevisiae*, *S. pastorianus*, *Yarrowia lipolytica*, *Zygosaccharomyces bisporus, and Zygosaccharomyces rouxii*. Finally *Enterobacter* sp., *Escherichia coli*, *Klebsiella* sp., *Proteus vulgaris*, and *Shigella* sp. are the Enterobacteriaceae isolated from *nunu* showing that its quality needs to be improved (Akabanda et al., 2010, 2013; Abdoul-Latif et al., 2013).

5.3.17 OBIOLOR

Acidity, taste, and aroma of *obiolor* are due to the action of *Lb. plantarum* and *Lc. Lactis*. During the fermentation stage *Bacillus* species presence have been reported (Achi, 1990; Ajiboye et al., 2014, 2016).

5.3.18 OGI

Lb. plantarum is reported as the predominant microorganism during fermentation. *Lb. fermentum* and *Lb. brevis* are also involved in the *ogi* production. Yeast species are *S. cerevisiae, C. mycoderma, C. krusei, C. tropicalis, Geotrichum candidum, G. fermentans,* and *Rhodotorula graminis* (Osungbaro, 2009b; Kohajdová, 2010, 2014; Omemu, 2011; Franz et al., 2014; Soni et al., 2014). *Corynebacterium* spp. hydrolyses the corn-starch and *Saccharomyces* and *Candida* contribute to the *ogi* flavor.

5.3.19 PITO

Candida spp., *C. tropicalis, Kluyveromyces* spp., *Hansenula anomala, Kloeckera apiculata, Kluyveromyces africanus, Schizosaccharomyces pombe, Saccharomyces cerevisiae, Torulaspora delbrueckii*are yeast species involved in the *pito* production. LAB are also involved in the fermentation of *pito*: those strains are *Lb. fermentum, Lb. delbrueckii ssp. delbrueckii, Lb. delbrueckii ssp. bulgaricus,* and *Pediococcus acidilactici* (Demuyakor and Ohta, 1991; Sefa-Dedeh et al., 1999; Orji et al., 2003; Glover et al., 2005; Sawadogo-Lingani et al., 2007; Yao et al., 2009).

5.3.20 TCHAPALO AND TCHOUKOUTOU

Clavispora lusitaniae, S. cerevisiae, Candida krusei and *C. rugosa* in *tchoukoutou* (Kayodé et al., 2007a, 2011; Greppi et al., 2013a, 2013b). Kayodé et al. (2007b) had reported that only *Lactobacillus* spp. are involved in the production of *tchoukoutou* in Bénin. *Candida tropicalis, Saccharomyces cerevisiaeLb. fermenteum, Lb. cellobiosus, Lb. brevis, Lb. coprophilus and lb. plantarum, and Leuconostoc* spp. are isolated from *tchapalo* (Aka et al., 2008; Dje et al., 2009; N'guessan et al., 2010).

5.4 ANTIMICROBIAL ACTIVITY OF MICROORGANISMS INVOLVED IN WEST AFRICAN INDIGENOUS FOOD AND FOOD PRODUCTS

Several West Africa researchers have proved that LAB isolated from fermented food are able to produce bacteriocins (Mohammed and Ijah, 2013), but up to date, few bacteriocins were characterized and properly named (Banwoet al., 2013). The results obtained by Adebayo et al. (2014) showed that bacteriocins from African food, especially Nigerian fermented foods, could be used as an effective control for pathogenic microorganisms. Indeed, the bacteriocin produced by *Lb. fermenteum* and *Lb. casei* were active against collected pathogenic clinical isolates such as Enterohaemorrhagic *E. coli* type 3, Enteropathogenic *E. coli* type 1, Enteropathogenic *E. coli* type 2, *Klebsiella pneumoniae*, *Salmonella typhi*, *Shigella dysenteriae*, *Shigella flexneri*, *Staphylococcus aureus*, and *Streptococcus pneumoniae*. In spite of lack of information's, we could find some partially results for some foods such as:

5.4.1 AKAMU

Ekwem (2014) reported that *Lb. acidophilus*, *Lb. delbrueckii* and *Lb. lactis* produced bacteriocin which showed inhibitory activity against *Staphylococcus aureus*, *Escherichia coli*, *Proteus* sp., *Bacillus subtilis* and *Salmonella sp*. Another study conducted by Obinna-Echem et al. (2014a) showed that *Lb. plantarum* inhibits the growth of five relevant foodborne pathogens: *Salmonella enterica serovar Enteritidis* NCTC 5188, *E. coli* NCTC 11560, *B. cereus* NCIMB 11925, *S aureus* NCTC 3750, and *L. monocytogenes* NCTC 7973.

5.4.2 CHEESE

Bacteriocinogenic LAB like *Lb. plantarum*, *Lb. brevis*, *Lb. fermentum*, *Lb. Lactis*, and *Streptococcus lactis* isolated from *wara* inhibit the growth of *B. subtilis*, *Listeria monocytogenes*, and *E. coli* (Adetunji and Adegoke, 2007). The inhibited compounds are characterized as bacteriocins.

5.4.3 DOKLU OR KOM

Lactobacillus fermentum, *Lb. plantarum*, *Ped. acidilactici*, *Ped. pentosaceus* and *Weissella cibaria* showed a positive antibacterial activity against *Lb. delbrueckii* F31 with inhibition diameters ranging from 11 to 20 mm (Assohounet al., 2013). Although *Lactobacillus delbrueckii* is a LAB commonly found in fermented foods, it is also a bacteria of alteration of certain food products, including citrus juices and those preserved in acid conditions (Guiraud, 2003; Koussémon et al., 2003).

5.4.4 FERMENTED MILK

Lactobacillus spp., *Pediococcus* spp., *Ln. mesenteroides* subsp. *mesenterooides*, and *Lactococcus* spp. isolated from Fulani fermented milk (Burkina Faso) and yogurt (Nigeria) were found to produce antimicrobial compounds known as bacteriocin because this antimicrobial activity were lost after treatment with all the proteolytic enzymes. The synthesized bacteriocins were active against *E. faecalis*, *B. cereus*, *Staph. aureus*, *Shigella* spp., *Salmonellatyphi*, and *E. coli* (Savadogo et al., 2004a; Adejumo, 2014). *Pediococcus* species isolated from Nigerian fermented cow and sheep milk are bacteriocin-producing strains. These bacteriocins PedA and PedB about 27 kDa had relative heat stability, high anti-listerial activity and a good spectrum of activity against some pathogenic microorganisms such as *E. coli* 0157:H7, *B. subtilis* ATCC 6633, *B. cereus* DSM 2301, *Staph. aureus* ATCC 27702, *Micrococcus luteus*, *Clostridium perfringens* DSM 456, *Listeria monocytogenes*, and *Lactobacillus casei* CMCC 1.539 (Banwo et al., 2013). *Klebsiella species*, *Shigella species*, *Salmonella* species, *Staphylococcus, and E. coli* are sensitive to bacteriocin produced by *Lactobacillus* spp. isolated from *nunu* (Obiazi et al., 2016).

5.4.5 FURA

Many works (Afolabi et al., 2008; Wakil and Osamwonyi, 2012; Okunye Olufemi et al., 2015) reported the ability of *Lb. plantarum*, *Lb. fermentum*, *Ln. meseteriodies*, *Ln. brevis*, *Ped. acidilactici* isolated from *fura* to inhibit pathogenic bacteria such as *Salmonella*, *Pseudomonas aeruginosa*,

Escherichia coli, Staphylococcus aureus, Bacillus cereus, B. licheniformis, Salmonella spp., *Pseud. flourescens, Pseud. aeruginosa, Proteus* spp., and *Serratia* spp.

5.4.6 KOKO

Microbes isolated from *koko* are recognized as producing antimicrobial substances. Those microbes tolerate oxgall bile at a concentration of 0.3% and grew at pH 2.5 (Lei et al., 2006). All these properties could make them as good probiotics strains.

5.4.7 KUNUN-ZAKI

Isolates of *Lb. plantarum, Lb. brevis, Lb. delbrueckii, Lb.* fermentum, *Ln.mesenteroides, Pediococcus* spp., and *Streptococcus thermophillus* produced active compounds which inhibited the growth of *Bacillus cereus, B. subtilis, Staphylococcus aureus, E. coli, Pseud. aeruginosa* and *Enterococcus faecalis* 29212 (Omemu and Faniran, 2011; Okereke et al., 2012; Oluwajoba et al., 2013; Okpara et al., 2014). Proteolytic enzymes, trypsin, and pepsin completely inactivated antagonistic activity of those compounds demonstrating their proteinaceous nature. Then the antimicrobial compounds could be designed as BLIS or bacteriocins.

5.4.8 OGI

Several works (Ogunremi and Sanni, 2011; Omemu and Faniran, 2011; Adebolu et al., 2012; Onwuakor et al., 2014b; Princewill-Ogbonna and Ojimelukwe, 2014) showed that the Lactobacilli species specially *Lb. plantarum, Lb. delbrueckii, Lb. brevis, Lb. casei,* and *Lb. Fermentum* synthesized bacteriocins who inhibit the growth of various microorganisms such as *B. subtilis, Staphylococcus aureus, Proteus* spp., *Klebsiella* spp., *E. coli, Salmonella* spp., *Pseud. aeruginosa, Haemophilus influenza, Listeria* spp., *S. cerevisiae,* and *Shigella dysenteriae.* pH changes and heat treatment up to 100°C had no effect on the activity of most of the produced bacteriocins.

5.5 ANTIFUNGAL ACTIVITY OF MICROORGANISMS INVOLVED IN WEST AFRICAN INDIGENOUS FOOD AND FOOD PRODUCTS

a. Inhibitory compound produced by *Lb. delbrueckii* affected fungi species *Candida albicans* and *Candida krusei* growth. So *Lb. delbrueckii* isolate from *Kunu-zaki* had an antifungal activity (Okpara et al., 2014).
b. The antifungal potential of LAB isolated during *doklu* fermentation was reported. *Lb. fermentum*, *Lb. plantarum*, *Ped. acidilactici*, *Ped. pentosaceus* and *Weissella cibaria* showed positive antifungal activity against at least one of the following fungal strains chosen as indicator strains: *Eurotium repens*, *Penicillium corylophilum*, *Aspergillus niger*, *Wallemia sebi* and *Cladosporium sphaerospremium* (Assohoun et al., 2013).
c. *Lactobacillus lactis*, *Lb. bulgaricus*, and *Streptococcus thermophilus* isolated from home-made yogurts in Nigeria were found to produce antifungal compounds active against *Candida albicans*, a pathogenic organism causing Candidiasis (Ayangbenro, 2015).
d. Bacteriocin-producing LAB isolated from *ogi* by Adebayo and Aderiye (2010) possessed antifungal activity because their produce compounds were active against *Penicillium citrinum*, *Aspergillus niger*, and *A. flavus*. They suggested the use of these antifungal substances in bio-control.

5.6 CONCLUSION

West Africa countries have a rich repertoire of indigenous dairy and food. The process is complex and simple in the same time because no modern education is needed to understand the technology. This technology is transmitted from generation to generation, contributing in the waste of information. The predominant microbes are LAB followed by yeasts and molds. In consideration of their popular demand, researchers from West African Universities and research institutes need to put together their knowledge and proposed an adequate technologies which must resolve their artisanal character and self-life problems. They have also increased their energy density. The foods industries from West Africa must also prefer transforming local foods.

5.7 SUMMARY

Several traditional fermented foods have been documented in African countries and include beverages (non-alcoholic and alcoholic), breads, pancakes, porridges cheeses, and milks. In West Africa countries, since antiquity, several traditional fermented food and non-fermented foods exist and are homemade with artisanal technology. In spite of their traditional character, making them as insecure foods, they are widely consumed by all type of ages and any social class. This work is aim to focus on some popular traditional food and food products of West Africa countries like Burkina Faso, Benin, Côte d'Ivoire, Ghana, Nigeria, and Togo. Thus, data were collected by consulting online available published works of researchers from those West Africa countries. Our recorded information reveals that the popular food and food products widely consumed in those West Africa countries are *Agbélima, Attiéké, Dégué, Doklu, Gari, Gowé, fura, Kenkey, Koko, Kunun drink, lafun, Mawè, Nunu, Obiolor, Ogi, Pito, Tchapalo, Tchoukoutou, Wagashie,* and *Wara*. From those foods, the LAB is reported to be the most predominant microorganisms involved in the fermentation process. Apart from their fermentation action, those microorganisms are able to control spoilage and pathogenic microbes by producing antimicrobial compounds such as bacteriocins. The investigated work reveals a lake of information's about, either the type of indigenous LAB bacteriocins or their mode and mechanism of action. So, further works need to be done in order to bring responses. Antifungal activity of bacteriocins is less documented. Finding bacteriocins active against fungi species will help to solve self-life of our traditional food and food products.

KEYWORDS

- ***agbélima***
- ***akamu***
- **antifungal**
- **antimicrobial**
- **cassava**
- ***cereal***

- ***gari***
- ***kenkey***
- **millet**
- **sorghum**
- **traditional fermented foods**

REFERENCES

Abdoul-latif, F. M., Bassolé, I. H. N., & Dicko, M. H., (2013). Proximate composition of traditional local sorghum beer "dolo" manufactured in Ouagadougou. *African Journal of Biotechnology, 12*(13), 1517–1522.

Abimbola, A. N., Adeniyi, R. O., & Faparusi, F., (2013). Microbiological, physicochemical and sensory assessment of improved Kunun-Zaki beverage made from millet and stored under different storage conditions. *International Journal of Food Safety, Nutrition, Public Health and Technology,* 5(2), 8–13.

Aboh, M. I., & Oladosu, P., (2014). Microbiological assessment of Kunun-zaki marketed in Abuja Municipal Area Council (AMAC) in the Federal Capital Territory (FCT), Nigeria. *African Journal of Microbiology Research, 8*(15), 1633–1637.

Abriouel, H., Omar, N. B., López, R. L., Martínez-Cañamero, M., Keleke, S., & Gálveza, A., (2006). Culture-independent analysis of the microbial composition of the African traditional fermented foods poto poto and dégué by using three different DNA extraction methods. *International Journal of Food Microbiology, 111*, 228–233.

Achi, O. K., & Ukwuru, M., (2015). Cereal-based fermented foods of africa as functional foods. *International Journal of Microbiology and Application, 2*(4), 71–83.

Achi, O. K., (1990). Microbiology of 'obiolor': A Nigerian fermented non-alcoholic beverage. *Journal of Applied Bacteriology, 69*, 321–325.

Adams, M. R., (1998). Fermented weaning foods. In: Wood, B. J. B., (ed.), *Microbiology of Fermented Foods* (pp. 790–811). BlackieAcademic, London.

Adebayo, C. O., & Aderiye, B. I., (2010). Antifungal activity of bacteriocins of lactic acid bacteria from some Nigerian fermented foods. *Research Journal of Microbiology, 5*(11), 1070–1082.

Adebayo, F. A., Afolabi, O. R., & Akintokun, A. K., (2014). Antimicrobial properties of purified bacteriocins produced from *lactobacillus casei* and *lactobacillus fermentum* against selected pathogenic microorganisms. *British Journal of Medicine & Medical Research, 4*(18), 3415–3431.

Adebayo-Tayo, B. C., & Onilude, A. A., (2008). Screening of lactic acid bacteria strains isolated from some Nigerian fermented foods for EPS production. *World Applied Science Journal, 4*, 741–747.

Adebolu, T. T., Ihunweze, B. C., & Onifade, A. K., (2012). Antibacterial activity of microorganisms isolated from the liquor of fermented maize Ogi on selected diarrheal bacteria. *Journal of Medicine and Medical Sciences, 3*(6), 371–374.

Adejumo, T. O., (2014). Antimicrobial activity of lactic acid bacteria isolated from fermented milk products. *African Journal of Science, 8*(10), 490–496.

Adelekan, A. O., & Oyewole, O. B., (2010). Production of ogi from germinated sorghum supplemented with soybeans. *African Journal of Biotechnology, 9*(42), 7114–7721.

Adelekan, A. O., Alamu, A. E., Arisa, N. U., Adebayo, Y. O., & Dosa, A. S., (2013). Nutritional, microbiological and sensory characteristics of malted soy-kunu zaki: An improved traditional beverage. *Advances in Microbiology, 3*, 389–397.

Adeleke, R. O., & Abiodun, O. A., (2010). Physico-chemical properties of commercial local beverages in Osun State, Nigeria. *Pakistan Journal of Nutrition, 9*(9), 853–855.

Adesokan, I. A., Avarenren, E. R., Salami, T. R., Akinlosotu, I. O., & Olayiwola, D. T., (2008). Management of spoilage and pathogenic organisms during fermentation of none an indigenous fermented milk product in Nigeria. *Journal of Applied Biosciences, 11*, 564–569.

Adetunji, V. O., & Arigbede, M. I., (2011). Occurrence of *E. coli* 0157:H7 and *listeria monocytogenes* and identification of hazard analysis critical control points (HACCPs) in production operation of a typical cheese "wara" and yoghurt. *Pakistan Journal of Nutrition, 10*(8), 796–804.

Adetunji, V. O., & Babalobi, O. O., (2011). A comparative assessment of the nutritional content of "wara" a West African soft cheese using *Calotropis procera* and *Cymbopogon citratus* as coagulants. *African Journal of Food, Agriculture, Nutrition and Development, 11*(7), 5573–5585.

Adetunji, V., & Adegoke, G., (2007). Bacteriocin and cellulose production by lactic acid bacteria isolated from West African soft cheese. *African Journal of Biotechnology, 6*(22), 2616–2619.

Adetunjia, V. O., Alongea, D. O., Singhb, R. K., & Chenc, J., (2008). Production of wara, a West African soft cheese using lemon juice as a coagulant. *LWT, 41*, 331–336.

Adeyeye, S. A. O., (2017). Safety issues in traditional West African foods: A critical review. *Journal of Culinary Science & Technology, 15*(2), 101–125.

Adimpong, D. B., Nielsen, D. S., Sørensen, K. I., Derkx, P. M. F., & Jespersen, L., (2012). Genotypic characterization and safety assessment of lactic acid bacteria from indigenous African fermented food products. *BMC Microbiology, 12*(75), 1–13. doi: 10.1186/1471–2180–12–75. Adinsi, L., Akissoe, N., Amoa-Awua, W., Awad, S., Dalode, G., Hounhouigan, D. J., Mestres, C., Oduro-Yeboah, C., & Sacca, C., (2013). *Results of Sampling and Determination of Biochemical and Nutritional Quality for Group 1*. African food tradition revisited by research, a project of CIRAD (Centre de coopération internationale en recherche agronomique pour le développement), URL: https://www.after-fp7.eu/ (accessed on 6 September 2019).

Adinsi, L., Vieira-Dalodé, G., Akissoé, N., Anihouvi, V., Mestres, C., Jacobs, A., Dlamini, N., Pallet, D., & Hounhouigan, D. J., (2014). Processing and quality attributes of gowe: A malted and fermented cereal-based beverage from Benin. *Food Chain, 4*(2), 171–183.

Afolabi, O. R., Bankole, O. M., & Olaitan, O. J., (2008). Production and characterization of antimicrobial agents by lactic acid bacteria isolated from fermented foods. *The Internet J. Microbiol., 4*(2), 1–7.

Afolayan, M. O., Afolayan, M., & Abuah, J. N., (2010). An investigation into sorghum-based ogi (ogi-baba) storage characteristics. *Advance Journal of Food Science and Technology, 2*(1), 72–78.

Aïssi, V. M., Soumanou, M. M., Bankolé, H., Toukourou, F., & De Souza, C. A., (2009). Evaluation of hygienic and mycological quality of local cheese marketed in Benin. *Aust. J. Basic Appl. Sci., 3*(3), 2397–2404.

Ajiboye, T. O., Iliasu, G. A., Adeleye, A. O., Abdussalam, F. A., Akinpelu, S. A., & Ogunbode, S. M., (2014). Nutritional and antioxidant dispositions of sorghum/millet-based beverages indigenous to Nigeria. *Food Science and Nutrition, 2*(5), 597–604.

Ajiboye, T. O., Iliasu, G. A., Adeleye, A. O., Ojewuyi, O. B., Kolawole, F. L., Bellod, S. A., & Mohammed, A. O., (2016). A fermented sorghum/millet-based beverage, *Obiolor*, extenuates high-fat diet-induced dyslipidaemia and redox imbalance in the livers of rats. *J. Sci. Food Agric., 96*, 791–797.

Aka, S., Djeni, N. D. T., N'Guessan, K. F., Yao, K. C., & Dje, K. M., (2008). Variability of the physicochemical properties and enumeration of the fermentative flora of chapalo, a traditional sorghum beer in Côte d'Ivoire (French title). *Afrique Science, 4*(2), 274–286.

Akabanda, F., & Owusu-Kwarteng, J. K., Glover, R. L., & Tano-Debrah, K., (2010). Microbiological characteristics of ghanaian traditional fermented milk product, Nunu. *Nature and Science, 8*(9), 178–187.

Akabanda, F., Owusu-Kwarteng, J., Tano-Debrah, K., Glover, R. L. K., Nielsen, D. S., & Jespersen, L., (2013). Taxonomic and molecular characterization of lactic acid bacteria and yeasts in nunu, a Ghanaian fermented milk product. *Food Microbiology, 34*, 277–283.

Akingbala, J. O., Oguntimehin, G. B., & Abass, A. B., (1991). Effect of processing method on quality and acceptability of 'fufu' from low cyanide cassava. *Journal of the Science of Food and Agriculture, 57*, 151–154.

Amoa-Awua, W. K. A., Appoh, F. E., & Jakobsen, M., (1996). Lactic acid fermentation of cassava dough into agbelima. *International Journal of Food Microbiology, 31*, 87–98.

Amusa, N. A., & Ashaye, O. A., (2009). Effect of processing on nutritional, microbiological and sensory properties of Kunun-Zaki (a sorghum-based non-alcoholic beverage) widely consumed in Nigeria. *Pakistan Journal of Nutrition, 8*, 288–292.

Amusa, N. A., & Odunbaku, O. A., (2009). Microbiological and nutritional quality of hawked kunun (a sorghum-based non-alcoholic beverage) widely consumed in Nigeria. *Pakistan Journal of Nutrition, 8*(1), 20–25.

Annan, T., (2013). *Development of Starter Culture for the Fermentation of Dehulled Maize into Nisho (white kenkey),* (pp. 1–106). M.Phil. (Food Science), Degree University of Ghana, Ghana.

Annan-Prah, A., & Agyeman, J. A., (1997). Nutrient content and survival of selected pathogenic bacteria in kenkey used as a weaning food in Ghana. *Acta Tropica, 65*, 33–42.

Assamoi, A. A., Regina, K. E., Ehon, A. F., Amani, N. G., Lamine, S. N., & Thonart, P., (2016). Isolation and screening of Weissella strains for their potential use as starter during attiéké production. *Biotechnol. Agron. Soc. Environ., 20*(3), 355–362.

Assanvo, J. B., Agbo, G. N., Behi, Y. E. N., Coulin, P., & Farah, Z., (2006). Microflora of traditional starter made from cassava for "attiéké" production in Dabou (Côte d' Ivoire). *Food Control, 17*, 37–41.

Assohoun, N. C. M., Djeni, T. N., Koussémon-Camara, M., & Brou, K., (2013). Effect of fermentation process on nutritional composition and aflatoxins concentration of Doklu, a fermented maize based food. *Food and Nutrition Sciences, 4*, 1120–1127.

Ayangbenro, A. S., (2015). Antimicrobial activity of bacteriocin-producing lactic acid bacteria isolated from yogurts against Candida albicans. *International Journal of Microbiology and Application, 2*(5), 84–87.

Ayo, J. A., Onuoha, G., Ikuomola, D. S., Esan, Y. O., & Ayo, V. A., (2011). Obetta production and evaluation of starter culture for Kunun zaki: A millet based beverage drink. *Nigerian Journal of Microbiology, 25*, 2457–2469.

Ayo, J. A., Onuoha, O. G., Ikuomola, D. S., Esan, Y. O., Ayo, V. A., & Oigiangbe, I. G., (2010). Nutritional evaluation of millet-beniseed composite based Kunun-zaki. *Pakistan Journal of Nutrition, 9*(10), 1034–1038.

Banwo, K., Sanni, A. I., Tan, H., & Tian, Y., (2012). Phenotypic and genotypic characterization of lactic acid bacteria isolated from some Nigerian traditional fermented foods. *Food Biotechnology, 26*, 124–142.

Banwo, K., Sanni, A., & Tan, H., (2013). Functional properties of *Pediococcus* species isolated from traditional fermented cereal gruel and milk in Nigeria. *Food Biotechnology, 27*, 14–38.

Belewu, M. A., Belewu, K. Y., & Nkwunonwo, C. C., (2005). Effect of biological and chemical preservatives on the shelf life of West African soft cheese. *African Journal of Biotechnology, 4*(10), 1076–1079.

Blansherd, A. F. J., Dahniya, M. T., Poulter, N. H., & Taylor, A. J., (1994). Fermentation of cassava into 'foofoo.' Effect of time and temperature of processing and storage quality. *Journal of the Science of Food and Agriculture, 66*, 485–492.

Cereda, M. P., & Mattos, M. C. Y., (1996). Linamarin: The toxic compound of cassava. *J. Venom. Anim. Toxins, 2*, 6–12.

Chelule, P. K., Mbongwa, H. P., Carries, S., & Gqaleni, N., (2010). Lactic acid fermentation improves the quality of amahewu, a traditional South African maize-based porridge. *Food Chem., 122*, 656–661.

Coulin, P., Farah, Z., Assanvo, J., Spillmann, H., & Puhan, Z., (2006). Characterization of the microflora of attieke a fermented cassava product, during traditional small scale preparation. *International Journal of Food Microbiology, 106*, 131–136.

Demuyakor, B., & Ohta, Y., (1991). Characteristics of pito yeast from Ghana. *Food Microbiology, 8*, 183–193.

Dje, K. M., Aka, S., Nanga, Y. Z., Yao, K. C., & Loukou, Y. G., (2009). Predominant lactic acid bacteria involved in the spontaneous fermentation step of tchapalo process, a traditional sorghum beer of Côte d'Ivoire. *Research Journal of Biological Sciences, 4*(7), 789–795.

Djeni, N. T., N'Guessan, K. F., Toka, D. M., Kouame, K. A., & Dje, K. M., (2011). Quality of attieke (a fermented cassava product) from the three main processing zones in Côte d'Ivoire. *Food Research International, 44*, 410–416.

Ekwem, O. H., (2014). Isolation of antimicrobial producing lactobacilli from akamu (a Nigerian fermented cereal gruel). *African Journal of Microbiology Research, 8*(7), 718–720.

Esther, L., Charles, A. O., Adeoye, O. S., & Toyin, O. A., (2013). Effects of drying method on selected properties of Ogi (Gruel) prepared from sorghum (*Sorghum vulgare*), millet (*Pennisetum glaucum*) and maize (*Zea mays*). *Journal of Food Processing & Technology, 4*(7), 1–4.

Evans, E., Musa, A., Abubakar, Y., & Mainuna, B., (2013). Nigerian indigenous fermented foods: Processes and prospects. In: *Mycotoxin and Food Safety in Developing Countries* (pp. 153–180). InTech. doi: 10.5772/52877.

Felber, C., Azouma, Y. O., & Reppich, M., (2016). Evaluation of analytical methods for the determination of the physicochemical properties of fermented, granulated, and roasted cassava pulp-gari. *Food Science & Nutrition Published,* 1–8.

Fortin, J., Desmarais, G., Asovié, O., & Diallo, M., (1998). Attiéké, couscous of Ivory Coast. In: Latour, J., (ed.), *Horizon Francophonie* (pp. 22–24). Quebec, Canada: The Food World.

Franz, C. M. A. P., Huch, M., Mathara, J. M., Abriouel, H., Benomar, N., Reid, G., Galvez, A., & Holzapfel, W. H., (2014). African fermented foods and probiotics. *International Journal of Food Microbiology, 190*, 84–96.

Glover, R. L. K., Abaidoo, R. C., Jakobsen, M., & Jespersen, L., (2005). Biodiversity of *Saccharomyces cerevisiae* isolated from a survey of pito production sites in various parts of Ghana. *Systematic and Applied Microbiology, 28*, 755–761.

Greppi, A., Rantisou, K., Padonou, W., Hounhouigan, J., Jespersen, L., Jakobsen, M., & Cocolin, L., (2013a). Yeast dynamics during spontaneous fermentation of mawè and tchoukoutou, two traditional products from Benin. *International Journal of Food Microbiology, 165*, 200–207.

Greppi, A., Rantsiou, K., Padonou, W., Hounhouigan, J., Jespersen, L., Jakobsen, M., & Cocolin, L., (2013b). Determination of yeast diversity in ogi, mawè, gowé and tchoukoutou by using culture-dependent and-independent methods. *International Journal of Food Microbiology, 165*, 84–88.

Guiraud, J. P., (2003). *Microbiologie Alimentaire* (p. 696). Dunod – RIA, France.

Halm, M., Amoa-Awua, W. K., & Jakobsen, M., (2004). Kenkey, an African fermented maize product. In: Hui, Y. H., Meunier-Goddik, L., Hansen, Å. S., Josephsen, J., Nip, W. K., Stanfield, P. S., & Toldra, F., (eds.), *Handbook on Fermented Foods and Beverage Science and Technology* (pp. 799–818). Marcel Dekker, New York.

Hama, F., Savadogo, A., Ouattara, C. A. T., & Traoré, A. S., (2009). Biochemical, microbial and processing study of dègué a fermented food from pearl millet dough) from Burkina Faso. *Pakistan Journal of Nutrition, 8*, 759–764.

Hayologlu, A. A., Guven, M., & Fox, P. F., (2002). Microbiological biochemical and technological properties of Turkish white cheese Beyaz peynir. *International Dairy Journal, 12*, 635–648.

Hounhouigan, D. J., Kayode, A. P., Bricas, N., & Nago, M. C., (2003). The culinary characteristics of yams sought in urban Benin. *Annals of Agricultural Sciences of Benin, 2,* 143–160.

Hounhouigan, D. J., Nout, M. J. R., Nago, C. M., Houben, J. H., & Rombouts, F. M., (1993a). Characterization and frequency distribution of species of lactic acid bacteria involved in the processing of mawe,' a fermented maize dough from Benin. *International Journal of Food Microbiology, 18*, 279–287.

Hounhouigan, D. J., Nout, M. J. R., Nago, C. M., Houben, J. H., & Rombouts, F. M., (1993b). Composition and microbiological and physical attributes of mawè, a fermented maize dough from Benin. *International Journal of Food Science and Technology, 28*, 513–517.

Idowu, M. A., & Akindele, S. A., (1994). Effect of storage of cassava roots on the chemical composition and sensory qualities of 'gari' and 'fufu.' *Food Chemistry, 51*, 421–424.

Ijeoma, I. O., Okerentugba, P. O., & Oranusi, N. A., (2015). Biotechnological characterization of lactic acid bacteria isolated from Ogi-a cereal fermented food product. *New York Science Journal, 8*(12), 21–26.

Ikpoh, I. S., Lennox, J. A., Ekpo, I. A., Agbo, B. E., Henshaw, E. E., & Udoekong, N. S., (2013). Microbial quality assessment of kunu beverage locally prepared and hawked in Calabar, Cross River State, Nigeria. *Journal of Current Research in Science, 1*(1), 20–23.

Irtwange, S. V., & Achimba, O., (2009). Effect of the duration of fermentation on the quality of gari. *Current Research Journal of Biological Sciences, 1*(3), 150–154.

Jideani, V. A., Nkama, I., Agbo, E. B., & Jideani, I. A., (2001). Fura production in some northern states of Nigeria-A survey. *Plant Foods for Human Nutrition, 56*, 23–36.

Kalui, C. M., Mathara, J. M., Kutima, P. M., Kiiyukia, C., & Wongo, L. E., (2009). Functional characteristics of *Lactobacillus plantarum* and *Lactobacillus rhamnosus* from ikii, a Kenyan traditional fermented maize porridge. *African Journal of Biotechnology, 8*(17), 4363–4373.

Kayodé, A. P. P., Adégbidi, A., Linnement, A. R., Nout, M. J. R., & Hounhouigan, D. J., (2005). Quality of farmer's varieties of sorghum and derived food as perceived by consumers in Benin. *Ecology of Food and Nutrition, 44*, 271–294.

Kayodé, A. P. P., Hounhouigan, D. J., Nout, M. J. R., & Niehof, A., (2007a). Household production of sorghum beer in Benin: technological and socio-economic aspects. *International Journal of Consumer Studies, 31*(3), 258–264.

Kayodé, A. P. P., Hounhouigan, J. D., & Nout, M. J. R., (2007b). Impact of brewing process operations on phytates, phenolic compounds and in vitro of iron and zinc in opaque sorghum beer. *Lebensm.-Wiss. Technol., 40*, 834–841.

Kayodé, A. P. P., Vieira-Dalodé, G., Linnemann, A. R., Kotchoni, S. O., Hounhouigan, A. J. D., Van Boekel, M. A. J. S., & Nout, M. J. R., (2011). Diversity of yeasts involved in the fermentation of tchoukoutou, an opaque sorghum beer from Benin. *African Journal of Microbiology Research, 5*(18), 2737–2742.

Kèkè, M., Yèhouénou, B., Dahouénon, E., Dossou, J., & Sohounhloué, D. C. K., (2008). Contribution to the improvement of the production and preservation technology of Peulhwaragashi cheese by injection of *Lactobacillus plantarum. Annals of Agronomy Sciences, Iques du Benin., 10*(1), 73–86.

Kogno, E., (2016). *Screening of Microorganisms with Amylolytic Potential Isolated from Fermented Foods Produced in Togo: Cases of Agbélima, Eblima and Epoma.* MSc Dissertation, Department of Plant Biology, University of Abomey-Calavi, Abomey-Calavi, Benin. (In French).

Kogno, E., Soncy, K., Taale, E., Anani, K., Karou, S., & Ameyapoh, Y., (2017). Molecular characterization of lactic acid bacteria involved in the fermentation of Togolese traditional cereal foods. *International Journal of Recent Advances in Multidisciplinary Research, 4*(2), 2308–2312.

Kohajdová, Z., (2010). Fermented cereal products. In: Pandey, A., Soccol, C. R., Gnansounou, E., Larroche, C., Neenigam, P. S., & Dussap, C. G., (eds.), *Comprehensive Food Fermentation Biotechnology* (pp. 57–82). Asiatech Publishers Inc., New Delhi, India.

Kohajdová, Z., (2014). Fermented cereal products. In: Ray, R. C., & Montet, D., (eds.), *Microorganisms and Fermentation of Traditional Foods* (pp. 78–107). CRC Press, Boca Raton.

Kohajdová, Z., (2017). Fermented cereal products. In: *Current Developments in Biotechnology and Bioengineering: Food and Beverages Industry* (91–117). Elsevier, Amsterdam, Netherlands.

Kordylasi, J. M., (1990). *Processing and Preservation of Tropical and Sub-Tropical Foods* (pp. 402–406). MacMillan, London.

Koussémon, M., Thammavongs, B., Gueguen, M., & Panoff, J. M., (2003). Cryotolerance in lactic acid bacteria: The case of Lactobacillus delbrueckii subsp. bulgaricus PAL D lb 18. *Sahelian Studies and Research,* 193–198. (In French).

Krabi, E. R., Assamoi, A. A., Ehon, A. F., Amani, N. G. G., Niamké, L. S., Cnockaert, M., Aerts, M., & Vandamme, P., (2016). Biochemical properties of three lactic acid bacteria strains isolated from traditional cassava starters used for attieke preparation. *African Journal of Food Science, 10*(11), 271–277.

Latunde-Dada, G. O., (1997). Fermented foods and cottage industries in Nigeria. *Food Nutr. Bull., 18*(4), 102.

Lei, V., & Jakobsen, M., (2004). Microbiological characterization and probiotic potential of koko and koko sour water, African spontaneously fermented millet porridge and drink. *Journal of Applied Microbiology, 96*, 384–397.

Lei, V., Friis, H., & Michaelsen, K. F., (2006). Spontaneously fermented millet product as a natural probiotic treatment for diarrhea in young children: An intervention study in Northern Ghana. *International Journal of Food Microbiology, 110*, 246–253.

Mbachu, A. E., Etok, C. A., Agu, K. C., Okafor, O. I., Awah, N. S., et al. (2014). Microbial quality of kunu drink sold in Calabar, Cross River State, Nigeria. *Journal of Global Biosciences, 3*(2), 511–515.

Mohammed, S. S. D., & Ijah, U. J. J., (2013). Isolation and screening of lactic acid bacteria from fermented milk products for bacteriocin production. *Annals. Food Science and Technology, 14*(1), 122–128.

Mokoena, M. P., Mutanda, T., & Olaniran, A. O., (2016). Perspectives on the probiotic potential of lactic acid bacteria from African traditional fermented foods and beverages. *Food&Nutrition Research, 60*, 29630 doi: 10.3402/fnr.v60.29630.

N'Guessan, F. K., N'Dri, D. Y., Camara, F., & Dje, M. K., (2010). Saccharomyces cerevisiae and Candida tropicalis as starter cultures for the alcoholic fermentation of tchapalo, a traditional sorghum beer. *World J. Microbiol. Biotechnol., 26*, 693–699.

N'Tcha, C., (2016). *Study of the Microbiological and Probiotic Properties of Traditional Tchoukoutou and Chakpalo Beers Produced in Benin.* PhD Dissertation, Department of Biochemistry and Cell Biology, Faculty of Sciences and Techniques (FAST), University of Abomey-Calavi, Abomey-Calavi, Benin [In French].

N'Tcha, C., Kayodé, A. P. P., Adjanohoun, A., Sina, H., Tanmakpi, G. R., Savadogo, A., Dicko, H. M., & Baba-Moussa, L., (2016). Diversity of lactic acid bacteria isolated from "kpètè-kpètè" a ferment of traditional beer "tchoukoutou" produced in Benin. *African Journal of Microbiology Research, 10*(16), 552–564.

Nahemiah, D., Bankole, O. S., Tswako, M. A., Nma-Usman, K. I., H., H., & Fati, K. I., (2014). Hazard analysis critical control points (HACCP) in the production of soy-kununzaki: A traditional cereal-based fermented beverage of Nigeria. *American Journal of Food Science and Technology, 2*(6), 196–202.

Nout, M. J. R., (2009). Rich nutrition from the poorest-cereal fermentations in Africa and Asia. *Food Microbiology, 26*(7), 685–692.

Nwachukwu, E., Achi, O. K., & Ijeoma, I. O., (2010). Lactic acid bacteria in fermentation of cereals for the production of indigenous Nigerian foods. *African Journal of Food Science and Technology, 1*(2), 21–26.

NWFASUAC (National Workshop of Faculty of Agronomic Sciences, University of Abomey-Calavi), (2006). *Milk Production and Processing Cheese Feed in Benin Good Practice Guide*. National Workshop of 14th July 2006, Faculty of Agronomic Sciences, University of Abomey-Calavi, Abomey-Calavi, Benin (In French).

Nwokoro, O., & Chukwu, B. C., (2012). Studies on Akamu, a traditional fermented maize food. *Rev. Chil. Nutr., 39*(4), 180–184.

Nyanzi, R., & Jooste, P. J., (2012). Cereal based functional foods. In: Rigobelo, E. C., (ed.), *Probiotics* (pp. 161–197). InTech, Rijeka.

Obiazi, H. A. K., Ugwuoke, H. C., Ebadan, M. I., & Osula, O. T., (2016). Inhibition effects and bacitriocin factors produced by Lactobacilli against enteric pathogenic bacteria. *American Journal of Research Communication, 4*(9), 90–102.

Obinna-Echem, P. C., Kuri, V., & Beal, J., (2014a). Evaluation of the microbial community, acidity and proximate composition of akamu, a fermented maize food. *Journal of the Science of Food and Agriculture, 94*(2), 331–340.

Obinna-Echem, P. C., Kuri, V., & Beal, J., (2014b). Fermentation and antimicrobial characteristics of *Lactobacillus plantarum* and Candida tropicalis from Nigerian fermented maize (akamu). *International Journal of Food Studies, 3*, 186–202.

Obiri-Danso, K., Ellis, W. O., Simpson, B. K., & Smith, J. P., (1997). Suitability of high lysine maize, Obantanpa for kenkey production. *Food Control, 8*, 125–129.

Obodai, M., Oduro-Yeboah, C., Amoa-Awua, W., Anyebuno, G., Ofori, H., Annan, T., Mestres, C., & Pallet, D., (2014). Kenkey production, vending and consumption practices. *Food Chain, 4*(30), 275–288.

Ogbonna, I. O., Opobiyi, M. Y., Katuka, B., & Waba, J. T., (2011). Microbial evaluation and proximate composition of Kunu zaki, an indigenous fermented food drink consumed predominantly in northern Nigeria. *Internet Journal of Food Safety, 13*, 93–97.

Ogunremi, O. R., & Sanni, A. I., (2011). Occurrence of amylolytic and/or bacteriocin-producing lactic acid bacteria in ogi and fufu. *Annals. Food Science and Technology, 12*(1), 71–77.

Oguntoyinbo, F. A., & Dodd, C. E. R., (2010). Bacterial dynamics during the spontaneous fermentation of cassava dough in gari production. *Food Control, 21*, 306–312.

Oguntoyinbo, F. A., & Narbad, A., (2012). Molecular characterization of lactic acid bacteria and *in situ* amylase expression during traditional fermentation of cereal foods. *Food Microbiology, 31*, 254–262.

Oguntoyinbo, F. A., Tourlomousis, P., Gasson, M. J., & Narbad, A., (2011). Analysis of bacterial communities of traditional fermented West African cereal foods using culture-independent methods. *International Journal of Food Microbiology, 145*, 205–210.

Ojo, R. O., Olatunde, O. A., & Akindotu, A. E., (2016). Isolation and biochemical characterization of lactic acid bacteria in local cheese "Wara" in Owo, Nigeria. *Achievers Journal of Scientific Research, 1*(1), 65–70.

Okeke, C. A., Ezekiel, C. N., Nwangburuka, C. C., Sulyok, M., Ezeamagu, C. O., Adeleke, R. A., Dike, S. K., & Krska, R., (2015). Bacterial diversity and mycotoxin reduction during maize fermentation (Steeping) for Ogi production. *Frontiers in Microbiology, 6*(Article1402), 1–12.

Okereke, H. C., Achi, O. K., Ekwenye, U. N., & Orji, F. A., (2012). Antimicrobial properties of probiotic bacteria from various sources. *African Journal of Biotechnology, 11*(39), 9416–9421.

Okpara, A. N., Okolo, B. N., & Ugwuanyi, J. O., (2014). Antimicrobial activities of lactic acid bacteria isolated from akamu and kunun-zaki (cereal-based non-alcoholic beverages) in Nigeria. *African Journal of Biotechnology, 13*(29), 2977–2984.

Okunye, O. L., Odeleye, F. O., & Abiodun, O. O. S., (2015). The effect of Fura De Nunu on selected clinical isolates of bacteria. *International Journal of Recent Research in Life Sciences, 2*(2), 50–53.

Ola, F., (2010). *Contribution to the Establishment of the Microbiological Typology of Fermented Products from West Africa: Cases of Togo, Benin, Burkina Faso and Nigeria.* MSc Dissertation, Department of School of Biological and Food Technology, University of Lomé, Lomé, Togo. (In French).

Oluseyi, A. K., Oluwafunmilola, A., Oluwasegun, S. T., Abimbola, A. A., Oluwatoyin, A. C., & Olubusola, O., (2013). Dietary fortification of sorghum-Ogi using crayfish (*Paranephrops planifrons*) as supplements in infancy. *Food Science and Quality Management, 15*, 1–9.

Oluwajoba, S. O., Akinyosoye, F. A., & Oyetayo, V. O., (2013). *In vitro* screening and selection of probiotic lactic acid bacteria isolated from spontaneously fermenting Kunu-Zaki. *Advances in Microbiology, 3*, 309–316.

Omafuvbe, B. O., Adigun, A. R., Ogunsuyi, J. L., & Asunmo, A. M., (2007). Microbial diversity in ready-to-eat fufu and Lafun-fermented cassava products sold in Ile-Ife, Nigeria. *Research Journal of Microbiology, 2*(11), 831–837.

Omemu, A. M., & Faniran, O. W., (2011). Assessment of the antimicrobial activity of lactic acid bacteria isolated from two fermented maize products - ogi and kunnu-zaki. *Malaysian Journal of Microbiology, 7*(3), 124–128.

Omemu, A. M., (2011). Fermentation dynamics during production of ogi, a Nigerian fermented cereal porridge. *Report and Opinion, 3*(4), 8–17.

Onwuakor, C. E., Nwaugo, V. O., Nnadi, C. J., & Emetole, J. M., (2014a). Effect of varied culture conditions on bacteriocin production of four *Lactobacillus* species isolated from locally fermented maize (Ogi). *Global Journal of Medical Research: Microbiology and Pathology,* 14(4 Version 1.0). http://pubs.sciepub.com/ajmr/2/5/1/ (accessed on 7 September 2019).

Onwuakor, C. E., Nwaugo, V. O., Nnadi, C. J., & Emetole, J. M., (2014b). Effect of varied culture conditions on crude supernatant (bacteriocin) production from four *Lactobacillus* species isolated from locally fermented maize (Ogi). *American Journal of Microbiological Research, 2*(5), 125–130.

Onyekwere, O. O., Akinrele, I. A., & Koleoso, O. A., (1989). Industrialisation of ogi fermentation. In: Steinkraus, K. H., (ed.), *Industrialization of Indigenous Fermented Foods* (pp. 329–362). Marcel Dekker.

Oranusi, S. U., Umoha, V. J., & Kwaga, J. K. P., (2003). Hazards and critical control points of kunun-zaki, a non-alcoholic beverage in Northern Nigeria. *Food Microbiology, 20*, 127–132.

Orji, M. U., Mbata, T. I., Aniche, G. N., & Ahonkhai, I., (2003). The use of starter cultures to produce 'Pito,' a Nigerian fermented alcoholic beverage. *World Journal of Microbiology & Biotechnology, 19*, 733–736.

Orutugu, L. A., Izah, S. C., & Aseibai, E. R., (2015). Microbiological quality of Kunun drink sold in some major markets of Yenagoa Metropolis, Nigeria. *Continental J. Biomedical Sciences, 9*(1), 9–16.

Osseyi, E. G., Tagba, P., Karou, S. D., Ketevi, A. P., & Lamboni, C. R., (2011). Stabilization of the traditional sorghum beer, "tchoukoutou" using rustic wine-making method. *Adv. J. Food Sci. Technol., 3*, 254–258.

Osungbaro, D. K., (2009a). "*Lactobacillus* in human health with reference to locally fermented foods." *Journal of Tropical Medicine, 2*(4), 28–36.

Osungbaro, T. O., (2009b). Physical and nutritive properties of fermented cereal foods. *African Journal of Food Science, 3*, 23–27.

Ouattara, C. A. T., Somda, M. K., Moyen, R., & Traoré, A. S., (2015). Isolation and identification of lactic acid and non-acid lactic bacteria from "dègue" of Western Africa traditional fermented millet-based food. *African Journal of Microbiology Research, 9*(36), 2001–2005.

Owusu-Kwarteng, J., Akabanda, F., Neilsen, D. S., Tano-Debrah, K., Glover, R. L., & Jespersen, L., (2012). Identification of lactic acid bacteria isolated during traditional fura processing in Ghana. *Food Microbiology, 32*(1), 72–78.

Owusu-Kwarteng, J., Tano-Debrah, K., Glover, R. L., & Akabanda, F., (2010). Process characteristics and microbiology of Fura produced in Ghana. *Nature and Science, 8*(8), 41–51.

Oyedeji, O., Ogunbanwo, S. T., & Onilude, A. A., (2013). Predominant lactic acid bacteria involved in the traditional fermentation of fufu and ogi, two Nigerian fermented food products. *Food Nutr. Sci., 4*, 40–46.

Padonou, S. W., Hounhouigan, J. D., & Nago, M. C., (2009a). Physical, chemical and microbiological characteristics of lafun produced in Benin. *African Journal of Biotechnology, 8*(14), 3320–3325.

Padonou, S. W., Nielsen, D. S., Hounhouigan, J. D., Thorsen, L., Nago, M. C., & Jakobsen, M., (2009b). The microbiota of lafun, an African traditional cassava food product. *International Journal of Food Microbiology, 133*, 22–30.

Panda, S. K., & Ray, R. C., (2016). Foods fermented and beverages from tropical roots and tubers. In: Sharma, H. K., Njintang, N. Y., Singhal, R. S., & Kaushal, P., (eds.), *Tropical Roots and Tubers: Production, Processing and Technology* (1st edn., pp. 225–252). Wiley-Blackwell, New Jersey, USA. doi: 10.1002/9781118992739.ch5.

Pedersen, L. L., Owusu-Kwarteng, J., Thorsen, L., & Jespersen, L., (2012). Biodiversity and probiotic potential of yeasts isolated from Fura, a West African spontaneously fermented cereal. *International Journal of Food Microbiology, 159*, 144–151.

Princewill-Ogbonna, I. L., & Ojimelukwe, P. C., (2014). Bacteriocins from lactic acid bacteria inhibit food borne pathogens. *IOSR Journal of Environmental Science, Toxicology And Food Technology (IOSR-JESTFT), 8*(1), 50–56.

Rühe, A., (1938). Die zubereitung und verwendung des käse in Africa. The preparation and use of the cheese in Africa. A contribution to the cultural history of the African milk industry. In: Plisscke, H. H., (ed.), *Göttingen Völker-Customer Studies*. (pp. 168–191). Leipzig.

Sangoyomi, T. E., Owoseni, A. A., & Okerokun, O., (2010). Prevalence of enteropathogenic and lactic acid bacteria species in wara: A local cheese from Nigeria. *African Journal of Microbiology Research, 4*(15), 1624–1630.

Sanni, L. A., Odukogbe, O. O., & Faborode, M. O., (2016). Some quality characteristics of Gari as influenced by roasting methods. *Agric. Eng. Int., 18*(2), 388–394.

Savadogo, A., Ouattara, C. A. T., Savadogo, P. W., Ouattara, A. S., Barro, N., & Traore, A. S., (2004b). Microorganisms involved in Fulani traditional fermented milk in Burkina Faso. *Pakistan Journal of Nutrition, 3*(2), 134–139.

Savadogo, A., Ouattara, C. A., Bassole, I. H., & Traore, A. S., (2004a). Antimicrobial activities of lactic acid bacteria strains isolated from Burkina Faso fermented milk. *Pakistan Journal of Nutrition, 3*(3), 174–179.

Savadogo, A., Tapi, A., Chollet, M., Wathelet, B., Traoré, A., & Jacques, P., (2011). Identification of surfactin producing strains in Soumbala and Bikalga fermented condiments using polymerase chain reaction and matrix-assisted laser desorption/ionization-mass spectrometry methods. *International Journal of Food Microbiology, 151*(3), 299–306.

Sawadogo-Lingani, H., Lei, V., Diawara, B., Nielsen, D. S., Møller, P. L., Traoré, A. S., & Jakobsen, M., (2007). The biodiversity of predominant lactic acid bacteria in dolo and pito wort for the production of sorghum beer. *Journal of Applied Microbiology, 103*, 765–777.

Schutte, L. M., (2013). *Isolation and Identification of the Microbial Consortium Present in Fermented Milks from Sub-Saharan African* (pp. 1–100). MSc Dissertation, Department of Food Science, Faculty of AgriSciences, University of Stellenbosch, South Africa.

Sefa-Dadeh, S., (1989). Effects of particle size on some physicochemical characteristics of agbelima (cassava dough) and corn dough. *Tropical Science, 29*, 21–32.

Sefa-Dedeh, S., Sanni, A. I., Tetteh, G., & Sakyi-Dawson, E., (1999). Yeasts in the traditional brewing of pito in Ghana. *World Journal of Microbiology & Biotechnology, 15*, 593–597.

Sessou, P., Farougou, S., Azokpota, P., Youssao, I., Yehouenou, B., Ahounou, S., & Sohounhloué, D. C. K., (2013a). Inventory and analysis of endogenous conservation practices of wagashi, a traditional cheese produced in Benin. *International Journal of Biological and Chemical Sciences, 7*(3), 938–952.

Sessou, P., Farougou, S., Azokpota, P., Youssao, I., Yèhouenou, B., Ahounou, S., & Sohounhloué, D. C. K., (2013b). Endogenous methods for preservation of wagashi, a Beninese traditional cheese. *African Journal of Agricultural Research, 8*(31), 4254–4261.

Shetty, P., & Jespersen, L., (2006). *Saccharomyces cerevisiae* and lactic acid bacteria as potential mycotoxin decontaminating agents. *Trends in Food Science & Technology, 17*(2), 48–55.

Soni, S. K., Soni, R., & Janveja, C., (2014). Production of fermented foods. In: Panesar, P. S., & Marwaha, S. S., (eds.), *Biotechnology in Agriculture and Food Processing: Opportunities and Challenges* (245–253). CRC Press/Taylor and Francis Group, Boca Raton.

Steinkraus, K. H., (2002). Fermentations in world food processing. *Comprehensive Reviews in Food Science and Food Safety, 1*, 23–32.

Taale, E., Savadogo, A., Sina, H., Zongo, C., Karou, S. D., Baba-Moussa, L., & Traore, A. S., (2016a). Searching for Bacteriocin pln loci from *Lactobacillus* spp. isolated from fermented food in Burkina Faso by molecular methods. *International Journal of Applied Biology and Pharmaceutical Technology, 7*(3), 86–94.

Taale, E., Savadogo, A., Zongo, C., Somda, M. K., Sereme, S. S., Karou, S. D., Soulama, I., & Traore, A. S., (2015). Characterization of *bacillus* species producing bacteriocin-like inhibitory substances (BLIS) isolated from fermented food in Burkina Faso. *International Journal of Advanced Research in Biological Sciences, 2*(4), 279–290.

Taale, E., Savadogo, A., Zongo, C., Tapsoba, F., Karou, S. D., & Traore, A. S., (2016b). Microbial Antimicrobial Peptide Review: The Case of Bacteriocins. *International Journal of Biological and Chemical Sciences, 10*(1), 384–399.

Taiwo, O. O., (2009). Physical and nutritive properties of fermented cereal foods. *African Journal of Food Science, 3*(2), 023–027.

Tchekessi, C., Bokossa, Y., Banon, J., Agbangla, C., Adeoti, K., Dossou-Yovo, P., & Assogba, E., (2013). Physico-chemical and microbiological characterizations of a traditional "gowé" pulp made from corn in Benin. *Journal of Scientific Research of the University of Lome, 15*(2), 91–101. (In French).

Teniola, O. D., & Odunfa, S. A., (2001). The effects of processing methods on the levels of lysine, methionine and the general acceptability of ogi processed using starter cultures. *International Journal of Food Microbiology, 63*, 1–9.

Teniola, O. D., & Odunfa, S. A., (2002). Microbial assessment and quality evaluation of ogi during spoilage. *World J. Microbiol. Biotechnol., 18*, 731–737.

Teniola, O. D., Holzapfel, W. H., & Odunfa, S. A., (2005). Comparative assessment of fermentation techniques useful in the processing of ogi. *World Journal of Microbiology and Biotechnology, 21*, 39–43.

Todorov, S. D., & Holzapfel, W. H., (2015). Traditional cereal fermented foods as sources of functional microorganisms. In: Holzapfel, W., (ed.), *Advances in Fermented Foods and Beverages* (pp. 123–153). Elsevier, Amsterdam, Netherlands.

Tohibu, A. S., (2009). *Sensory and Chemical Stability of Vaccum Packaged Wagashie* (pp. 1–77). MSc (Food Science) Dissertation, Department of Biochemistry, Faculty of Biosciences and Biotechnology, College of Science, Kwame Nkrumah University of Science and Technology, Kumasi, Ghana.

Umaru, G. A., Tukur, I. S., Akensire, U. A., Adamu, Z., Olumuyiwa, A. B., Shawulu, A. H. B., Audu, M., Sunkani, J. B., Adamu, S. B. G., & Adamu, N. B., (2014). Microflora of Kunun-Zaki and Sobo drinks in relation to public health in Jalingo Metropolis, North-Eastern Nigeria. *International Journal of Food Research, 1*, 16–21.

Vieira-Dalodé, G., Akissoé, N., Annan, N., Hounhouigan, D. J., & Jakobsen, M., (2016). Aroma profile of gowe, a traditional malted fermented sorghum beverage from Benin. *African Journal of Food Science, 10*(2), 17–24.

Vieira-Dalodé, G., Jespersen, L., Hounhouigan, J., Moller, P. L., Nago, C. M., & Jakobsen, M., (2007). Lactic acid bacteria and yeasts associated with gowé production from sorghum in Bénin. *Journal of Applied Microbiology, 103*, 342–349.

Wakil, S. M., & Osamwonyi, U. O., (2012). Isolation and screening of antimicrobial producing lactic acid bacteria from fermentating millet gruel. *International Research Journal of Microbiology, 3*(2), 072–079.

Westby, A., (1991). Importance of fermentation in cassava processing. In: Ofori, F., & Hahn, S. K., (eds.), *Proceedings of 9th Symposium of the International Society for Tropical Root Crops* (pp. 249–255). ISTRC, Accra, Ghana.

Yao, A. A., Egounlety, M., Kouame, L. P., & Thonart, P., (2009). Lactic acid bacteria in starchy and fermented foods or beverages from West Africa: their current use. *Ann. Med. Vet.*, (French title), *153*, 54–65.

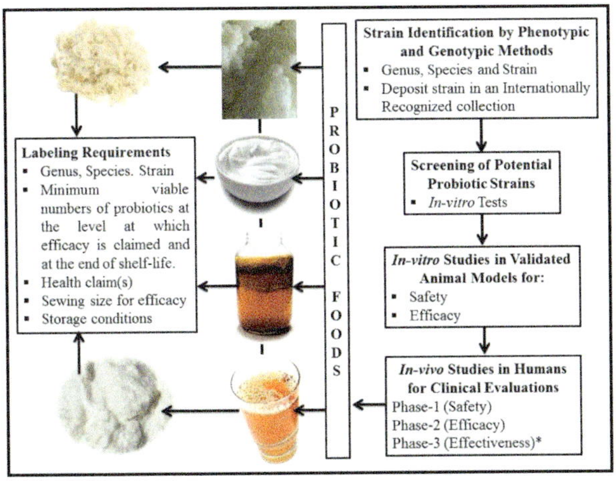

FIGURE 1.3 Guidelines for evaluation of candidate probiotic strains.
Note: *required only when a specific health claim is made] [Adopted from Ganguly, N. K., Bhattacharya, S. K., Sesikeran, B., Nair, G. B., Ramakrishna, B. S., Sachdev, H. P. S., et al. (2011). ICMR-DBT guidelines for evaluation of probiotics in food. *Indian Journal of Medical Research*, *134*(1), 22–25 with modification.

FIGURE 4.1 (A) Juçara (*Euterpe edulis Martius*), (B) Caraguata (*Bromelia antiacantha*), and (C) Okra (*Abelmoschus esculentus L*).

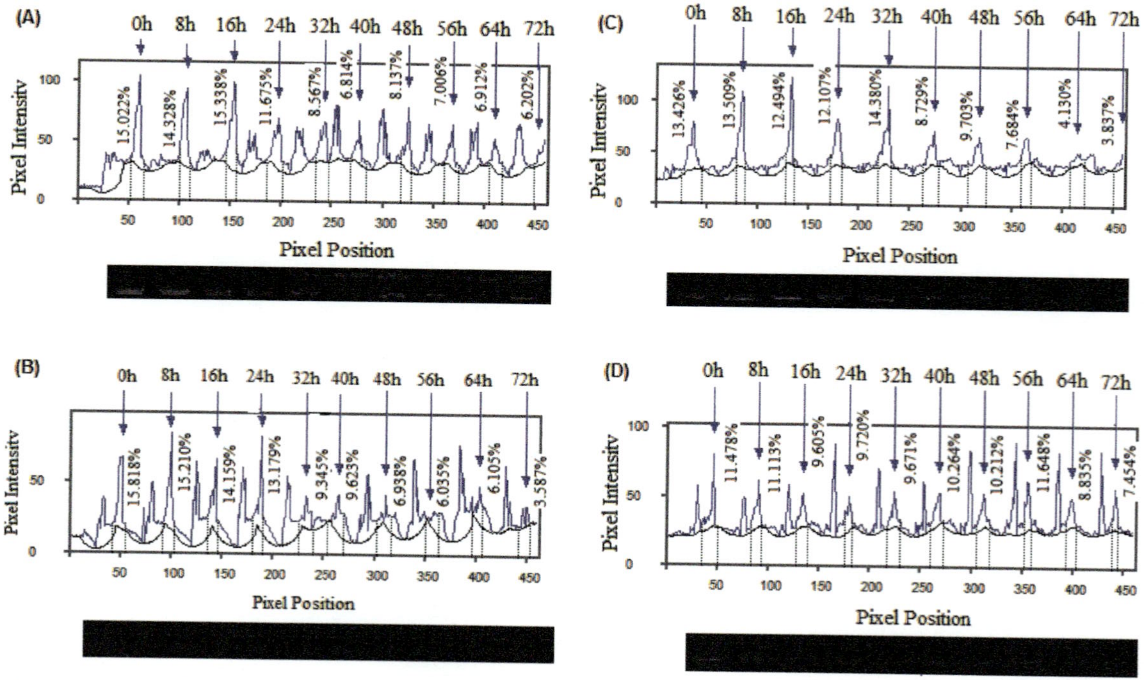

FIGURE 4.4 Electrophoresis pattern on agarose gel (1%), DNA incubated with Fenton reagent at 0, 8, 16, 24, 32, 40, 48, 56, 64 and 72 h times. They are; (A) Fermented milk by *L. helveticus* (29°C), (B) By *L. plantarum* (27°C) with green banana flour (1.08% m.v^{-1}), (C) Fermented soybean extract by *L. acidophilus* (37°C), and (D) *L. plantarum* (37°C) with okara (3% m.v^{-1}).

Note: Image processing by João Paulo Andrade de Araujo. The quantification of plasmid DNA bands was done using LabImage 1D software version 3.3.2 (Kapeplan Bio-imaging Solutions, 2006), represented in Brazil by the company Loccus Biotecnologia.

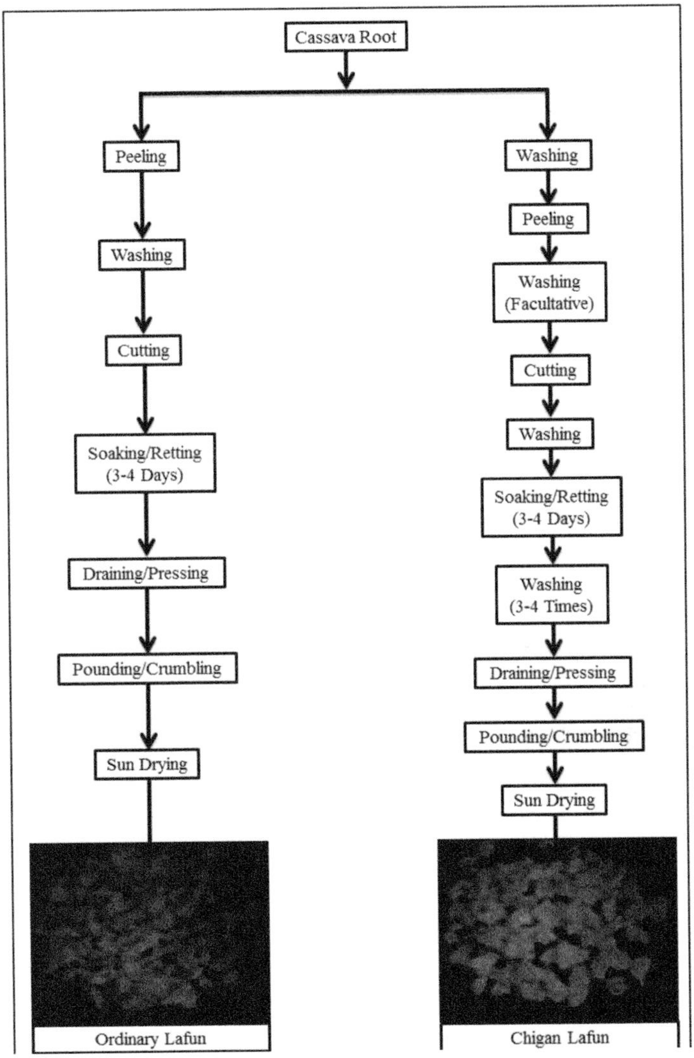

FIGURE 5.4 Schematic diagram for process and product technology of ordinary *lafun* and *chigan lafun* production (Padonouet al., 2009a).

[Reprint from: Padonou, S. W., Hounhouigan, J. D., & Nago, M. C., (2009a). Physical, chemical, and microbiological characteristics of lafun produced in Benin. *African Journal of Biotechnology, 8*(14), 3320–3325. No permission is needed as the article is open access].

FIGURE 5.5 Traditional *tchoukoutou* beer aspect. (Reprinted from N'tcha, 2016. With permission.)

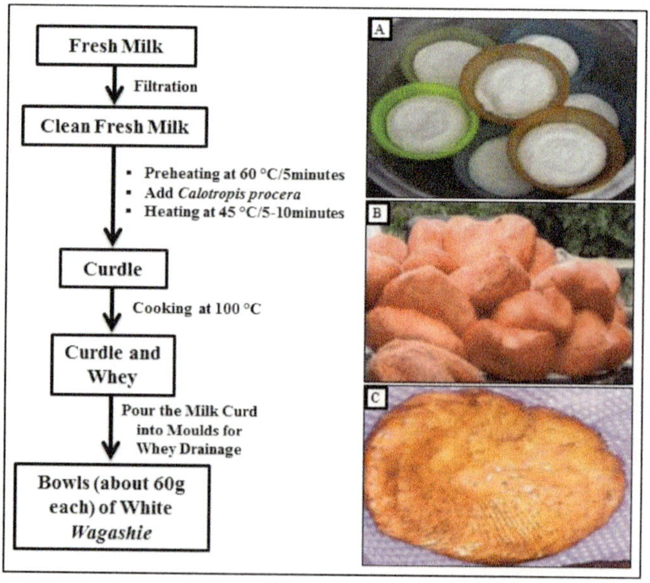

FIGURE 5.6 Schematic diagram for process and product technology of white *wagashie* production (NWFASUAC, 2006). A) White *wagashie*, B) Red *wagashie*, and C) Light red *wagashie*.

[Reprint from NWFASUAC (National Workshop of Faculty of Agronomic Sciences, University of Abomey-Calavi), (2006). Milk Production and Processing Cheese Feed in Benin Good Practice Guide. National Workshop of 14th July 2006, Faculty of Agronomic Sciences, University of Abomey-Calavi, Abomey-Calavi, Benin. (In French) Permission is not needed as it is a workshop report].

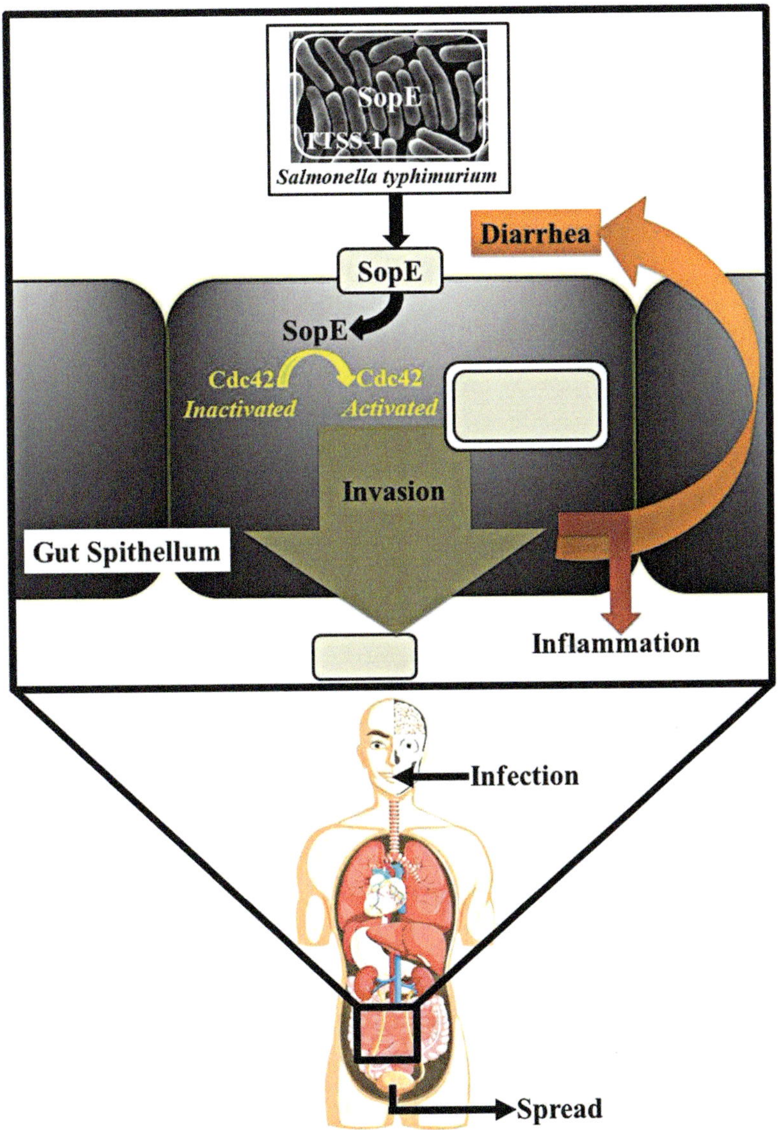

FIGURE 7.1 Role of the *Salmonella typhimurium* TTSS-1 in the induction of enterocolitis.
Source: Adapted from http://www.micro.biol.ethz.ch/research/hardt.html

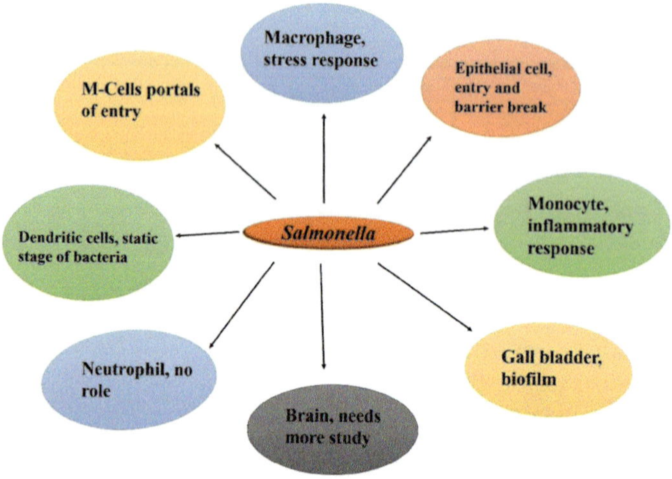

FIGURE 7.2 *Salmonella* infection in different cells and organs. (Reprinted with permission from Lahiri, et al. 2010. © 2010 Elsevier.)

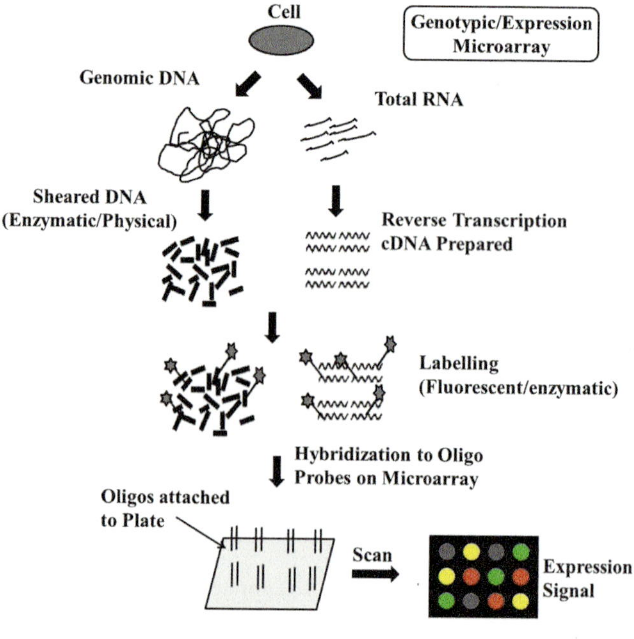

FIGURE 7.4 Schematic diagrams of genotyping and gene expression microarray experiments. *Source:* Adapted from Josefsen, et al., 2012.

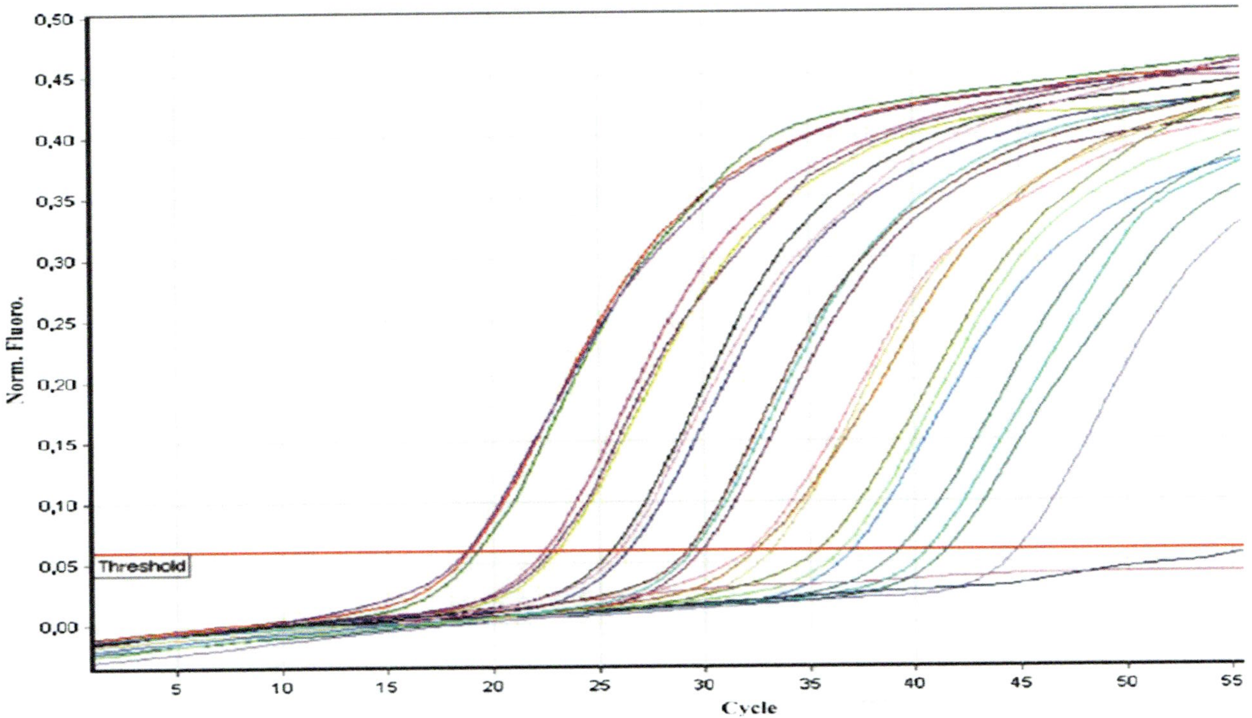

FIGURE 7.5 An amplification plot obtained by real-time PCR. The y-axis corresponds to the fluorescence emitted from either the labeled probe or DNA binding dye, which rises in each amplification cycle (x-axis). The CT is the amplification cycle where the fluorescence exceeds the background fluorescence. (Reprinted with permission from Hornyák et al., 2012. © 2012 Elsevier.).

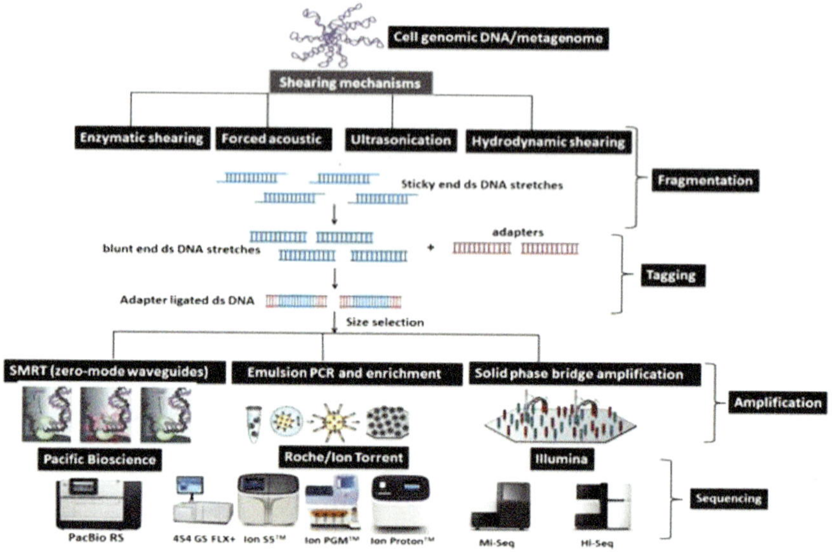

FIGURE 7.6 Workflow showing the NGS based DNA sequencing and sample preparation.

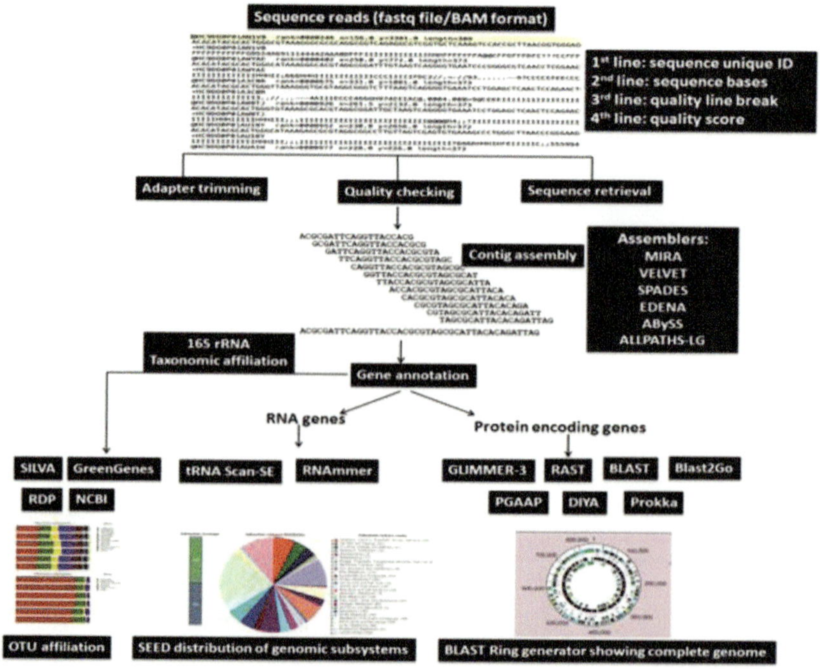

FIGURE 7.7 Detail pipeline for microbial genome and metagenome sequence analysis.

PART III

Innovative Microbiological Approaches and Technologies in the Food Industry

CHAPTER 6

Microbiological Approach for Environmentally Friendly Dairy Industry Waste Utilization

GEMILANG LARA UTAMA,[1*] FAYSA UTBA,[1] WIDIA DWI LESTARI,[1] HERLINA,[1] and ROOSTITA BALIA[2]

[1]*Faculty of Agro-Industrial Technology, Universitas Padjadjaran, Sumedang–45363, West Java, Indonesia,*

[2]*Faculty of Animal Husbandry, Universitas Padjadjaran, Sumedang–45363, West Java, Indonesia*

Corresponding author. E-mail: lugemilang@gmail.com

6.1 INTRODUCTION

The dairy industry is one of the fast-growing businesses in all over the world. With 2% production increased from 2014, there are still found imbalance supply and demand in 2015. According to Annual Dairy Situation Report by The International Dairy Federation, total world milk production in 2015 was estimated at 818 million tonnes, while the world total consumption was 111.3 kg per capita and refer to the OECD and FAO, this number should increase 12.5% by 2025 (IDF, 2017).

The number had shown the potential of pollution load that can be generated by dairy industries activities. Of the milk volume processed, dairy industries generate one to three times effluent or exactly 0.2–10 L of effluent with an average 2.5L of wastewater per L of milk processed (Raghunath et al., 2016; Vignesh et al., 2016). The effluent released by the dairy industries still has high organic matter which shown by COD of 468 mg/L and BOD (biological oxygen demand) of 210 mg/L, suspended solids (TSS of 942 mg/L and TDS of 680 mg/L), oil, and grease of 240 ml/L, chloride of 136 mg/L, pH of 7.34–7.38, fats, nitrogen, phosphorus, and calcium (Kharbanda and Prasanna, 2016; Pathak et al., 2016).

The disposal of dairy wastes directly to the stream or land can cause environmental problems. High pollution load of dairy wastewater that disposed directly to the stream can causing eutrophication that reduce oxygen in the water (Utama et al., 2016a). Meanwhile there a possibility of nitrate contamination to the groundwater that usually used for drinking water if the dairy wastewater disposed to the ground and if it stagnant, wastewater will decompose rapidly then produce undesirable odor (Raghunath et al., 2016).

Dairy wastes contain organic compound that mainly consists of protein, fat, and lactose with small amount of fiber, other solids, and minerals. In semisolid form, dairy sludge contains 34.98% protein, 11.04% fat, 9.77% fiber and4.42% lactose, 2.33% calcium, 1.05% phosphor and 0.4% magnesium based on dry matter (Marlina et al., 2012). In liquid form such as cheese-making wastewater, based on dry matter it contains 49.52% lactose, 34.7% protein, 6.3% fat, and 8.36% other solids (Utama, 2016).

The organic compound contained in dairy wastes should be utilized so it can provide benefit and able to reduce the environmental problems. Utilization of dairy wastes have been widely done in making food, feed, plastics, bioenergy, and biofuels, pharmaceuticals, and many other products which aims to reduce environmental pollution and the damage caused by the dairy manufacturing processes (De Jesus et al., 2015).

Dairy wastes can be utilized in many approaches. In simple ways, dairy wastes can be used as feed or fertilizer (Marlina et al., 2012; Santos et al., 2016). Dairy wastewater (whey) can be utilized for energy drink products with the addition of sodium, potassium, calcium, chloride, citrate, and other contents which similar with oral rehydration solution (Singh and Singh, 2012). Dairy wastewater also can be purified into lactose or processed further into other derivatives such as single cell protein (SCP), poly-hydroxylalkanoate (PHA), bio-polyester, lactic acid, vinegar, antibiotic, and sophorolipid (Koller et al., 2012). However, biotechnological approach with the use of microorganisms still the most widely used methods in dairy wastes utilization to reduce environmental problems (De Jesus et al., 2015).

6.2 DAIRY INDUSTRY WASTES CHARACTERISTICS

6.2.1 PHYSICAL-CHEMICAL CHARACTERISTICS OF DAIRY INDUSTRY WASTES

Raw milk processed into various products like butter, cheese, yogurt, condensed milk and milk powder involving processes such as chilling, pasteurization, and homogenization so that resulting various physical-chemical characteristics on the wastes (Pathak et al., 2016). Water used in every stage of dairy processes was only 1 from 10 parts of milk; this resulted in dairy wastes containing a high concentration of organic materials (Qasim and Mane, 2013). All of the processes will contribute towards high values of BOD, inflated rates of chemical oxygen demand (COD), and all of the physical-chemical characteristics shown in Table 6.1.

TABLE 6.1 Physical-Chemical Characteristics of Dairy Industry Wastes

Physical-Chemical Parameters	**Reported Values**		
	Rao and Datta (2012)	**Qasim and Mane (2013)**	**Pathak et al. (2016)**
Biological oxygen demand (BOD)	1240 mg/L	442 mg/L	210 mg/L
Chemical oxygen demand (COD)	84 mg/L	8960 mg/L	468 mg/L
Oil and Grease	290 mg/L	n/a	240 mg/L
Chloride	105 mg/L	78.3 mg/L	136 mg/L
Alkalinity	600 mg/L $CaCO_3$	n/a	462.5 mg/L $CaCO_3$
pH	7.2	7.10 (±0.12)	7.34–7.38
TSS	760 mg/L	253.6 mg/L	942 mg/L
TDS	1060 mg/L	543.4 mg/L	680 mg/L
Turbidity	n/a	49.5 mg/L	n/a
Total N	84 mg/L	120.1 mg/L	n/a
Sodium	n/a	125.4 mg/L	n/a
Potassium	n/a	78.3 mg/L	n/a
Phosphorus	11.7 mg/L	186.4 mg/L Whitish	n/a
Color	n/a	Unpleasant	n/a
Odor	n/a		n/a

The dairy wastes physical-chemical characteristics were affected by the passed processes. Casein decomposition is resulting in whitish color with heavy black sludge and strong butyric acid unpleasant odors (Shete and Shinkar, 2013). Fermentation of lactose to lactic acid has decreased the pH from slight alkaline to acidic wastewater quite rapidly; meanwhile, other contents such as cream or whey has mainly affected the COD of dairy wastes (Pathak et al., 2016).

6.2.2 MICROBIOLOGICAL CHARACTERISTICS OF DAIRY INDUSTRY WASTES

Microflora of dairy industry wastes will depend on the biochemical properties and the physical-chemical conditions, which affect the optimum metabolic activity of the microorganisms (Shivsharan et al., 2013). Carbon and organic nitrogen compounds such as lactose, casein, and other derivatives (lactic and amino acid) that contained in the dairy wastewater have a correlation with bacteria, yeast, and mold culture diversity (Palela et al., 2008). The increase of temperature would raise the species diversity and microorganisms count, while distribution of microorganisms also determined by light, temperature, pH, phosphates, chloride nitrate and sulfate (Prakashveni and Jagadeesan, 2012).

The diversity of microorganisms in dairy wastes has shown consists of gram-negative bacteria as predominant microorganisms. Identification of bacterial isolate from dairy wastes are *Lactobacillus*, *Bacillus*, *Pseudomonas*, *Azotobacter*, *Arthrobacter*, *Zoogloea*, *Microbacterium*, *Staphylococcus*, *Micrococcus*, *Cardiobacterium*, *Bifidobacterium*, *Pasteurella*, *Escherichia*, and *Eikenella* (Shivsharan et al., 2013). Species diversity of bacteria in diary wastes are *Alcaligens faecalis*, *Leuconostoc lactis*, *Meniscus*, *Silfidobacillus*, *Sparolacto bacillus*, *Trichococcus*, *Syntrophospara*, *Lactococcus*, *Pediococcus*, *Sprillum*, *Aquasprillum*, *Methylococcus*, *Psychrobacter*, *Campylobacter*, *Monococcus*, *Phalylobacterium*, and *Ocnosprillum* (Prakashveni and Jagadeesan, 2012). Besides the bacteria which dominated by genus *Bacillus*, *Pseudomonas*, and *Escherichia*, microbiological tests showed that there are also yeasts which belong to the genus *Saccharomyces*, *Kluyveromyces*, and *Torulopsis* and molds belong to the genus *Aspergillus*, and *Geotrichum* has found in dairy wastes (Palela et al., 2008).

6.3 MICROBIOLOGICAL UTILIZATION OF WHEY AS COMMON DAIRY INDUSTRY WASTES

Whey is one of the dairy industry wastes which considered had a major problem on their disposal, while on the other hand, whey has a great potential to utilize. Whey is very rich in nutrients, such as lactose, soluble proteins, lipids, and mineral salts (Guha et al., 2013). Whey can be classified into two types based on the technology process that it went through, namely sweet whey and acid whey. Sweet whey is derived from rennet-induced coagulation of casein, while acid whey results from fermentation-induced coagulation or addition of organic or mineral acids. Both sweet and acid whey contains lactose for approximately 70–72% of total solids, 8–10% whey proteins, and 12–15% minerals (Panesar et al., 2007).

Great amount of whey cheese is a potential that needs to be considered, especially if the treatment will be undertaken. In 2000, the EU produces 4.04×10^7 tons of cheese whey from cheese production as much as 6.38×10^6 tonnes and 67% whey is used as a raw material in the production of health food and beverage as well as animal feed. The remaining amount of cheese whey cannot be processed and it is a serious problem for cheese manufacturer or dairy industry (Koller et al., 2008). Globally, the number of cheese whey can be reached $1.15–1.40 \times 10^8$ tons per year (Koller et al., 2012).

Whey can be utilized into various products. Heating at high temperatures can produce condensed whey, complete acid or sweet-whey powders. Ultrafiltration and diafiltration can increase the amount of protein and minerals to produce whey protein concentrate free, so it can be used as baby food as well as functional foods (Dalev, 1994). Meanwhile, the use of microorganisms could diversify the utilization of whey into biotechnological products such as SCP and bioactive peptide or renewable energy such as methane and bioethanol (Chatzipaschali and Stamatis, 2012).

Moreover, lactose as whey main substrate can be used as sweetener or utilized into various lactose derivative products through the help of microorganisms. Whey lactose can be converted directly into polyhydroxyalkanoates by *Escherichia coli* recombinant, *Hydrogenophaga psedoflava* and *Azotobacter* sp., hydrolyzed into lactic acid by *Lactobacilli*, fermented to vinegar by *K. marxianus* and *Acetobacter pasteurans* or bioethanol by *Saccharomyces cerevisiae, Klyveromyces lactis, Klyveromyces marxianus, Candida pseudotropicalis* (Koller et al., 2012).

6.3.1 BIOACTIVE COMPOUND

6.3.1.1 ORGANIC ACID PRODUCTION

Organic acid dominates the third-largest category in the global market of fermentation products. It is very versatile in food and beverages industry. Organic acid can be produced by chemical synthesis or through direct fermentation using carbohydrate as substrates Fermentation is somehow preferred since all organic acids of tricarboxylic acid cycle can be produced in high yields by microorganisms (Soccol et al., 2007; Hatzinikolaou and Wang, 1992). Organic acid might be produced as either the end products or intermediate product of a fermentation process. The structure and properties of the organic produced are based on the type of microorganisms and the biochemical pathway that the microorganism can provide. Based on the biochemical pathway, there are 4 categories of organic acids. Type I is produced through the citric acid cycle (tricarboxylic acid pathway) under aerobic condition. Type II is produced under anaerobic condition through the metabolic pathway of butyric acid bacteria like *Clostridia*. *Clostridia* can convert many sugars to many variation of organic acids such as acetic, butyric, lactic, and formic acids. Type III organic acids consist of important compounds related to pharmaceutical uses such as antibiotics, vitamins, and amino acids, that are produced by specific microorganisms through more complicated biochemical pathway. Type IV, include fatty acids, are produced through hydrolysis (Hatzinikolaou and Wang, 1992). Whey can be used as a base for organic acid fermentation (Colomban et al., 1993). For example, there has been increasing interest in producing propionic acid from whey lactose as a cheap carbon source by using propionibacteria (Lewis and Yang, 1992).

Organic acid might be produced as either the end products or intermediate product of a fermentation process. The most used organic acids are citric acid, acetic acid, tartaric acid, malic acid, gluconic acid, propionic acid, fumaric acid, and also lactic acid, which has been discussed above. Citric acid and acetic acid are the most popular food acidulants (Soccol et al., 2007). In the food industry, propionic acid and its salts are used as preservator that inhibits growth of mold or food spoilage in many products like bread, cake, meat, fruit, and vegetables and is considered as GRAS (Colomban et al., 1993). Succinic acid is an important precursor molecule in the synthesis of biodegradable polyester resins, dyestuffs, and pharmaceuticals, and it also used in food industry as additive (Kurzrock and Weuster-Botz, 2010).

Bacteria, fungi, and yeasts are all capable of producing organic acid like *Arthrobacter paraffinens, Bacillus licheniformis, Corynebacterium* sp., *Lactobacillus casei, L. helveticus, L. paracasei, Streptococcus thermophiles* (bacteria); *Aspergillus niger, A. aculeatus, A. carbonarius, A. awamori, A. foetidus, A. fonsecaeus, A. phoenicis, Rhizopus oryzae, Penicillium janthinellum* (fungi); and *Candida tropicalis, C. oleophila, C. guilliermondii, C. citroformans, Hansenula anomala, Yarrowia lipolytica* (yeasts). *A. niger* is the most frequently used for the production of citric acid (Soccol et al., 2007). Several strains are proven to be able to ferment succinic acids, such as *Anaerobiospirillum succiniciproducens, Actinobacillus succinogenes*, and a recombinant *of Escherichia coli* strains and *Corynebacterium glutamicum* called *Mannheimia succinici-producens*. All of those bacteria are able to produce succinic acid with the presence of glucose and CO_2 (Kurzrock and Weuster-Botz, 2010) *Actinobacillus succinogenes* and *Manheimia succiniciproducens* are facultative anaerobic bacteria, proven to be effective succinic acid producers for its ability to endure high glucose osmotic and produce great amounts of products with a high productivity (Huh et al., 2006).

There are some problems in production of organic acids. Most organic acid fermentations are inhibited by the acidic pHs and also its major fermentation product (e.g., propionic acid fermentation is inhibited by propionic acid itself), making the fermentation rate and concentration low. Also, for heterogeneous fermentation, it produces many byproducts and this results in expensive purification process and low yields. This problem can be overcome by using integrated fermentation-separation system, like extractive fermentation system. This system can remove the acidic product to provide pH control and higher reaction rate. It will also result in pure and concentrated forms of product, making the recovery and purification costs lower. This extractive fermentation is used widely in homofermentation of lactic, acetic, and citric acid. But, research has shown that it is also effective for heterofermentation of propionic acid (Lewis and Yang, 1992). Production of organic acids also requires purification process due to the many impurities caused by produced acids, carbon sources, and salts (Huh et al., 2006). The first step is to separate debris and other components. Separation is usually done by combination of two or more methods to gain better results. For example, most usual methods for succinic acid cell separation are centrifugal separation or microfiltration that is followed by ultrafiltration to eliminate other solid components like debris, protein, and polymers (Kurzrock and Weuster-Botz, 2010). The proposed recovery techniques mainly used are precipitation, chromatography, membrane separation, extraction, and distillation (Li et al., 2016).

Lactic acid is one of organic acids that have a wide range of application throughout many industrial fields. In food industry, lactic acid can be used as food preservatives or acidifying agents, for textile industry as monomer for polylactic acid polymers, cosmetic industry needs it for the manufacture of hygiene and esthetical products, meanwhile pharmaceutical industry uses it as supplement in the synthesis of drugs, and many more (Shah and Patel, 2013; Martinez et al., 2013).

The synthesized of lactic acid can be done by both chemical and fermentative way. Production of lactic acid through fermentation is relatively fast, has high yields, and it is likely to obtain different lactic acid isomers only by changing the fermentation condition (Martinez et al., 2013; Bernardo et al., 2016). Fermentation has the ability to produce optically pure L(+)- or D(–)-lactic acid which distinguishes it from chemical synthesis of lactic acid. Chemical synthesis produces racemic DL-lactic acid or both of the isomers of lactic acid (L(+)-lactic acid and D(–)-lactic acid) and it is almost impossible to obtain only one of the isomers. Meanwhile, the D(–)-lactic acid is less desirable because resulting in acidosis and decalcification which harmful to human metabolism (Wee et al., 2006).

Lactic acid fermentation is done by the activity of microorganisms that converts carbohydrate to lactic acid (Burgain et al., 2014). The presence of lactose as carbohydrate source in dairy wastes made it possible to produce lactic acid, since it role as the main substrate that will metabolize by microorganisms into lactic acid (Hitha et al., 2014).

Microorganisms involved in the fermentation process are the important material which will determine the optimum lactic acid production. Fermentation can be done by involving lactic acid bacteria (LAB) such as *L. delbrueckii, L. acidophilus, L. bulgaricus, L. casei,* and *L. helveticus* from *Lactobacillus* sp., *Lactococcus lactis* and *S. thermophilus*. Fungal fermentation is also possible to produce lactic acid, but it has low production rate.

It is important to provide sufficient nutrients on the media. Supplementation of nutrients is likely to be done, such as addition of whey as nitrogen source to support the microorganisms, but also to ensure the total utilization of whey lactose and make the fermentation more economically viable as well as reducing the number of nutrients left during fermentation (Wee et al., 2006).

Fermentation of lactic acid can be held by both homofermentative and heterofermentative way. It differs by the end product, where homofermentation will produce 80% lactic acid, while heterofermentation will produce not only lactic acid, but also some co-products like CO_2, ethanol, or acetic acid (Wu et al., 2015). Homofermentation occurred in two steps, the first one is glycolysis where glucose is transformed into pyruvic acid and the next step is the reduction of pyruvic acid to lactic acid. This fermentation, theoretically, will yield 2 mol of lactic acid per mole of glucose. Heterofermentation also takes place in two steps. The first step is the pentose phosphate pathway, where glucose is converted to glyceraldehyde 3-phosphate, acetyl-phosphate, and CO_2. Glyceraldehyde 3-phosphate then enters glycolysis pathway where it is converted to lactic acid. Acetyl-phosphate is also converted to acetic acid or ethanol. Galactose is also converted to lactic acid via homofermentation (Martinez et al., 2013).

The most frequently used methods for lactic acid fermentation are batch, fed-batch, repeated batch, and continuous fermentation. Higher lactic acid concentration can obtain by using batch and fed-batch cultures, but higher productivity is achieved by using continuous cultures (Wee et al., 2006). Newer technology is also applied to fermentation of lactic acid to obtain higher productivity, such as immobilized-cell system. This system allowed increasing yields and productivities by preventing the limits related to washout (Martinez et al., 2013). It also allows higher cell densities, increase stability, makes reutilization and continuous operation possible, and it avoids the separation of cells from substrate products following processing (Panesar et al., 2007). Gel entrapment is one of the most popular methods of immobilization, using materials like alginate gel, agar, or k-carrageenan. After fermentation is completed, lactic acid is recovered and purified. Some techniques are available for the recovery of lactic acid, such as extraction, adsorption, membrane separation, filtration, and electrodialysis.

6.3.1.2 ACETALDEHYDE PRODUCTION

Another utilization of whey is to produce acetaldehyde. Acetaldehyde is used as fragrance or flavor additive by food industry, but mostly is used for acetic acid synthesis (Moroz et al., 2000). Using yeast or bacteria, acetaldehyde can be produced through fermentation. As of the production of lactic acid, lactose content inside whey will act as the carbon source that will be fermented into acetaldehyde.

Yeast can produce acetaldehyde via ethanol or alcoholic fermentation. Yeast such as *Saccharomyces cerevisiae* and *Schizosaccharomyces pombe* are often used to conduct ethanol fermentation in winemaking. *S. cerevisiae* strains are the most popular because of its ability to produce high levels of acetaldehyde, ranging from 50 to 120 mg/l. Important thing to notice is that acetaldehyde is not the final product of the fermentation, it is actually an intermediate product. Lactose will have to be broken down to its monomers, facilitated by invertase, an enzyme secreted by the yeast. Invertase will break the glycosidic linkage between the monomers, thus breaking the lactose into galactose and glucose. Next, the glycolytic pathway takes place and converts glucose into pyruvate acid. Pyruvate acid is then converted to acetaldehyde, before the yeast use it further to produce ethanol. It is important to avoid the further conversion of acetaldehyde to ethanol by the activity of alcohol dehydrogenase (ADH) enzyme. One way is to control the optimum condition during fermentation, such as temperature. Research shows that there is a significant accumulation of acetaldehyde production at 30°C that could be due to the inhibitory effect of the temperature towards the activity of ADH. Also, the accumulation of acetaldehyde depends on the equilibrium between the enzymes alcohol, dehydrogenase, and ADH. Production is also dependent on many other factors such as anaerobic condition, nutrients available in medium, etc. Thus, like all microbiological fermentation, controlling condition during process is very vital and will affect greatly on quality and quantity of the end product (Romano et al., 1994). It is also available to produce acetaldehyde via bioconversion of ethanol. Using alcohol oxidase, an enzyme that is collected form methylotropic yeast such as *Hansenula polymorpha* that is capable of oxidizing short-chain alcohol (Moroz et al., 2000).

Bacteria also have the ability to convert lactose to acetaldehyde. LAB is able to produce acetaldehyde, though the pathway of acetaldehyde production is still unclear. The most widespread pathway in bacteria is a three-step pathway, where it starts from pyruvate decarboxylation by pyruvate ferredoxin oxidoreductase or pyruvate formate lyase into acetyl coenzyme A, which is further converted by CoA-dependent-acetylating acetaldehyde dehydrogenase (AcDH) to acetaldehyde (Eram and Ma, 2013). Degradation of thymidine by deoxyriboaldolase can also produce acetaldehyde and glyceraldehyde-3-phosphate. Threonine, unlike other amino acids that can be converted into acetaldehyde via pyruvate as intermediate metabolite, can be converted directly into acetaldehyde by the activity of threonine aldolase. Production of acetaldehyde by LAB is also strain dependent. Several researches show that *L. delbrueckii* subsp. *bulgaricus* can produce great amount of acetaldehyde and is better than *S. thermophilus* (Chaves et al., 2002). Many researches on engineering bacteria to produce acetaldehyde are also conducted nowadays. For example, engineering *E. coli* by deleting the original fermentation pathways and introducing an exogenous acetyl-coenzyme A (CoA) reductase/acetaldehyde dehydrogenase will allow the production of acetaldehyde with high yield. The research further shows, that adding 1 g of yeast extract/liter and lowering the pH to 6.0 can increase the production of acetaldehyde even more (Zhu et al., 2011).

6.3.1.3 BACTERIOCINS AND ANTIMICROBIALS PRODUCTION

Utilization of whey as an antimicrobial source is of great interests nowadays. One of the main focuses being bacteriocins, an antimicrobial produced by bacteria. Both gram-positive and negative bacteria produce proteins that possess antimicrobial activity, called bacteriocins, as primary metabolite since it is made during the primary phase of growth (Zacharof and Lovitt, 2012). Whey, rich in lactose, is a cheap carbon source for growing bacteriocin-producing microbes as well as the production of the bacteriocin itself. Research shows that cheese way is able to support the growth of and bacteriocin production from *E. faecium*. The research also shows that all tested strains of *E. faecium* are capable to use cheese whey as carbon and nitrogen source efficiently (Schirru et al., 2014). Whey can be used as feedstock of bacteriocin production with supplementation of other nutrients, such as yeast extract (as nitrogen source), KH_2PO_4, $MgSO_4$, amino acids serine, threonine, and cysteine that can stimulate bacteriocin production (Liu et al., 2003).

Although bacteriocin has antimicrobial activity, it differs from antibiotic. It is caused by the preferential effect, where it only attacks closely related or strains of the same species, even itself. It is to say that antimicrobial spectrum of bacteriocin is not as wide as antibiotics. There are several killing mechanisms of bacteriocin. In colicin, for instance, there are three different bactericidal mechanisms once it enters bacteria, first by forming a pore inside inner membrane that can cause leakage of cytoplasmic compound, destruction of an electrochemical gradient, loss of ion, and cell death. Second, the activity of nuclease that can digest DNA and RNA of targeted bacteria. The last one is by the activity of peptidoglycanase that digest peptidoglycan precursor causing failure in peptidoglycan synthesis that leads to bacterial death (Yang et al., 2014).

Bacteriocins are considered as GRAS because it is degraded by proteolytic enzymes inside the gastrointestinal tract, so there is no residual to be concerned. Thus, bacteriocin is utilized by food industry for food preservator on egg, dairy, meat products, etc. (Zacharof and Lovitt, 2012). Bacteriocin is also used in pharmaceutical as "designer drugs" that only target specific pathogenic bacteria due to its preferential effect. This serves as an alternative towards the use of antibiotic that lately causes multi-resistance bacteria (Riley and Wertz, 2002).

Many strains and species of bacteria is able to produce bacteriocins. *L. plantarum, Pediococcus, Carnobacteria, Clostridia,* and *Propionibacteria,* is proven to be able to produce plantaricin, a type of bacteriocin. *L. fermentum* that has the ability to ferment sugar from galactose, glucose, and lactose, is able to produce bacteriocin inside whey as media culture (Udhayashree et al., 2012). *Bifidobacterium* sp. is also capable of producing bacteriocin, like Bifidocin B and Bifidi I that can kill foodborne pathogenic microbes and both gram-positive and negative bacteria (Balciunas et al., 2016). Bacteria strain and species, therefore, will determine the types of bacteriocin production. For example, nisin is produced by *Lactococcus lactis,* nukacin by *Streptococcus simulans,* hyicin from *Staphylococcus hycius*, mersacidin from *Bacillus* sp., Enterocin (food biopreservative) from *Enterococcus faecalis* (Bastos et al., 2015).

One of the most important bacteriocin is Nisin. Nisin is proven to be very efficient for its high activity, even at nanomolar concentrations, against gram-positive bacteria including *S. aureus* and *L. monocytogenes* which often cause food spoilage (Zacharof and Lovittb, 2012). Production of nisin from whey has also been conducted using mixed cultures of *Lactococcus lactis* and *Saccharomyces cerevisiae*, production of nisin resulted in 85% greater amount than only using pure culture of *Lactococcus lactis*. This is due to the consumption of lactic acid by *S. cerevisiae.* Lactic acid, simultaneously produced during the nisin biosynthesis by LAB, inhibits the optimum nisin production. An alternative way to separate lactic acid and improve nisin production is by involving *S. cerevisiae* that will consume lactic acid. The yeast itself will not harm the nisin-producing bacteria, since it doesn't interfere with lactose which is the major carbohydrate in whey that used for bacterial growth and nisin production (Liu et al., 2006).

Bacteriocins are produced in ribosom as inactive prepeptides. These prepeptides are then to be modified by proteins or amino acid encoded by the bacteriocin gene cluster before exit. It is important that the bacteria producing bacteriocin is capable of protecting itself from their own bacteriocin. During production, the bacteria will also produce specific immunity proteins to protect itself from its own bacteriocin. (Zacharof and Lovitt, 2012). Bacteriocin production is influenced by many factors such as growth temperature, time of incubation, pH, whey concentration, and the chemical composition of the culture medium (Schirru et al., 2014). Research shows that the maximum inhibition zone of bacteriocin originated from *L. fermentum* was at 37°C and at pH 2 (Udhayashree et al., 2012).

Bacteriocins are secreted into the culture from the bacteria, therefore, isolation and purification are crucial. There are many methods for isolation and purification of bacteriocin, that is usually be based on their affinity to organic solvents, solubility in concentrated salt solutions and at a given pH value. Among the most often used methods are salting out, solvent extraction, ultrafiltration, adsorption-desortion, ion-exchange, and size exclusion chromatography (Parada et al., 2007). The secretion of bacteriocin from the bacteria is based on types, sources, and amounts of sugar, inorganic minerals, and organic nitrogen (Schirru et al., 2014).

Production of bacteriocin by *Bacillus licheniformis* in cheese whey is dependent on whey concentration, where increased concentration of whey resulted in increased bacteriocin production (Cladera-Olivera et al., 2004). However, bacteriocin is not the only antimicrobial that can be, indirectly, derived from whey. Current research shows that hydrolysis of bovine whey protein to peptide, such as α-lactalbumin and β-lactoglobulin, lactoferin, and lysozyme, is found to have antimicrobial activity (Esmaeilpour et al., 2017). Antimicrobial activity of these peptides are dependent on the enzyme that hydrolyze it. α-lactalbumin hydrolyzed by trypsin and chymotrypsin shows inhibitory effect towards *E. coli* while hydrolyzation by al calase does not. These peptides can act in various ways to kill bacteria which is based on amino acid composition as well as structural and physicochemical characteristics (Akalin, 2014).

6.3.1.4 BIOACTIVE PEPTIDES PRODUCTION

Most of the bioactive functional compounds can be derived from milk such as lipids, minerals, vitamins, lactose, casein, and whey protein. Bioactivity of milk components have been categorized as four major areas: (1) gastrointestinal development, activity, and function; (2) immunological development and function; (3) infant development; and (4) microbial activity, including antibiotic and probiotic action (Gobbetti et al., 2007). The biologically active compounds in milk can guard human body against illnesses such as bioactive peptide. Bioactive peptides have been defined as specific protein fragments that have positive influence on physiological and metabolic functions or condition of the body and may have ultimate beneficial effects on human health (Park and Nam, 2015).

Bioactive peptides can also be derived from dairy wastes such as whey. Scotta is "Ricotta" cheese whey that has a peptide fraction of 3000 Da and highlighted a wide presence of potential bioactive peptides with health benefits that potentially utilized in nutraceutical formulations and functional foods (Sommella et al., 2016). Whey protein are predominated by α-lactalbumin and β-lactoglobulin, then the other are bovine serum albumin, immunoglobulins, lactoferrin, and lactoperoxidase (Dziuba and Dziuba, 2014). These bioactive peptides can be generated by the proteolytic activity of LAB (Korhonen and Pihlanto, 2006).

The generation of bioactive peptides can be done by the bacteria used in the manufacture of fermented dairy products. LAB, e.g., *Lactococcus lactis, Lactobacillus helveticus, and Lb. delbrueckii ssp. bulgaricus* has proteolytic systems that consist of a cell wall-bound proteinase and a number of distinct intracellular peptidases, including endopeptidases, aminopeptidases, tripeptidases, and dipeptidases which possible to manipulate the formation of peptides (Christensen et al., 1999; Williams et al., 2002).

Most of microbial bioactive peptides release from whey reported as ACE-inhibitory or antihypertensive peptides, immunomodulatory, antioxidative, and antimicrobial peptides. *Lb. helveticus* strains that widely used as starter for fermented milk has capability in releasing ACE-inhibitory peptides such as Val-Pro-Pro (VPP) and Ile-Pro-Pro (IPP) (Nakamura et al., 1995; Sipola et al., 2002). Antioxidative activities shown by *Lb. delbrueckii subsp. bulgaricus* IFO13953 through releasing peptides such as Ala-Arg-His-Pro-His-Pro-His-Leu-Ser-Phe-Met and *Lb. rhamnosus* with the peptides of Val-Lys-Glu-Ala-Met-Ala-Pro-Lys (Kudoh et al., 2001; Ashar and Chand, 2004).

6.3.2 ENZYME PRODUCTION

6.3.2.1 LIPOLYTIC ENZYME

Breakdown of lipid or fat content in the treatment of dairy wastewater is the importance of anaerobic digestion which involves enzyme activities (Soundarya et al., 2016). Lipolytic enzyme or lipases catalyze the hydrolysis of ester bonds of triacylglycerols into diacylglycerides, monoacylglycerides, glycerol, and fatty acids. Lipases break the emulsions of esters, glycerides, and long-chain fatty acids, which may be produced by microorganisms such as bacteria, fungi, and yeasts (Pastore et al., 2003).

Microorganisms have a role in the degradation of lipid or fat in dairy wastewater through enzymatic approach (Brooksbank et al., 2007). There are several microorganisms known as good producers of lipase, are *Penicillium chrysogenum, Rhizopus oligosporus, Aspergillus niger, and the genus Pseudomonas, Bacillus, and Burkholderia* (Hermes et al., 2013). Predominant microorganism in dairy wastewater is *Bacillus sp.* and *Pseudomonas* which can produce lipase enzyme. Lipase enzyme production by *Pseudomonas* sp. had the optimum range at the pH of 6–7, temperature 37°C–45°C and treatment period 48 hrs–72 hrs (Afroz et al., 2015).

The stages of lipolytic enzyme isolation is screening for microorganism with lipolytic activity, plate assay for bacterial lipases, culture conditions for lipolytic enzyme production, activity assay, morphological, and physiological characterization, 16S rDNA amplification and sequencing, phylogenetic analysis, lipolytic enzyme purification, electrophoresis, and protein determination, and determination of the optimum pH, temperature, and NaCl concentration for lipolytic activity (de Mariana et al., 2008).

Hermes et al. (2013) mentioned that lipases activities of microorganisms isolated from dairy waste such as cheese whey may vary. Regarding the pH of 6.2–7 and 24h incubation time, the microorganisms that incubated at 35°C with 150 rpm shaker shown lipases production of 2.76–4.87 U/ml. However, the pH value of 6.2 shown higher lipase production, with 0.13 U/ml superior than the pH of 7.0.

6.3.2.2 PROTEOLYTIC ENZYME

Proteolytic enzyme, also called protease are a group of enzyme that break the long-chain molecule of protein to shorter fragments and/or their components, an amino acid. The Proteolytic enzyme or proteases have many applications in pharmaceutical, leather, food, and waste processing industries (Vijayalakshmi and Murali, 2015). Proteolytic enzymes can release antioxidative peptides from caseins, soybean, and gelatine by enzymatic hydrolysis (Park and Nam, 2015).

Proteases are naturally in all organism, where the sources of proteases include plants, animals, and microorganism. Microorganism serves as a source of these enzymes because of their rapid growth, the limited space required for their cultivation and the ease with which they can be genetically manipulated to generate new enzymes with altered properties that are used for various applications (Kocher and Mishra, 2009).

Production of a proteolytic enzyme from dairy waste can be generated by bacteria. Qureshi et al. (2015) mention that the proteolytic system of bacteria is essential for their growth in milk. Converting milk casein to the free amino acids and peptides are the complex proteolytic system of bacteria.

Vijayalaksmi and Murali (2015) found that *Bacillus subtilis* is one of the bacteria that can produce a proteolytic enzyme which can be isolated from dairy wastes. The highest activity of proteases by the isolated *Bacillus subtilis* generated at pH 7 and temperature 35°C with the specific activity of 0.4125–0.4128 U/ml/min. The results show that secreted protease enzyme is a neutral protease, and the optimization can be done the addition of the medium for *Bacillus subtilis* growth with a rich carbon and nitrogen source such as rice bran.

6.3.2.3 β-GLUCOSIDASE ENZYME

One of the by-products discharged from dairy industry is whey. As a by-product, whey could be utilized into something useful like lactic acid. Lactic acid is produced from the conversion of lactose through fermentation process by LAB (Patel and Samir, 2016). LAB are indigenous flora that happens to be natural contaminants in dairy products. It could be isolated using selective media under two conditions, which are aerobic and anaerobic (Khalil and Anwar, 2016).

LAB from lactic acid fermentation at industrial level has a tremendous application to produce metabolic end products or secondary metabolites (Patel et al., 2013). It can also be used as source of enzymes, such as malolactic, phenoloxidases, proteases, and peptidases, lipases, glycosidases, ureases, and polysaccharide-degrading enzymes (Matthews et al., 2004).

One of the glycosidases synthesized enzyme which can grow easily and also stable at 100°C for 85 hours long is β-glucosidase enzyme (Divakar, 2013). LAB species which connected with fermented products known to supply carbohydrates degrading enzymes are *Lactobacillus, Lactococcus, Pediococcus,* and *Bifidobacterium* (Patel et al., 2012).

β-glucosidases was mainly applicative to hydrolyzed lactose in milk or derived products such as cheese whey. However, recently transgalactosylation activities (i.e., which can oligomerise galactosides) of β-glucosidases enzyme have been extensively exploited for the production of functional galactosylated products (Oliveira et al., 2011). β-glucosidase enzyme (β-D-glucoside glucohydrolases) presumed to play a significant role which could produce plant secondary metabolites, as well as part of cellulase enzyme with endoglucanase and cellbiohydrolase which can specifically produce glucose and acts as β-cellobiose disaccharides (Kuhad et al., 2011; Michlmayr and Kneifel, 2013). β-glucosidases is potentially useful for biotechnological processing such as synthesizing oligosaccharides and alkyl-glycosides also take part on releasing aromas, flavors, and isoflavone aglycons. Activity of β-glucosidases could be influenced by the substrate and cellobiose configuration, which requires a change of structure (Singhania et al., 2013).

β-glucosidases are found in fungi, bacteria, plants, and animals (Krisch et al., 2010). In plants, β-glucosidases involved in the synthesis of β-glucan during cell wall development, fruit ripening, and pigment metabolism. In humans and mammals, β-glucosidases involved in glucosyl ceramides hydrolysis which is associated with the Gaucher's disease (Lieberman et al., 2007). β-glucosidases that found in fungi are from filamentous fungi, like *Aspergillus oryzae* (Riou et al., 1998), *Aspergillus niger* (Gunata and Vallier, 1999), *Penicillum decumbens* (Chen et al., 2010), *Penicillum brasilianum* (Krogh et al., 2010), *Paecilomyces sp* (Yang et al., 2009), and *Phanerochaete chrysosporium* (Tsukada et al., 2006).

β-glucosidases generated by *Aspergillus* sp., *Kluyveromyces* sp. and LAB are widely used in industry because generally recognized as safe (GRAS) for human consumption, which is critical for food-related applications (Kosseva et al., 2009). *Aspergillus* sp. produces extracellular β-galactosidase to the acidic medium with a pH of 2.5–5.4 with the highest optimum temperature of 50°C (Panesar et al., 2006). Meanwhile, in *Kluyveromyces* sp. generated intracellular β-galactosidase with the mechanisms of lactose transportation into the interior of the yeast cell by a permease and then hydrolyzed intracellularly to glucose and galactose, which follow the glycolytic pathway or the Leloir pathway, respectively (Domingues et al., 2010). Beside fungi and yeasts, LAB such as *lactococci, streptococci*, and *lactobacilli* and *bifidobacterium*, have been regarded as safe bacteria that generate β-galactosidases, especially for functional food applications (Husain, 2010).

6.3.3 ENERGY CHEMICAL COMPOUND

6.3.3.1 ETHANOL PRODUCTION

Production of ethanol from whey can be done by yeasts fermentation. Lactose as carbon source contained in whey, fermented by yeasts such as *S. cereviseae*, *Kluyveromyces* sp. and *C. pseudotropicalis*, then converted into ethanol and carbon dioxide (Guimarães et al., 2010). *Kluyveromyces lactis* is one of indigenous yeasts isolated from whey that has the ability to ferment lactose into ethanol, with the optimum fermentation conditions is at 30°C for 36 h with agitation (Tipteerasri et al., 2007).

Ethanol generated by yeast through stages bioconversion. At an early stage, lactose hydrolyzed into glucose and galactose by lactase enzyme (Parrondo et al., 2009). The monosaccharides hydrolyzed into pyruvic acid through the glycolytic pathway (Flores et al., 2000). Pyruvic acid is an important ramification in the metabolic pathways of yeast, because it will determine the ethanol production or respiration.

The role of indigenous yeasts is essential, considering the specificity of carbon source utilized. *Saccharomyces* sp. found in cheese whey microflora; however, the wild *Saccharomyces* sp. has lack ability to produce ethanol from lactose (Guimarães et al., 2010). *Candida* sp. isolated from cheese whey has the ability to hydrolyzed lactose to increase the levels of ethanol and have high resistance to extreme conditions such as high sugar and alcohol contents (El Dein, 1997; Nakayama et al., 2008; Jolly et al., 2014; Utama et al., 2016b).

6.3.3.2 METHANE PRODUCTION

Producing dairy products such as yogurt, cheese, milk, butter, ice cream, etc.) is the biggest source for wastewater production (Kothari et al., 2016).

The effluent from the dairy industry, mostly are biodegradable organic matter. Large amount of organic matter produced could generate bioenergy (Beszedes et al., 2010). Dairy manure contains lactose, as the prime contributors which cause significant oxygen demand. Lactose can be utilized in bioconversion process that could supply biomass and extracellular products (Mawson, 1994). Lactose as a waste product could be utilized by fermentation to produce biofuels. Biofuels that widely produced are biogas (methane) and alcohol fuels (ethanol). Ethanol is produced by the conversion of glucose through fermentation process (Kisielewska, 2012). Methane is biologically exerted by anaerobic digestion process (Casey et al., 2006; Grant et al., 2015). Methane is used as the alternative of petroleum-based fuels (Mawson, 1994) of its performance and cost. The utilization of methane could also reduce the emission of greenhouse gases (GHGs) (Kisielewska, 2012).

Anaerobic digestion is done by microorganism to decompose organic matter in oxygen-free condition. Decomposition of organic material by anaerobic bacteria produces both methane and carbon dioxide. During the process, organic compounds are converted to another derivative. Sulfur compounds to hydrogen sulfide, nitrogen compounds to ammonia, and inorganic matter to various advantageous products. The final outcome from anaerobic digestion is methane as natural gas that could be used for the production of energy, heat, rich-nutrient organic slurry, and various inorganic products (Burke, 2001).

Microorganisms that involved in anaerobic digestion are hydrolytic bacteria, acetogenic bacteria, homoacetogenic bacteria, and methanogenic bacteria. The microorganisms involved have equal important role in the production of methane. Hydrolytic bacteria breakdown organic matter into simpler compound, such as CO_2, alcohol, hydrogen, and organic acids. Acetogenic bacteria form acetate and hydrogen from organic acids and alcohol. Homoacetogenic bacteria convert hydrogen, CO_2, alcohols, organic acids, and carbohydrates to acetate. Methanogenic bacteria form methane from hydrogen, CO_2, and acetate. Organisms that produces methane are classified into Archaea, phylum II, Euryarchaeota (Tavares and Malcata, 2016).

Anaerobic digestion starts with hydrolysis, acidogenesis, followed by methanogenesis. Hydrolysis is the breakdown of polymerized and insoluble organic compounds, such as carbohydrates, proteins, and fats to soluble monomers and dimmers, such monosaccharides, amino acids and fatty acids. Hydrolysis is done by group of relative anaerobic bacteria, like *Streptococcus* and *Enterobacterium* (Smith, 1966; Bryant, 1979; Shah and Patel, 2013).

Acidogenesis is the hydrogenation and fermentation of organic constituent to form short-chain volatile fatty acids, hydrogen, and carbon dioxide. Fatty acids produced then degraded by acetogenic bacteria to form acetate, hydrogen, and carbon dioxide (Mawson, 1994). Bacteria used are *Methanobacterium suboxydans,* which is used to form propionic acid from pentanoic acid, whereas *Methanobacterium propionicum*, used to form acetic acid from propionic acid (Shah and Patel, 2013). Acetic acid is the major substrate for methanogens to produce methane (Mawson, 1994).

Methanogenesis is complex reduction and oxidation biochemical reaction that happen in anaerobic condition with symbiotic effects from a variety of anaerobic and relatively anaerobic bacteria to decompose multimolecular organic material into simple and chemically stabilized substances (Naik et al., 2010; Shah and Patel, 2013). Methanogenesis consist of liquefaction, hydrolysis of insoluble materials, gasification with the help of partial or complete mineralization, also humification of organic material (Lyberatos and Skiadas, 1999; Shah and Patel, 2013). 30% of methane produced from the reduction of CO_2 done by autotrophic methane bacteria. Even though only a few bacteria that can decompose acetic acid to form methane, heterotrophic bacteria have an important role in the methane digestion process to produce a vast of methane (Verstraete et al., 2002; Shah and Patel, 2013). Methanogens that decompose acetates are *Methanosarcina barkeri* and *Methanosarcina sp.* (Shah and Patel, 2013). The dominant species in degradation of sewage sludge among acetotrophic methanogens is *Methanosaeta concilii* (McMahon et al., 2004; Shah and Patel, 2013).

Obtained methane may be utilized in a variety fields of technological process, economy, power engineering purposes, such as production of heat and electrical energy in associated units (1m^3 methane could produce 2.1 kWh electrical energy and 2.9 kWh heat energy), fuel in motor-car engines, production of methanol, and spark ignition from electrical energy obtained (Shah and Patel, 2013). In the form of Compressed Natural Gas (CNG), methane is used for vehicle fuel which is more environmentally friendly than petrol and diesel (Cornell, 2008). Combustion of methane and oxygen could create a fire that is used as fuel for turbines, ovens, automobiles, water-heater, etc. (Weinschenk, 2015). Methane in liquid form could be used for rocket fuels, which is more advantageous than kerosene because of the lesser deposition of carbon and lesser difficulty of reusing boosters (Thunnissen et al., 2004).

6.3.4 FUNCTIONAL COMPOUND

6.3.4.1 SINGLE CELL PROTEIN (SCP)

Microbial biomass has been considered as an alternative to conventional sources of protein in food and feed. Yeast was the first microorganisms that recognized as an animal feed supplement in almost a century and since 1996, protein production from biomass originating from different microbial sources such as yeast, fungi, bacteria, and algae and start to be known as SCP (Nasseri et al., 2011). Various microorganisms used for SCP production are bacteria (*Cellulomonas, Alcaligenes, B. subtilis, etc.*), algae (*Spirulina, Chlorella, etc.*), molds (*Trichoderma, Fusarium, Rhizopus, Aspergillus, etc.*), and yeast (*Candida, Kluyveromyces, Saccharomyces, etc.*) (Somaye et al., 2008; Nasseri et al., 2011; Ashok et al., 2014; Yadav et al., 2016).

The production of SCP by utilizing many substrates such as lactose, glucose, cellulose, hemicellulose, etc. which can be found in agro-industrial wastes (Table 6.2). Cellulose and hemicellulose can easily found in food and vegetable wastes; meanwhile, lactose or glucose can be found in dairy wastes such as whey (Utama et al., 2016). The SCP production from whey has evolving the bioconversion process which turns low-value by-products into the products that has nutritional and market value (Somaye et al., 2008).

TABLE 6.2 Microorganism and Substrate Used for SCP Production

Microorganism	Substrate
	Bacteria
Acinetobacter calcoacenticus	Ethanol
Aeromobacter delvacvate	n-alkanes
Aeromonas hydrophylla	Lactose
Bacillus megaterium	Non-protein nitrogenous compounds
Bacillus subtilis,	Cellulose, Hemicellulose
Cellulomonas sp.,	Cellulose, Hemicellulose

TABLE 6.2 *(Continued)*

Microorganism	Substrate
Flavobacterium sp.,	Cellulose, Hemicellulose
Lactobacillus sp,	Glucose, Amylose, Maltose
Methylomonas clara	Methanol
Methylomonas methylotropus	Methanol
Pseudomonas fluorescences	Uric, acid, and other Non-protein nitrogenous compounds
Rhodopseudomonas capsulate	Glucose
Thermomonospora fusca	Cellulose, Hemicellulose
Fungi	
Aspergillus fumigatus	Maltose, Glucose
Aspergillus niger	Cellulose, Hemicellulose
Aspergillus oryzae	Cellulose, Hemicellulose
Chaetomium cellulolyticum	Cellulose, Hemicellulose
Chepalosporium eichhorniae	Cellulose, Hemicellulose
Eytalidium acidiphlium	Cellulose, Pentose
Penecillium cyclopium	Glucose, Lactose, Galactose
Rhizopus chinensis	Glucose, Maltose
Thricoderma alba	Cellulose, Pentose
Thricoderma viridae	Cellulose, Pentose
Yeast	
Amoco torula	Ethanol
Candida intermedia	Lactose
Candida novellas	n-alkanes
Candida tropicalis	Maltose, Glucose
Candida utilis	Glucose
Saccharomyces cerevisiae	Lactose, Pentose, Maltose
Algae	
Chlorella pyrenoidosa	*Chlorella pyrenoidosa*
Chlorella sorokiana	*Chlorella pyrenoidosa*
Chondrus crispus	*Chlorella pyrenoidosa*
Porphyrium sp.	*Chlorella pyrenoidosa*
Scenedesmus sp.	*Chlorella pyrenoidosa*
Spirulina sp.	*Chlorella pyrenoidosa*

Sources: Modified from Bhalla et al. (2007) and Nasseri et al. (2011).

There are so many alternative ways in producing SCP with using bacteria, yeast, or mold. *Bacillus subtilis* can be used to ferment whey and has been found to be the most efficient in for SCP production because of its fast cell division with the SCP production of 0.32 mg/ml (Ashok et al., 2014). *Kluyveromyces marxianus* has been choose as potential yeast in SCP production from whey because its ability to grow in the lactose-based substrate with 82% total protein produced in the first 18 h of fermentation (Simaye et al., 2008). Meanwhile mixed culture between *K. marxianus* and *S. cereviseae* resulting higher biomass yield of 0.31 g biomass for every g lactose consumed with productivity 0.33 g/L.h (Yadav, 2016). *Aspergillus oryzae* or *Rhizopus arrhizus* chosen as SCP producing mold because of their non-toxic nature with 20–70% true protein weight (Srividya et al., 2014).

6.3.5 FERTILIZER

At the beginning, the industry only could use the whey in a limited way. Most whey used as fed for poultry or livestock, and some of the approaches are not economically beneficial. One of the most ways done by the industry is to throw directly to the ground under the pretext of organic fertilizer (Utama, 2016). However, improper approaches can cause problems given the large volume and organic material gained therein. The use of whey in adequate proportions may have positives effects as fertilizer, while excessive use may present unbalance soil microflora which could affect the plants negatively (Sonnleitner et al., 2003a, 2003b; Demir and Ozrenk, 2009; Grosu et al., 2012).

Whey has been known to have physical-chemical characteristics that can be useful. Nitrogen, phosphorus, sulfur, calcium, sodium, and magnesium, lactose, and proteins from whey have a role in increasing crop productivity (Demir and Ozrenk, 2009; Stefan et al., 2009; Tsakali et al., 2010). Low pH of whey makes most macronutrient cations more available for plants grown. In sodic soil, the Ca^{2+}, Mg^{2+}, and K^{+} from whey tend to lower the soil solution pH which will speed the leaching of exchangeable Na when sufficient water is passed through the soil (Lehrsch et al., 1993a). Addition whey soluble salts to the soils will increase the flux density by both water and air which caused by proportion of larger soil pores (Lersch et al., 1993b).

Beside the physics-chemical characteristics, microorganisms also have role in increasing soil fertility. Indigenous microorganisms will stimulate whey aerobically into polysaccharide that will stabilize aggregates and improves soil structures (Lehrsch et al., 1993a). Microorganisms also have role in degrading lactose and proteins to produce CO_2 and organic acid that could decrease the pH that could increase Ca solubility (Robbins, 1985). Microorganisms have specific activities in increasing the number of free carbon and nitrogen from carbohydrates and proteins hydrolysis which can affects the C/N ratio of plant growth medium (Utama, 2016). The activities of microorganism in whey have the ability in soil structure improvements, solubility of nutrients, proper C/N ratio that will increase soil fertility for plant growth.

6.4 ENVIRONMENTAL ASPECT OF MICROBIOLOGICAL APPROACH TOWARDS DAIRY INDUSTRY WASTES UTILIZATION

6.4.1 ADVANTAGES AND DISADVANTAGES

Utilization of dairy waste is actually a work of bioremediation. Bioremediation is a naturally occurring process by which microorganisms either immobilize or transform environmental contaminants to harmless, even useful, end products. Bioremediation will help to preserve the environment since it utilizes naturally occurring biogeological process and eliminate contaminant rather than transfers it from one environmental medium to another. It will also lower some costs like capital expenditure (Thassitou and Arvanitoyannis, 2001). Waste reduction of dairy waste by microbial utilization can have a significant impact on product processing efficiency and improved financial returns.

Nowadays, 50% of worldwide's whey production has already been utilized and transformed into various products as mentioned above (Ghosh et al., 2016). Compared to chemical synthesis, fermentation has advantages like low cost of substrates, low production temperature, and low energy consumption (John et al., 2009). Cost of substrates can be minimized, especially when it comes to using byproducts as the source. Also, byproducts of dairy are renewable resources so it is guaranteed that the resources for fermentation process will always be available since the production of waste from dairy industry is inescapable. Meanwhile, chemical synthesis of succinic acid for instance, is produced through catalytic hydrogenation of crude oil as the substrate. Using crude oil is now avoided because of its availability is highly restricted since it is not a renewable resource and the prices keep increasing (Kurzrock and Weuster-Botz, 2010). Chemical synthesis also requires high temperature to catalyze and enable reactions to occur, thus making the energy consumption is also high, while fermentation doesn't. Fermentation needs only the appropriate temperature as well as other parameters to ensure the life of microorganisms involved. The temperature required is not as high as chemical process; in fact, the high temperature will kill the microorganisms involved.

Although it has many benefits, the microbial utilization of dairy wastewater are sometimes having trouble, related to the technicality. Fermentation process for industrial scale are lacking in engineering tools such as mathematical models and optimization techniques. One of the most important engineering challenges are scaling up the fermentation process. Scaling up is meant to transform optimum operating conditions found in pilot-scale to the production scale in order to reach an efficient production. One major problem related to the pilot-scale bioreactors, is that different experimental scales are very hard to compare to larger size industrial bioreactors (Formenti et al., 2014).

An often time, the downstream process of fermentation is what hampers the whole fermentation process such as isolation, recovery, or purification of the end product. For example, the purification process of organic acid fermentation costs for around 60–70% of the production cost in the fermentation process (Huh et al., 2006). This is due to the production of other byproducts during the fermentation process that causes the impurities of the wanted product. Therefore, the end product must be purified and this causes an awful lot of expenditure. Production of other byproducts sometimes also has inhibitory effect. For example, in production of nisin, where the production of nisin is coupled by the lactic acid produced that can inhibit the optimum production nisin (Liu et al., 2006). However, these problems have already been a subject of research, thus providing many alternative solutions to solve it. Though it is viewed as expensive, fermentation process is still more cost effective, and therefore is more likely used, for example in lactic acid production where 90% of all lactic acid worldwide is produced through bacterial fermentation (Coelho et al., 2011).

6.4.2 CARBON EMISSION AVOIDANCE

Dairy industry contributes in 3.739 and 11.217 million m^3 effluent per year (Tikariha and Sahu, 2014; Monroy et al., 1995). Waste produced from dairy industry generated in the manufacturing of dairy products, milk processing units, such as pasteurization, homogenization, etc., and the cleaning of milk processing units using chemical solution (Tikariha and Sahu, 2014; Thompson and George, 1998). Even though dairy factory waste contains low amount of heavy metal material, high amount of organic constituents from dairy industry could cause pollution (Tikariha and Sahu, 2014). The organic constituents usually act as nutritional content, which are lactose, caseins, *maclura*, etc. (De Jesus et al., 2015).

Dairy waste could be utilized by anaerobic digestion process to create renewable energy and to minimize the emission of GHGs (Gloy, 2010). GHG could cause greenhouse effect that lead to global warming. One of the GHG compound is carbon dioxide (CO_2) (Mishra et al., 2013). CO_2 could cause irreversible climate change which is a great impact towards environment (Solomon et al., 2008). In high concentration, CO_2 exposure may also cause health problems. CO_2 exposures in 1–5% may cause dyspnea, tremor, modified breathing, acidosis, headaches, reduced fertility, visual impairment, increased blood pressure, and intercostals pain (Bierwith, 2016).

Carbon dioxide exerted from whey as dairy waste could be utilized to obtain poly-hydroxybutyrate (PHB) through microbial fermentation process produce biodegradable food packaging material. Food packaging manufacturers usually use petrochemical-based polymers that could degrade ecosystems environment. The material used is not only not reusable but also not foreseen for reuse. Because of its non-biodegradability, petrol-based plastic highly contributes in the emission of carbon dioxide equivalent. By using whey to form PHB-based food packaging, could reduce: 1) 35% of carbon emission than using PP-based food packaging; 2) 75% of BOD than the original whey (before treatment); and 3) 50% of COD than the original whey (before treatment) (Alborch, 2014).

Alternative energy that is renewable is needed because of the deficits of energy supply and the elevation of energy demands, which could be achieved with the help of microorganisms. Energy in form of electricity could be produced with the help of microbial fuel cell. It could be an important form for bio-energy because of the efficiency, recyclability, and green way to generate electricity. Rather than energy from fossil fuels which has a big contribution of the carbon dioxide emission (Aurora, 2012). Microbiological approach towards dairy industry waste might be a profitable option because it could generate energy to utilize CO_2, treat wastewater, and also avoid carbon emissions (Chinnasamy et al., 2012).

6.5 FUTURE PROSPECTIVE

Trends of environmentally friendly dairy industries are highly escalated lately. Support to various efforts in realizing environmentally sound dairy processing can be achieved through microbiological approach in utilizing wastes resulted from the dairy processing. Microbiological approach resulting metabolites that are highly useful and some of them also available to develop as functional compound that has high value. The technology offered is very possible to be applied widely considering the ease and also potentially to develop in vary levels. At sophisticated levels, the technology can be increased to determine more specific metabolites and very possible to combine it with nano-technology. Meanwhile, at a simpler level, the approach can be done in simple and easy batch methods that focus more on the results obtained. The approach will result high reduction of wastes volume, potential of soil and water pollution, pathogenic risk and also carbon emissions. Besides gives tangible benefit from the income of bioconversions product sales, the approach also gives intangible benefit such as better corporate image as environmental friendly dairy industry.

6.6 SUMMARY

There are so many ways to utilized wastes generated by the dairy industries. Whey as common dairy industry wastes can be used through the physical, chemical, and microbiological approach. Microbiological approach is one of the best ways to ensure the dairy wastes utilization was done in the environmental friendly way. High regeneration rate and broad metabolism ability make the microorganisms are potential for dairy industry wastes utilization, meanwhile harmless residue and degradability characteristics makes it environmentally sound. Strain selection of the microorganisms become the important thing to determine the resulted metabolites from dairy wastes bioconversions and also specify post-treatment that should be done to ensure there are no pathogenic risk or harmful residues resulted. At the application level, the use of microorganisms is widely known by the community; this will simplify the implementation at various level of utilization, especially if it expects the sustainability.

KEYWORDS

- **bioactive peptides**
- **bio-energy**
- **bioremediation**
- **carbon emission**
- **dairy effluent**
- **functional food**
- **greenhouse gasses**
- **microbiological utilization**
- **organic acid**
- **renewable energy**
- **single-cell protein**
- **β-galactosidase**
- **β-glucosidase**

REFERENCES

Afroz, Q. M., Khan, K. A., & Pervez, A. U. S., (2015). Enzyme used in dairy industries. *International Journal of Applied Research, 1*(10), 523–527.

Akalin, A. S., (2014). Dairy-derived antimicrobial peptides: Action mechanisms, pharmaceutical uses and production proposals. *Trends in Food Science & Technology, 36*, 79–95.

Ashar, M. N., & Chand, R., (2004). Fermented milk containing ACE-inhibitory peptides reduces blood pressure in middle aged hypertensive subjects. *Milchwissenschaft, 59*, 363–366.

Ashok, V. G., Minakshi, A. P., & Pranita, A. G., (2014). Liquid whey: A potential substrate for single cell protein production from *bacillus subtilis* NCIM 2010. *International Journal of Advanced Life Sciences, 2*(2), 119–123.

Balciunas, E. M., Arni, S. A., Converti, A., Leblanc, J. G., De Souza, O. R. P., (2016). Production of bacteriocin-like inhibitory substances (BLIS) by *Bifidobacterium lactis* using whey as a substrate. *International Journal of Dairy Technology, 69*(2), 236–242.

Bastos, M. D. C. D. F., Coelho, M. L. V., & Santos, O. C. D. S., (2015). Resistance to bacteriocins produced by Gram-positive bacteria. *Microbiology, 161*, 683–700.

Bernardo, M. P., Coelho, L. F., Sass, D. C., & Contiero, J., (2016). L-(+)-Lactic acid production by *Lactobacillus rhamnosus* B103 from dairy industry waste. *Brazilian Journal of Microbiology, 47*, 640–646.

Bhalla, T. C., Sharma, N. N., & Sharma, M., (2007). *Production of Metabolites, Industrial Enzymes, Amino Acid, Organic Acids, Antibiotics, Vitamins and Single Cell Proteins*. National Science Digital Library, India.

Brooksbank, A. M., Latchford, J. W., & Mudge, S. M., (2007). Degradation and modification of fats, oils and grease by commercial microbial supplements. *World Journal of Microbiology and Biotechnology, 23*(7), 977–985.

Burgain, J., Scher, J., Francis, G., Borges, F., Corgneau, M., Revol-Junelles, A. M., Cailliez-Grimal, C., & Gaiani, C., (2014). Lactic acid bacteria in dairy food: Surface characterization and interactions with food matrix components. *Advances in Colloid and Interface Science, 213*(2014), 1–15.

Chatzipaschali, A. A., & Stamatis, A. G., (2012). Biotechnological utilization with a focus on anaerobic treatment of cheese whey: Current status and prospects. *Energies, 5*, 3492–3525.

Chaves, A. C. S. D., Fernandez, M., Lerayer, A. L. S., Merau, I., Kleerebezem, M., & Hugenholtz, J., (2002). Metabolic engineering of acetaldehyde production by streptococcus thermophilus. *Applied and Environmental Microbiology, 68*(11), 5656–5662.

Christensen, J. E., Dudley, E. G., Pederson, J. A., & Steele, J. L., (1999). Peptidase and amino acid catabolism in lactic acid bacteria. *Antoine Van Leeuwenhoek, 76*, 217–246.

Cladera-Olivera, F., Caron, G., & Brandelli, A., (2004). Bacteriocin production by *Bacillus licheniformis* strain P40 in cheese whey using response surface methodology. *Biochemical Engineering Journal, 21*, 53–58.

Dalev, P. G., (1994). Utilization of waste whey as a protein source for production of iron proteinate: an antianemic preparation. *Bioresource Technology, 48*, 75–77.

De Jesus, C. S. A., Ruth, V. G. E., Daniel, S. F. R., & Sharma, A., (2015). Biotechnological alternatives for the utilization of dairy industry waste products. *Advances in Bioscience and Biotechnology, 6*, 223–235.

De Mariana, M. N., Vargas, V. A., Antezana, H., & Svoboda, M., (2008). Lipolytic enzyme production by halophilic/halotolerant microorganisms isolated from Laguna Verde, Bolivia. *Bolivian Journal of Chemistry, 25*(25), 14–23.

Demir, S., & Ozrenk, E., (2009). Effects of whey on the colonization and sporulation of arbuscular mychorizal fungus, glomus intraradices, in lentil (*Lens orientalis*). *African Journal of Biotechnology, 8*(10), 2151–2156.

Domingues, L., Guimarães, P., & Oliveira, C., (2010). Metabolic engineering of *Saccharomyces cereviseae* for lactose/whey fermentation. *Bioengineered Bugs, 1*, 164–171.

Dziuba, B., & Dziuba, M., (2014). Milk proteins-derived bioactive peptides in dairy products: Molecular, biological and methodological aspects. *Food Sciences Technology, 13*(1), 5–25.

El Dein, M., (1997). Role of lipids in ethanol tolerance of Candida lambica. *Indian Journal of Experimental Biology, 35*(1), 89–91.

Eram, M. S., & Ma, K., (2013). Decarboxylation of pyruvate to acetaldehyde for ethanol production by hyperthermophiles. *Biomolecules, 3*, 578–596.

Esmaeilpour, M., Ehsani, M. R., Aminlari, M., Shekarforoushd, S., & Hoseini, E., (2017). Antimicrobial peptides derived from goat's milk whey proteins obtained by enzymatic hydrolysis. *Journal of Food Biosciences and Technology, 7*(1), 65–72.

Flores, C., Rodriguez, C., Petit, T., & Gancedo, C., (2000). Carbohydrate and energy-yielding metabolisme in non-conventional yeasts. *FEMS Microbiology Review, 24*, 507–529.

Gobbetti, M., Minervini, F., & Rizzello, C. G., (2007). Bioactive peptides in dairy products. In: Hui, Y. H., (ed.), *Handbook of Food Products Manufacturing* (pp. 489–517). John Wiley & Sons, New Jersey.

Grosu, L., Fernandez, B., Grigoras, C. G., Patriciu, O. I., Grig-Alexa, I., Nicuta, D., Ciobanu, D., Gavrilla, L., & Finaru, A. L., (2012). Valorization of whey from dairy industry for agricultural use as fertilizer: Effects on plant germination and growth. *Environmental Engineering and Management Journal, 11*(12), 2203–2210.

Guha, A., Banerje, S., & Bera, D., (2013). Production of lactic acid from sweetmeat industry waste by *lactobacillus delbruki. International Journal of Research in Engineering and Technology, 02*(04), 630–634.

Guimarães, P. M. R., Teixeira, J. A., & Domingues, L., (2010). Fermentation of lactose to bioethanol by yeasts as part of integrated solutions for the valorization of cheese whey. *Biotechnology Advances, 28*, 375–384.

Hermes, E., Da Rocha, D. C., Orssatto, F., Lucas, J. F. R., Gomes, S. D., & Sene, L., (2013). Isolation of microorganisms of cheese whey with lipolytic activity for removal of COD. *Engenharia Agrícola, 33*(2), 379–387.

Hitha, C., Hima, C. S., Yogesh, B. J., Bharathi, S., & Sekar, K. V., (2014). Microbial utilization of dairy waste for lactic acid production by immobilized bacterial isolates on sodium alginate beads. *International Journal of Pure and Applied Bioscience, 2*(4), 55–60.

Husain, Q., (2010). Beta galactosidases and their potential applications: A review. *Critical Reviews in Biotechnology, 30*, 41–62.

IDF (International Dairy Federation), (2016). The world dairy situation. *Bulletin of the International Dairy Federation 485/2016.* URL: http://www.idfa.org/docs/default-source/d-news/world-dairy-situationsample.pdf (accessed on 7 September 2019).

Jolly, N. P., Varela, C., & Pretorius, I. S., (2014). Not your ordinary yeast: Non-*saccharomyces* yeasts in wine production uncovered. *FEMS Yeast Research, 14*(2), 215–237.

Kharbanda, A., & Prasanna, K., (2016). Extraction of nutrients from dairy wastewater in the form of MAP (magnesium ammonium phospate) and HAP (Hidroxyapatite). *Rasayan Journal of Chemistry, 9*(2), 215–221.

Kocher, G. S., & Mishra, S., (2009). Immobilization of *bacillus circulans* MTCC 7906 for enhanced production of alkaline protease under batch and packed bed fermentation conditions. *Internetional Journal of Microbiology, 7*(2), 1–6.

Koller, M, Salerno, A., Muhr, A., Reiterer, A., Chiellini, E., Casella, S., Horvat, P., & Braugnegg, G., (2012). Whey lactose as raw material for microbial production of biodegradable polyesters. In: Saleh, H. E. D., (ed.), *Polyester.* InTech, doi: https://www.intechopen.com/books/polyester/whey-lactose-as-a-raw-material-for-microbial-production-of-biodegradable-polyesters (accessed on 7 September 2019).

Koller, M., Atlic, A., Gonzalez-Garcia, Y., Kutschera, C., & Braunegg, G., (2008). Polyhydroxyalkanoate (PHA) biosynthesis from whey lactose. *Macromolecular Symposia, 272*(1), 87–92.

Korhonen, H., & Pihlanto, A., (2006). Bioactive peptides: Production and functionality. *International Dairy Journal, 16*, 945–960.

Kosseva, M., Panesar, P., Kaur, G., & Kennedy, J., (2009). Use of immobilized biocatalysts in the processing of cheese whey. *International Journal of Biological Macromolecules, 45*, 437–447.

Kudoh, Y., Matsuda, S., Igoshi, K., & Oki, T., (2001). Antioxidative peptide from milk fermented with *Lactobacillus delbrueckii* subsp. bulgaricus IFO13953. *Journal of the Japanese Society of Food Science, 48*, 44–50.

Lersch, G. A., Robbins, C. W., & Hansen, C. L., (1993a). Cottage cheese (acid) whey effects on sodic soil aggregate stability. *Acid Soil Research and Rehabilitation, 8*, 19–31.

Lersch, G. A., Sojka, R. E., Carter, D. L., & Jolley, P. M., (1993b). Freezing effects on aggregate stability of soils amended with lime and gypsum. In: Poesen, J. W. A., & Nearing, M. A., (eds.), *Soil Surface Sealing and Crusting* (pp. 115–127). Catena, Cremlingen.

Liu, C., Hu, B., Liu, Y., & Chen, S., (2006). Stimulation of nisin production from whey by a mixed culture of *lactococcus lactis* and *saccharomyces cerevisiae*. *Applied Biochemistry and Biotechnology, 129*(132), 751–761.

Liu, C., Liu, Y., Liao, W., Wen, Z., & Chen, S., (2003). Application of statistically-based experimental designs for the optimization of nisin production from whey. *Biotechnology Letters, 25*, 877–882.

Marlina, E. T., Balia, R. L., & Hidayati, Y. A., (2012). *Acid and Total Bacteria of Dairy Wastewater Solid and Cassava Waste Flour Mixture Fermented by Aspergillus Niger as Broiler Feed.* URL: http://repository.unpad.ac.id/20666/ (accessed on 7 September 2019).

Martinez, F. A. C., Balciunas, E. M., Salgado, J. M., Gonzalez, J. M. D., Converti, A., & Oliveira, R. P. S., (2013). Lactic acid properties, applications and production: A review. *Trends in Food Science and Technology, 30*, 70–83.

Moroz, O. M., Gonchar, M. V., & Sibirny, A. A., (2000). Efficient bioconversion of ethanol to acetaldehyde using a novel mutant strain of the methylotrophic yeast hansenula polymorpha. *Biotechnology and Bioengineering, 68*(1), 44–51.

Nakamura, Y., Yamamoto, M., Sakai, K., Okubo, A., Yamazaki, S., & Takano, T., (1995). Purification and characterization of angiotensin I- converting enzyme inhibitors from sour milk. *Journal of Dairy Science, 78*, 777–783.

Nakayama, S., Morita, T., Negishi, H., Ikegami, T., Sakaki, K., & Kitamoto, (2008). *Candida krusei* produces ethanol without production succinic acid, a potential advantage for ethanol recovery by pervaporation membrane separation. *FEMS Yeast Research, 8*, 706–714.

Nasseri, A. T., Rasoul-Amini, S., Morowvat, M. H., & Ghasemi, Y., (2011). Single cell protein: Production and process. *American Journal of Food and Technology, 6*(2), 103–116.

Oliveira, C., Guimarães, P. M. R., & Domingues, L., (2011). Recombinant microbial systems for improved β-galactosidase production and biotechnological applications. *Biotechnology Advances, 29*, 600–609.

Palela, M., Ifrim, G., & Bahrim, G., (2008). Microbiological and biochemical characterization of dairy and brewery wastewater microbiota. *The Annals of the University Dunarea de Jos of Galati, Fascicle VI – Food Technology, New Series, 2*(31), 23–30.

Panesar, P. S., Kennedy, J. F., Gandhi, D. N., & Kataryzma, B., (2007). Bioutilization of whey for lactic acid production. *Food Chemistry, 105*, 1–14.

Panesar, P. S., Panesar, R., Singh, R. S., Kennedy, J. F., & Kumar, H., (2006). Microbial production, immobilization and application of beta-D-glucosidase. *Journal of Chemical Technology and Biotechnology, 81*, 530–543.

Parada, J. L., Caron, C. R., Medeiros, A. B. P., & Ricardo, C., (2007). Bacteriocins from lactic acid bacteria: Purification, properties and use as biopreservatives. *Brazilian Archives of Biology and Technology, 50*(3), 521–542.

Park, Y. W., & Nam, M. S., (2015). Bioactive peptides in milk and dairy products: A review. *Korean Journal Food Science, 35*(6), 831–840.

Parrondo, J., Garcia, L. A., & Diaz, M., (2009). Nutrient balance and metabolic analysis in Kluyveromyces marxianus fermentation with lactose-added whey. *Brazilian Journal of Chemical Engineering, 26*(3), 445–456.

Pastore, G., Costa, V., & Koblitz, M., (2003). Partial purification and biochemical characterization of extracellular lipase produced by new *Rhizopus* sp. *Food Science and Technology, Campinas, 23*, 135–140.

Pathak, U., Das, P., Banerjee, P., & Datta, S., (2016). Treatment of wastewater from a dairy industry using rice husk as adsorbent: Treatment efficiency thermodynamics, and kinetics modeling. *Journal of Thermodynamics,* 1–7.

Prakashveni, R., & Jagadeesan, M., (2012). Isolation identification and distribution of bacteria in dairy effluent. *Pelagia Research Library Advances in Applied Science Research, 3*(3), 1316–1318.

Qasim, W., & Mane, A. V., (2013). Characterization and treatment of selected food industrial effluents by coagulation and adsorption technique. *Water Resources and Industry, 4*, 1–12.

Raghunath, B. V., Punnagaiarasi, A., Rajarajan, G., Irshad, A., Elango, A., & Kumar, G. M., (2016). Impact of dairy effluent on environment: A review. In: Prashanthi, M., & Sundaram, R. (eds.), *Integrated Waste Management in India, Environmental Science and Engineering* (pp. 239–249). Springer International Publishing, Switzerland.

Rao, M., & Datta, A., (2012). *Waste Water Treatment* (pp. 1–383). Oxford & IBH Publishing, Delhi.

Riley, A. M., & Wertz, E. J., (2002). Bacteriocins: Evolution, ecology, and application. *Annual Review of Microbiology, 56*(37), 117–137.

Robbins, C. W., (1985). The $CaCO_3$-CO_2-H_2O system in soils. *Journal of Agronomics Education, 14*, 3–7.

Romano, P. S. G., Turbanti, L, & Polsinelli, M., (1994). Acetaldehyde production in *Saccharomyces cerevisiae* wine yeasts. *FEMS Microbiology Letters, 118*, 213–218.

Santos, P. M., Silva, J. E. C., Santos, A. C., Santos, J. G. D., Araújo, A. S., & Rodrigues, M. O. D., (2016). Liquid dairy waste as fertilizer for 'Mombaça' grass. *Communicata Scientiae, 7*(2), 251–261.

Schirru, S., Favaro, L., Mangia, N. P., Basaglia, M., Casella, S., Comunian, R., Fancello, F., Gombossy, D. M. F. B. D., De Souza, O. R. P., & Todorov, S. D., (2014). Comparison of bacteriocins production from *enterococcus faecium* strains in cheese whey and optimized commercial MRS medium. *Annals of Microbiology, 64*, 321–331.

Shah, N., & Patel, A., (2013). Lactic acid bacteria in the treatment of dairy waste and formation of by-products-a promising approach. *Indian Food Industry Mag., 32*(4), 45–49.

Shete, B. S., & Shinkar, N. P., (2013). Dairy industry wastewater sources characteristics and its effects on environment. *International Journal of Current Engineering and Technology, 3*, 1611–1615.

Shivsharan, V. S., Wani, M. P., & Kulkarani, S. W., (2013). Isolation of microorganism from dairy effluent for activated sludge treatment. *International Journal of Computational Engineering Research, 3*(3), 161–167.

Singh, A. K., & Singh, K., (2012). Study on hydrolysis of lactose in whey by use of immobilized enzyme technology for production of instant energy drink. *Advance Journal of Food Science and Technology, 4*(2), 84–90.

Sipola, M., Finckenberg, P., Korpela, R., Vapaatalo, H., & Nurminen, M. L., (2002). Effect of long-term intake of milk products on blood pressure in hypertensive rats. *Journal of Dairy Research, 69*, 103–111.

Somaye, F., Marzich, M., & Lale, N., (2008). Single cell protein (SCP) production from UF cheese whey by *kluyveromyces marxianus*. *Proceeding of 18th National Congress on Food Technology* (pp. 1–6).

Sommella, E., Pepe, G., Ventre, G., Pagano, F., Conte, G. M., Ostacolo, C., et al. (2016). Detailed peptide profiling of "Scotta": From dairy waste to a source of potential health-promoting compounds. *Dairy Science and Technology, 96*(5), 763–771.

Sonnleitner, R., Lorbeer, E., & Schinner, F., (2003a). Effects of straw, vegetable oil and whey on physical and microbiological properties of chernozem. *Applied Soil Ecology, 22*, 195–204.

Sonnleitner, R., Lorbeer, E., & Schinner, F., (2003b). Monitoring of changer in physical and microbiological properties of chernozem amended with different organic substrates. *Plant and Soil, 253*, 391–402.

Soundarya, R. V., Justus Reymond, D., & Karthikeyan, V., (2016). Biodegradation of fat, oil, and grease by using an up flow anaerobic sludge blanket reactor for dairy wastewater. *International Journal of Innovative Research in Science, Engineering and Technology, 5*(5), 7854–7861.

Srividya, A. R., Vishnuvarthan, V. J., Murugappan, M., & Dahake, P. G., (2014). Single cell protein: A review. *International Journal of Pharmaceutical Research Scholars, 2*(4), 472–485.

Stefan, D. S., Stefan, M., & Penu, A. I., (2009). Treatment and management of distillery fermented fruit spent marc. II. Processing by composting. *Environmental Engineering and Management Journal, 9*, 971–976.

Tipteerasri, T., Hanmoungjai, W., & Hanmoungjai, P., (2007). *Ethanol Production from Crude Whey by Kluyveromyces marxianus TISTR 5695*. Chiangmai University, Thailand.

Tsakali, E., Perotos, K., D'Allesandro, A. G., & Goulas, P., (2010). A review on whey composition and the method used for its utilization for food and pharmaceutical products. In: Cadavez, V., & Thiel, D., (eds.), *6th International Conference on Simulation and Modeling in the Food and Bio-Industry (FOODISM 2010)*. CIMO, Braganca, Portugal.

Udhayashree, N., Duraisamy, S., & Senthilkumar, B., (2012). Production of bacteriocin and their application in food products. *Asian Pacific Journal of Tropical Biomedicine*, S406–S410.

Utama, G. L., (2016). *Analysis of the Utilization of Indigenous Yeasts Consortium in Bioconversion Cheese Whey and Vegetable Wastes into Bioethanol and Social Economic and Environmental Benefits.* MSc Dissertation, Faculty of Agro-Industrial Technology, Universitas Padjadjaran, Bandung.

Utama, G. L., Balia, R. L., Kurnani, T. B. A., & Sunardi, S., (2015). Cost benefit analysis of bioconversion Neufchatel whey into rectified ethanol and organic liquid fertilizer in semi pilot scale. *Agrolife Scientific Journal, 4*(1), 192–196.

Utama, G. L., Kurnani, T. B. A., Sunardi, & Balia, R. L., (2016b). The isolation and identification of stress tolerance ethanol-fermenting yeasts from mozzarella cheese whey. *International Journal on Advanced Science, Engineering and Information Technology, 6*(2), 252–257.

Utama, G. L., Kurnani, T. B. A., Sunardi, S., & Balia, R. L., (2016a). Waste minimization of cheese-making by-product disposal through ethanol fermentation and the utilization of distillery wastes for fertilizer. *Second International Conference on Science, Engineering and Environment*, Osaka.

Vignesh, R. A., Meenakshi, P., Padma, A., Sahaja, N., Aarthi, N., Rajkumar, J., & Sridhar, D., (2016). Effect of dairy industry effluent on living organisms. *International Journal of Applied Research, 2*(3), 229–232.

Vijayalakshmi, T. M., & Murali, R., (2015). Isolation and screening of bacillus substilis isolated from the dairy effluent for the production of protease. *International Journal of Current Microbiology and Applied Sciences, 4*(12), 820–827.

Wee, Y. J., Kim, J. N., & Ryu, H. W., (2006). Biotechnological production of lactic acid and its recent applications. *Food Technology and Biotechnology, 44*(2), 163–172.

Williams, A. G., Noble, J., Tammam, J., Lloyd, D., & Banks, J. M., (2002). Factors affecting the activity of enzymes involved in peptide and amino acid catabolism in nob-starter lactic acid bacteria isolated from cheddar. *International Dairy Journal, 12*, 841–852.

Wu, Y., Ma, H., Zheng, M., & Wang, K., (2015). Lactic acid production from acidogenic fermentation of fruit and vegetable wastes. *Bioresource Technology, 191*, 53–58.

Yadav, J. S. S., Yan, S., Ajila, C. M., Bezawada, J., Tyagi, R. D., & Surampalli, R. Y., (2016). Food-grade single cell protein production, characterization and ultrafiltration recovery of residual fermented whey proteins from whey. *Food and Bioproducts Processing, 99*, 156–165.

Yang, S. C., Lin, C. H., Sung, C. T., & Fang, J. Y., (2014). Antibacterial activities of bacteriocins: Application in foods and pharmaceuticals. *Frontiers in Microbiology, 5*(241), 1–10.

Zacharof, M., & Lovitt, R. W., (2012). Bacteriocins produced by lactic acid bacteria: A review article. *APCBEE Procedia, 2*, 50–56.

Zhu, H., Gonzalez, R., & Bobik, T. A., (2011). Coproduction of acetaldehyde and hydrogen during glucose fermentation by *Escherichia coli*. *Applied and Environmental Microbiology, 77*(18), 6441–6450.

CHAPTER 7

Molecular Techniques for Detection of Foodborne Pathogens: Salmonella and *Bacillus cereus*

DEEPAK KUMAR VERMA,[1*] BALARAM MOHAPATRA,[2*] CHANCHAL KUMAR,[3] NEHA BAJWA,[4] KUMARI SHANTI KIRAN,[2] G. KIMMY,[5] ASHISH BALDI,[4] AMI R. PATEL,[6] BARKHA SINGHAL,[7] ALAA KAREEM NIAMAH,[8] and PREM PRAKASH SRIVASTAV[1]

[1]*Agricultural and Food Engineering Department, Indian Institute of Technology, Kharagpur, West Bengal, India,*

[2]*Department of Biotechnology, Indian Institute of Technology, Kharagpur, West Bengal, India,*

[3]*Department of Microbiology, Vallabhbhai Patel Chest Institute, University of Delhi, New Delhi, India,*

[4]*Department of Pharmaceutical Sciences & Technology, Maharaja Ranjit Singh Punjab Technical University, Bathinda, Punjab, India,*

[5]*Department of Food Engineering & Technology, Sant Longowal Institute of Engineering and Technology Longowal, Sangrur, Punjab, India*

[6]*Division of Dairy and Food Microbiology, Mansinhbhai Institute of Dairy & Food Technology-MIDFT, Dudhsagar Dairy Campus, Mehsana, Gujarat, India*

[7]*School of Biotechnology, Gautam Buddha University, Greater Noida, Gautam Budh Nagar, U.P., India,*

[8]*Department of Food Science, College of Agriculture, University of Basrah, Basra City, Iraq*

[*]*Corresponding author. E-mail: deepak.verma@agfe.iitkgp.ernet.in; rajadkv@rediffmail.com (D.K. Verma); balarammohapatra09@gmail.com (B. Mohapatra)*

7.1 INTRODUCTION

Contaminated food items and its consumption are found to be the one of the major cause of illness worldwide, where almost one in 10 people become ill every year. Amongst all, food-borne diarrheal diseases are the most common with 550 million cases and 230,000 deaths every year (WHO, 2016). In the European Union (EU), around 5,648 food-borne outbreaks affecting 69,553 people were reported in 2011 (EFSA, 2013). The effect is more dangerous in case of children under five years of age with potentially health threats, most notably caused by both *Campylobacter jejuni* and *Salmonella enterica* or enteropathogenic coliforms causing 80 million infections every year (WHO, 2016). Contaminated food and drinking water are the major root cause of diarrheal diseases (www.who.int/mediacentre/factsheets/fs237/en/), and in 2005, 1.8 million people died of the disease in the underdeveloped world. An estimated total of 76 million people with 325,000 hospitalizations and 5000 deaths occurred in the US per year (Mead et al., 1999). The magnitude of the disease prevalence becomes more complicated due to the associated chronic infections that cause end-stage disease (e.g., cirrhosis or malignancy). The ever increasing trend towards travel and trade, infectious disease are expected to emerge and cause global threats. There is a lag between the understanding and mechanism of microbial cause of intestinal illness and its comprehensive diagnosis, where 50% of the causative agents are unidentified (Tompkins et al., 1999). The food-borne intoxications by *Bacillus cereus*, coliforms like *Escherichia coli* and *S. enteric* are underestimated because of inadequate diagnostic tools. The rapid population, demographic shift, change in food habit, demand of global food market and its improper safety considerations, improper transport logistics, spread over of human population with infected intestinal flora, increasing proportions of immunologically compromised individuals, changing farming practices (production of cheaper food), climate change and other associated environmental factors are the major contributor of these food-borne illness among human populations (Diane et al., 2010).

Most food-borne pathogens are transmitted from the environment or animals through food products to humans. To reduce the mortality rate by food-borne pathogens, in the food production chain require increasing the food safety, knowledge of the pathogens and detection methods. Owing to the large scale disease prevalence and severity of food-borne infections, a total of 86 countries have initiated a worldwide program called "PulseNet International," a global laboratory network dedicated for food-borne disease surveillance. The laboratory networks of many countries like Europe, Africa, Asia Pacific, Canada, Latin America, the United States (US), etc. are connected to identify and subtype food-borne bacterial pathogens through standardized whole genome sequencing (WGS) and real-time surveillance of the disease outbreaks (Nadon et al., 2017). The main aim of the program is enabling the protection and improvement of public health with respect to food-borne disease. The molecular diagnostic methods (including recent whole genome, transcriptome, and proteome) for detection of the causative agents and their factors are found to be with greater specificity and sensitivity (Weile and Knabbe, 2009; Dwivedi and Jaykus, 2011) and could resolve the causes of mortality.

7.2 FOOD-BORNE PATHOGENS AT A GLANCE

On an average, 48 million people get sick, 128,000 hospitalizations, and 3,000 deaths happen every year indicating the alarming situation of food-borne illness for public health as per Centers for Disease Control and Prevention (CDCP) (CDC, 2012). Also estimates and provides the accurate exposure of which foodborne bacteria, viruses, microbes (pathogens) are causing the most illnesses in the US. Documented studies say that the sources of many food poisoning cases caused by the bacteria, viruses, and parasites (Table 7.1), usually due to improper food handling. Contaminated food may not look, taste or smell any different from foods that are safe to eat. Symptoms of food poisoning vary individuals and may develop as quickly as 30 minutes to as long as several days after eating food that's been infected. As identified *Salmonella* species *and Bacillus cereus* are the most common food-borne pathogen accounted for the majority of food-borne illness. In the US, approximately 400 deaths occur annually (Lowther et al., 2011). The outbreak of *S. enterica* serotype Newport infection in 2004 caused a total of 130 hospitalization across Northern Ireland (Irvine et al., 2009). Many of the reports thereafter suggested the contamination of *S.* pathovars (*S. senftenberg*) (Pezzoli et al., 2008) and other food poisonings related to *Bacillus cereus* (Drobniewski, 1993, Dierick et al., 2005). As *Bacillus cereus* is commonly associated taxa from soil, its dispersal to foods and to the human intestinal tract becomes easy for its host invasion (Stenfors et al., 2008). The trouble started when certain bacteria and other harmful pathogens replicate and spread, which can happen when food is mishandled.

TABLE 7.1 Common Pathogenic Microorganism and Associated Food Products

Microorganism	Related Foods	References
Bacillus cereus/ subtilis	Water, vegetables, soup, meat, milk, and milk products, uncooked, and poultry	Chon et al., 2015; Dierick et al., 2005; Tewari et al., 2015
Campylobacter jejuni	Raw milk, undercooked meat, poultry, and shellfish	Magistrado et al., 2001; Wysok et al., 2011; Pergola et al., 2017
Clostridium botulin	Improperly canned foods, vacuum packaged and tightly wrapped foods	Moorhead et al., 1999; Rasooly et al., 2010
Escherichia coli	Raw/undercooked eggs, meat, raw milk, and dairy products	Awadallah et al., 2016; Stromberg et al., 2017
Salmonella spp.	Undercooked food, Eggs, poultry, meat, raw milk and dairy products, chocolate, salad, and spices	Abdi et al., 2017; Kusunoki et al., 1997; Vignaud et al., 2017

7.2.1 SALMONELLA: SYSTEMATICS AND BIOLOGY

The genus *Salmonella* (Lignieres, 1900) is phylogenetically placed in the family *Enterobacteriaceae*, belonging to the order *Enterobacteriales* of class *Gammaproteobacteria*. The major two validly proposed classification systems are recommended for the member species of this genus are *S. enterica* and *S. bongori* and five species system with *S. bongori*, *S. choleraesuis*, *S. enteritidis*, *S. typhi*, and *S. typhimurium* are described and differentiated (Crosa et al., 1973, Hernan et al., 1997). Among the five distinct species members, the DNA-DNA homology (DDH) was found to be high. Based on the latest taxonomic opinion (Judicial opinion 80), *S. enterica* declared as the only representative type species of the genus *Salmonella* with strain LT2 as type strain (= ATCC 43971= CIP 60.62 = NCIMB 11450 = NCTC 12416). All the members are Gram-stain-negative, rod-shaped and motile rods with an optimal temperature of 37°C, but can grow in a quite wide range (7–45°C). They are facultative anaerobic and non-spore forming (Popoff and Le Minor, 2005). Till date, there are 9 species and 14 sub-species of *Salmonella* have been described as valid members and standing in nomenclature (http://www.bacterio.net/salmonella.html). The multi serotyping includes *S. enterica* subsp. *enterica* (I); associated with warm blooded animals, *S. enterica* subsp. *salamae* (II), *S. enterica* subsp. *Arizonae* (IIIa), *S. enterica* subsp. *diarizonae* (IIIb), *S. enterica* subsp. *houtenae* (IV), *S. enterica* subsp. *indica* (VI), *S. bongori* as subspecies (V) (Tindall et al., 2005). Many species are found to be associated with cold-blooded animals as well as environment (Grimont and Weill, 2007).

The Kauffman-White scheme classified *Salmonella* on the basis of serovars (Grimont and Weill, 2007). Based on antigenicity members are differentiated to somatic (O), flagellar (H) antigens with over 2000 *Salmonella* serotypes. These serotypes are host-specific and infect many animals with capacity to cause invasive diseases, like gastroentritidis, typhloid, fever, etc. (Crosa et al., 1973). Currently, there are more than 2,500 known serovars of *Salmonella*, where most of the serotypes belong to *S. enterica*. For simplicity, the *Salmonella* strains mentioned in this thesis are abbreviated by *S.* followed by the serovar name, e.g., *S. enterica* subsp. *enterica* serovar Typhimurium is abbreviated as *S. typhimurium*.

Salmonella, isolated from natural environments such as water, soil, and plants, does not show significant growth. However, *Salmonella* is known to be quite resistant towards different environmental factors, it can survive for several weeks in water and several years in soil, as long as the conditions are favorable (Tietjen and Fung, 1995; Jensen et al., 2006). *Salmonella* growth in food has been observed at 2–4°C, although optimal growth temperature is 37°C, and can grow in a quite broad range like 7–45°C (Ferreira and Lund, 1987). *S. enterica* is a common agent of food-borne infections and the human infections like enteric fever (typhoidal fever) and (Fierer and Guiney, 2001). Under certain conditions, salmonellae can cause invasive disease in which it enters the bloodstream and spreads throughout the body. The severity of *Salmonella* infections depends on several factors such as infecting *Salmonella* serotype, virulence of the strain, dose of the infection, and host parameters. *Enteritidis* and *Typhimurium* (Primarily serovars) cause gastrointestinal infections (Jantsch et al., 2011). Usually, the *Salmonella* colonizes in the intestine by attaching to the epithelial cells with the help of fimbria, followed by intestinal mucosa invasion and multiply in the gut-associated lymphoid tissue (Galan and Curtiss, 1989; Gahring et al., 1990; Ginocchio et al., 1992). Based on 16S rRNA gene sequence analysis, there are more than 500 isolates have been reported in RDP II (https://rdp.cme.msu.edu/) to be associated with food-borne infections and shared closeness among each serotype and closely cladded with each other (Figure 7.1).

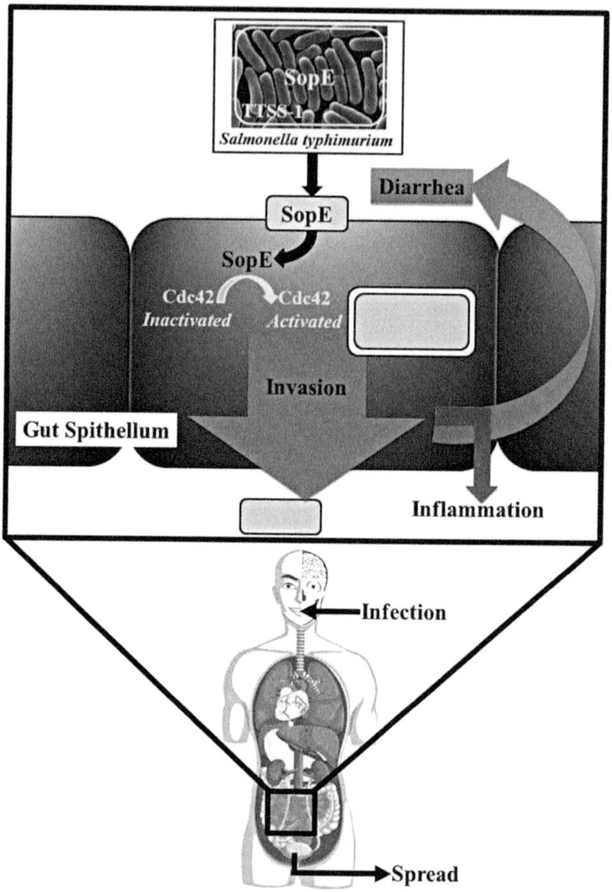

FIGURE 7.1 (See color insert.) Role of the *Salmonella typhimurium* TTSS-1 in the induction of enterocolitis.

Source: Adapted from http://www.micro.biol.ethz.ch/research/hardt.html

7.2.1.1 HISTORICAL RESUME

Salmonella accompanied humankind since ancient times. Evolutionary evidences, and phylogenetic analysis of *S. typhi* strains, indicate that a common *S. Typhi* ancestor existed 15,000 to 150,000 year ago. Typhi was not described as the etiologic agent of enteric fever until 1880 by Eberth; descriptions are recognized in Greek and Chinese texts dating back to hundreds of years BC (August et al., 1993; Cunha, 2004). The term typhoid fever, also known as enteric fever, derived from the Greek typhos, which translates as "putrid odor." Which is also proposed as the most likely cause of death among Athens indwells during the plague of Athens (Papagrigorakis et al., 2006). During the 20th century, enteric fever was a leading cause of mortality in several important American and European metropolitan areas such as Chicago, New York, London, among others. Death rates were as high as 174 per 100,000 inhabitants (Budd, 1873; Sedgwick et al., 1892). Latter on the reduction of mortality observed due to the availability of antibiotics as well as water supply and sanitation. New millstones updated with novel antibiotics development for the treatment of salmonella's infection.

7.2.1.2 EPIDEMIOLOGY

93.8 million cases of human gastroenteritis each year, and of these, 80.3 million cases are estimated from food-borne are effected by *Salmonella* (Majowicz et al., 2010). *Salmonella* borne infections were reported to the European Food Safety Authority (EFSA) in 2011 (EFSA, 2013) with 95,548 confirmed cases. The two serotypes, *S. enteritidis*, and *S. typhimurium*, are the most frequent reported serovars, together accounting for almost 70% of all human *Salmonella* infections in Europe. *S. enteritidis* cases which are associated with Humans mostly caused by the contaminated poultry meat and eggs, whereas *S. typhimurium* infectious cases commonly linked with the ingestion of contaminated pork, beef, and poultry meat (EFSA, 2013).

In 2012, CDC was reported 831 food-borne outbreaks. All of them caused by a variety of pathogens, and 106 cases were caused by *Salmonella. Salmonella* accounted for the most hospitalizations (64%) in outbreaks (CDC, 2012). In the recent largest outbreak, between March 2013 and July 2014, over 600 individuals in 29 states and Puerto Rico were infected with seven outbreak strains of *S. Heidelberg* (CDC, 2014). Transmitted routes of *Salmonella* infection to humans, by contact with animals and human-to-human contact including environments. In the case of industrialized countries, the main sources of infections are contaminated animal-derived food product, notably fresh eggs, and meat. A large proportion of contaminated vegetables and water sources are the primary sources of *Salmonella* pathogenesis in the developing countries (Wegener et al., 2003). In Denmark, domestic pork has been identified as the most important food source of salmonellosis. The second most important source of *Salmonella* infection is imported pork (Anonymous, 2012), affecting the various body and cellular functions (Figure 7.2).

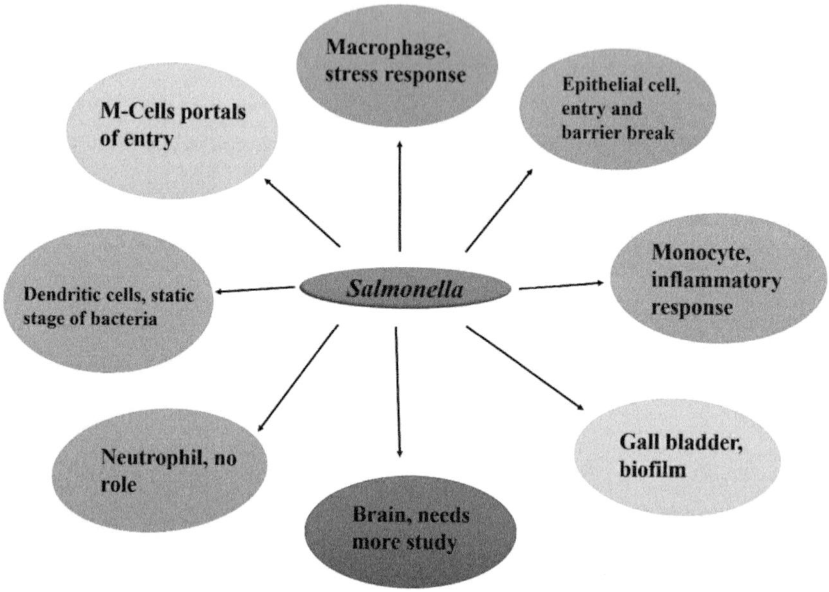

FIGURE 7.2 (See color insert.) *Salmonella* infection in different cells and organs. (Reprinted with permission from Lahiri, et al. 2010. © 2010 Elsevier.)

7.2.1.3 SALMONELLA IN FOOD PRODUCTION: CASE STUDY

The route of entry of *Salmonella* is a food chain, starts directly from the livestock feed, at the slaughterhouse or packing plant. Infection may also occur from manufacturing, processing, and be retailing of food, through catering and cooking at home (Lo Fo Wong et al., 2002). The slaughter line is an open process with many possibilities for cross-contamination of both *Salmonella* positive and negative carcasses (van Hoek et al., 2012). Infected pigs comprise the main reservoir for carcass contamination in slaughterhouses, and many of the carcasses harbor *Salmonella* on the skin or in the rectum (Hald et al., 2003; Visscher et al., 2011; van Hoek et al., 2012). Slaughter line carcasses can be contaminated from tools and equipment used for the slaughtering because the equipment can be very difficult to clean and sterilize, the *Salmonella* has the possibility to colonize and form biofilms (Zottola and Sasahara, 1994). Adaptation and persistence of *Salmonella* in food production chains have been suggested to be a consequence of bacterial attachment and surface colonization that increases the risk of contamination (Swanenburg et al., 2001; Vestby et al., 2009). *Salmonella* present in biofilm generally are more resistant towards environmental stresses, such as disinfection, organics acids and heat compared with their planktonic counterparts (Scher et al., 2005; Wong et al., 2010; Morild et al., 2011; Nielsen et al., 2013). *Salmonella* has the capability to form biofilms on different surfaces used in the food production chain, which includes plastics (Joseph et al., 2001), glass (Prouty and Gunn, 2003) and stainless steel (Hood and Zottola, 1997; Joseph et al., 2001). Apart from attaching to the materials used in the production line, *Salmonella* can also attach to the surface of the meat.

Attachment of the surface is influenced by different factors of *Salmonella*, such as the serovars, surface, contact time and the growth state (Hood and Zottola, 1997; Joseph et al., 2001; Oliveira et al., 2006). Also found that the longer *S. typhimurium* cells were allowed to attach to a pork meat surface. The impact of the contact time was further influenced by the growth state, i.e., the effect of the contact time seemed to be more pronounced for the planktonic bacteria compared with their immobilized counterparts. This suggests that is a liquid contamination might contribute to a higher occurrence of bacterial attachment to the meat surface. The *Salmonella* serovars also seem to impact the attachment ability, together with a variation between strains of the same serotype. In a study by Oliveira et al. (2006), it was found that the difference in adhesion ability of different *S. enteritidis* strains, on materials used in kitchens, where strongly strain-dependent (Oliveira et al., 2006). The adhesion also varied between the different contact surfaces, i.e., polyethylene, polypropylene, and granite.

7.2.2 BACILLUS CEREUS: CLASSIFICATION AND BIOLOGY

The genus Bacillus as a taxon is a phylogenetically incoherent, lacking a common ancestral history. Among most prokaryotic genera, the genus *Bacillus* is of great medical, economic, and historical importance (Logan and De Vos, 2009). 16S rRNA gene-based sequencing and identification of conserved signature indels (CSIs) are the most reliable differentiating methods to position members belonging to *Bacillus* spp. Most species of the genus *Bacillus* are of clinical significance like *Bacillus anthracis*, the causative agent of anthrax and *B. cereus*, a causative agent of food-poisoning (Bottone, 2010; Logan, 2012). Additionally, *Bacillus* members act as a model for Gram-positive organisms and in the understanding of stress-resistant endospore formation (Logan and De Vos, 2009; Zeigler, 2011). Despite a great amount of research focused on species of the genus, this group still remains disordered and often confusing. The genus *Bacillus* is a rod-shaped aerobic to anaerobic Gram-positive bacterial group producing endospores (Logan and De Vos, 2009). The *Bacillus cereus* group consists of seven closely related species; *Bacillus cereus*, *B. thuringiensis*, *B. anthracis*, *B. mycoides*, *B. pseudomycoides*, *B. weihen- stephanensis*, and *B. cytotoxicus* (Lapidus et al., 2008; Guinebretière et al., 2013). The strains of *B. cereus*, *B. anthracis*, and *B. thuringiensis* are closely related but share highly conserved chromosomes and differ in their plasmids virulence encoding (Rasko et al., 2005). Members of *B. cereus* group (*sensu-lato*) is accompanied as food poisoning bacterium that causes diarrhea and vomiting. The *B. anthracis*, is responsible of causing anthrax, can infect both animals and humans worldwide (Drobniewski, 1993). The three species have specific phenotypes and pathogenicity, which often are plasmid encoded, such as the insecticidal crystal proteins in *B. thuringiensis* (Berry et al., 2002), the cereulide synthesis of emetic *B. cereus* strains (Ehling-Schulz et al., 2006) also the toxins and capsule of *B. anthracis* (Keim et al., 2009). *B. cereus* group has been associated with food-poisoning illnesses in humans and other kinds of clinical infections (Bottone, 2010). As *B. cereus* is closely related to *B. anthracis*, and is sometimes used as a model organism for development and validation of new methods in order to minimize the risk arising from working with *B. anthracis* (Fricker et al., 2011).

The members of B. cereus groups (*sensu-lato* and *sensu-stricto*) are widespreadincluding soil, sediments, water, and plants (von Stetten et al., 1999; Kotiranta et al., 2000; Østensvik et al., 2004; Schoeni and Wong, 2005). *B. cereus* at 28–35°C shows optimal growth, but can grow over a wide temperature range from 45°C to 55°C. *B. cereus* can survive in harsh environments as it can form spores that are resistant to heat, dehydration, and other physical stresses (Stenfors et al., 2008). The general characteristics and properties of *Bacillus cereus* strains are presented in Table 7.2. *B. cereus* poisoning can be caused by either infection or intoxication that further leads to a diarrheal or an emetic type of disease. The diarrheal syndrome is caused by enterotoxins produced by vegetative cells during the growth in the small intestine after consumption of contaminated foods (Schoeni and Wong, 2005). The emetic syndrome is associated with the production of the cereulide toxin in foods before consumption (Ehling-Schulz et al., 2004; Stenfors et al., 2008). Both diseases are relative mild and not last more than 24 hours. For both types of food-poisoning, the contaminated foods have usually been heat-treated, which triggers the spore germination, i.e., transfer from spore to vegetative cell, and the absence of the other bacterial flora, gives the opportunity for *B. cereus* to grow well (Granum and Lund, 1997).

TABLE 7.2 General MIGS Characteristics Properties of Bacillus Cereus Members

MIGS ID	Properties	Terms
	Classification	Domain *Bacteria*
		Phylum *Firmicutes*
		Class *Firmibacteria*
		Order *Bacillales*
		Family *Bacillaceae*
		Genus *Bacillus*
		Species *Bacillus cereus*
	Gram stain	Positive
	Cell shape	Rod
	Motility	Motile with peritrichous flagella/some are non-flagellated
	Sporulation	Endospore-forming
	Temperature range(avg.)	5°C–50°C
	Optimum temperature	28–35°C
	pH range; Optimum	4.0–9.5, optimum of 6.0–8.0
	Carbon source	Wide spectrum (simple sugars, sugar acid, alcohols, fatty acids, hydroacrbons)
MIGS-6	Habitat	Ubiquitous (especially in soil/sediment/food)
MIGS-6.3	Salinity	0–8% NaCl (w/v)
MIGS-22	Oxygen requirement	Aerobe/Facultative anaerobes/anaerobes
MIGS-15	Biotic relationship	Free-living/some are host-associated like animal or plant rhizosphere

7.2.2.1 HISTORICAL RESUME

Bacillus cereus isolated in 1887, from the air in a cowshed by Frankland. Since 1950, many outbreaks from a variety of foods including vegetable, soups, milk, and ice cream and meat and cooked meat and poultry, fish, were described in Europe. The first well-characterized *B. cereus* outbreak in the USA was documented in 1969. Since 1971, a number of *B. cereus* poisonings of different type symptoms like, vomiting, were reported.

The *B. cereus sensu stricto*, also known as *B. cereus*, it is an egressing pathogen that causes gastroenteritis in humans. In 2008, the European community found 102 confirmed outbreaks caused by food-born *B. cereus*, corresponding to more than a thousand patients (Anonymous, 2009). After *Salmonella* and *Staphylococcus aureus*, the third most important food poisoning incidents caused by *B. cereus* in Europe (Anonymous, 2009). *B. cereus* causes food-borne illnesses, symptoms characterized by diarrhea, and the other called emetic toxin, by nausea and vomiting. Both symptoms are usually observed less than 24 h, but in older and debilitated persons have been observed bloody diarrhea and emetic poisoning (Kotiranta et al., 2000).

7.2.2.2 EPIDEMIOLOGY

According to a report from EFSA, *Bacillus* toxins were responsible for only 3.9% of food-borne outbreaks in the EU in 2011, which corresponds to 220 outbreaks out of which 47 were with strong evidence caused by *B. cereus* (EFSA, 2013). The two main sources of *B. cereus* infection, accounting for more than 42.5% of the outbreaks, were from mixed foods and cereal products. However, the low number of reported *B. cereus* outbreaks might be a result of underestimation. Many cases might not be identified due to mild and transient symptoms of the illness. In addition, *B. cereus* is not classified as a zoonosis by EFSA and has therefore received less attention in surveillance and reporting compared with, e.g., *Salmonella*, and *Campylobacter* (Ceuppens et al., 2013).

The members of *B. cereus* are important opportunistic pathogen and categorized as (1) gastrointestinal pathogens causing diarrheal symptoms and (2) systemic and local infections in the respiratory tracts of immunologically compromised patients (Carretto, 2000; Rasko et al., 2005). It has been known to contaminate diverse types of foods including meat, soups, vegetable dishes, dairy products, and seafood and has been known to produce several pathogenic compounds (Table 7.3) and virulence factors including spores, dodecadepsipeptide cereulide, enterotoxins, extracellular protein, dodecadepsipeptide cereulide associated with emesis after food ingestion contaminated by *B. cereus* and three different enterotoxins including hemolysin BL, nonhemolytic enterotoxin, and cytotoxin K, causing diarrhea (Bottone, 2010; Jeßberger et al., 2014). Environmental soil with spores of *B. cereus* is the primitive source of food contamination. As a result, various raw food ingredients such as vegetables, potatoes, herbs, milk, and spices are often contaminated with *B. cereus* spores (EFSA, 2005). The disease of *B. cereus* in humans arises as a direct result of the survival of the *B. cereus* spores under normal cooking and improper storage conditions after cooking, the spores germinate and the vegetative cells multiply (Logan, 2012). The two types of food-poisoning from *B. cereus* are often related to different types of food products. The diarrheal disease is often related to meat products, soups, vegetables, sauces, and dairy products, whereas the emetic type is related to mainly to rice and pasta (Schoeni and Wong, 2005; Abee et al., 2011); however, emetic strains have been isolated from milk, pudding, and ice cream (Messelhäusser et al., 2010). Due to the difference in food sources, different distributions between countries are observed for the emetic and diarrheal diseases. The emetic disease dominates in Japan and the UK, whereas in Northern Europe and North America, the diarrheal disease seems to be more prevalent (Kotiranta et al., 2000).

7.2.2.3 BACILLUS CEREUS IN FOOD PRODUCTION: CASE STUDY

The *B. cereus* has the capability to causing food-associated illness were recorded over 100 years ago by fledgling microbiologists (Giffel and Beumer, 1999). In 1950s, an understanding of *B. cereus* as a pathogen was not noted (Stenfors et al., 2008). Because of its abundance and the resistance of its spores, *B. cereus* can contaminate most agricultural products and play a major role in the contamination and spoilage of food products. *B. cereus* is also frequently present in the food production environments due the adhesion ability and resistance of the spores towards various environmental extremities like heat, desiccation, radiation, and capacity to resist wide range of metabolic inhibitors, surface sterilant, sanitizers as well as biofilm-forming abilities on food processing containers (Andersson and Rönner, 1998; Lindsay et al., 2000; Peng et al., 2002; Ryu and Beuchat,

2005; Stenfors et al., 2008). Studies have shown that *B. cereus* biofilms in the air-liquid interface contain up to 90% spores, which contributes to the spread of spores in food processing areas and production-line (Wijman et al., 2007). A strategy to reduce or even deplete *B. cereus* in food products are to promote the spore germination before implementing the food storage methods, as the vegetative cells are less resistant compared with the spores and thereby relatively easy to kill (Abee et al., 2011).

TABLE 7.3 Virulence Factor/Annotation of Genes for Pathogenesis by *B. Cereus*

Virulence Factors	
Lipase	
plcA	Phosphatidylcholine-preferring phospholipase C (PC-PLC)
piplc	Phosphatidylinositol-specific phospholipase C (PI-PLC)
Protease	
inhA	Immune inhibitor A metalloprotease
	Immune inhibitor A metalloprotease
Regulation	
plcR	PlcR
Toxin	
antO	Anthrolysin O
cytK	Cytotoxin K
hlyII	Hemolysin II
hlyIII	Hemolysin III
-	Hemolysin III homolog
hblC, hblD, hblB, hblA	Hemolytic enterotoxin HBL
nheC, nheB, nheA	Non-hemolytic enterotoxin NHE

Major concern in food processing methodology and human consumption are environmental endospores (Vilas-Boas et al., 2007). Endospores are present in soil and easily transferred into agricultural foods like vegetables, fruits, grains, and spices (Giffel and Beumer, 1999; Stenfors et al., 2008). Endospores have ability to freely migrate into dairy products from bovine udders through contact with field grain, dust, feces, or bedding (Giffel and Beumer, 1999; Stenfors et al., 2008; Bartoszewicz et al., 2008; Griffiths, 2010). These endospores may revert into vegetative bacterial state and have the potential to multiply invariably, or vice versa. Endospores can be survived in myriad conditions, pasteurization may not eliminate them. There are chances to transfer when a new food batch is processed on the same contaminated machinery. Meat products can become contaminated with *B. cereus* endospores because endospores spread through the air and contaminate non-mechanized areas (Juozaitis et al., 1994; Stenfors et al., 2008). There are possibilities to accidentally ingest contaminated water by a human then display *Bacillus* intoxication symptoms (Cheung et al., 1990).

B. cereus prevalence of food poisoning probably is much higher globally (Vilas-Boas et al., 2007; Stenfors et al., 2008; Griffiths, 2010). Because *B. cereus* intoxication usually seems flulike and short-lived. Sometimes most of the patients simply allow the infection to take a natural course instead of medical treatment (Stenfors et al., 2008). The symptoms of *B. cereus* intoxication mimic by emetic *S. aureus* and *C. perfringens* type A infection. These factors make obligation an accurate diagnosis of *B. cereus*. Vegetative *B. cereus* cells growing in contaminated foods and produced exotoxin, called cereulide (Stenfors et al., 2008; Griffiths, 2010). These cereulide toxins may influence the emetic syndrome, a classic intoxication scenario required between 0.5 to 6 hours to appear signs and symptoms (Stenfors et al., 2008). In some individuals, symptomatic nausea and emesis occur in 6 to 24 hours, can leads to severity and death.

There is a case study in August 2003, there is a family have five children. Children became sick after eating pasta salad. They all are going to the picnic on coming Saturday, so that the pasta salad was prepared on a Friday; remaining pasta had been stored in the refrigerator until the following next day evening. Next day pasta had unusual smell. They were served for first three children, boy 14 years (B14) and two girls 10 years (G10) and 9 years old (G9) ate only a small quantity. After 6 h the meal the youngest girl (G9), 9 years old, started vomiting. She complained breathing problems and was taken to the emergency department of a local hospital. Upon arrival, her brothers and sisters started vomiting as well. Because the clinical condition of two children (G10 and B14) deteriorated rapidly, they were intubated and mechanically ventilated. They were transported to the University Hospital in Leuven. During the transfer, G9 had severe pulmonary hemorrhage and needed continuous resuscitation. Upon arrival, she was moribund with coma, diffuse bleeding, and severe muscle cramps. She died within 20 min, at 13 h after the meal. After autopsy *Bacillus cereus* was detected in her gut and spleen. A postmortem liver biopsy showed microvascular and extensive coagulation necrosis.

All four other children were also affected, with different degrees (Dierick et al., 2005).

7.3 MOLECULAR TECHNIQUES FOR DETECTION OF PATHOGENS

Food-borne disease is caused by food poison produced by contaminant. Now food-borne pathogen is a major concern due to their ability to adapt and survive into different environmental conditions. The changes in the genes function or mutation help in the survival of pathogens. Certain antibodies, virulent genes or other complex defensive mechanisms also contribute to their adaptability and survival under various environmental conditions. For their surveillance of efficient and reliable characterization are primarily required detection (Adzitey and Nurul, 2011; Adzitey et al., 2011). Although, there are so many detection methods are available for food-borne pathogens. But still, there are required some best molecular based early stage detection methods. Emphasis of this study is to highlights, available molecular techniques and there usage to detect and characterize of food-borne pathogens.

The recent trend of testing food samples for the presence of *Salmonella* involves the detection of the pathogen, identification of the isolate as *Salmonella* and its specific serovar designation (presumably with Koch Postulates), and subtyping of the isolate for the cause of salmonellosis. Detection methods rely on traditional bacterial culture procedures on selective-differential agar plates specific for *Salmonella* and type organisms (Andrews et al., 2011). The process is time-dependent and tedious with a lot of error-prone result interpretation and hand expertise with a presumptive confirmation. Many traditional biochemical testing (like testing on blood agar, nutrient utilization on media can take at least 24 h for a confirmation of *Salmonella*. Here the molecular techniques like DNA fingerprinting, pulsed-field gel electrophoresis (PFGE), 16S rRNA gene and 16S-23S rRNA ITS typing, ribotyping, etc. have all been used to subtype *Salmonella* isolates. There are publicly available databases (NCBI, RDP-II, Green genes, SILVA, PubMLST, etc.) for verifying these fingerprints for species confirmation and strain identifications (Bailey et al., 2002). The use of molecular methods instead of the culture-based conventional methods, such as enzyme-linked immunosorbent assays (ELISA), DNA microarray, PCR, and real-time PCR to enable faster detection (Van der Zee and Huisint Veld, 2000). As many of the techniques have a lower detection level, the enrichment step can be either reduced in time or even omitted. Several published studies have been reported for the detection of *Salmonella* by molecular methods (Uyttendaele et al., 2003; Eriksson and Aspan, 2007; Malorny et al., 2007). Introduction of the molecular methods based on the detection of the organism's nucleic acids (e.g., DNA or RNA) such as, PCR, microarray, and sequencing have been made for detection and identification of pathogens faster along with greater specificity and sensitivity (Weile and Knabbe, 2009; Dwivedi and Jaykus, 2011). PCR and real-time PCR methods are mainly used for the detection of the different toxin genes of the emetic and diarrheal strains of *B. cereus*. For detection and quantification of *B. cereus* several PCR methods have been published, for some examples see Fricker et al. (2007), Martínez-Blanch et al. (2009), and Reekmans et al. (2009).

7.3.1 TECHNIQUES FOR DETECTION OF PATHOGENS

7.3.1.1 DEOXYRIBONUCLEIC ACID (DNA) SEQUENCING

The arrangement of nucleic acids in polynucleotide chains ultimately contains the information for the hereditary and biochemical properties of life. By the decoding, the arrangement of nucleotides gives information about any gene function and genetic variability. Deoxyribonucleic acid (DNA) sequencing technologies used to determine the order of the nucleotide bases, like Adenine (A), Cytosine (C), Guanine (G) and Thymine (T) in a DNA strand. Currently, DNA sequencing is widely used for the identification of novel pathogens. The conventional DNA sequencing approach was introduced in 1975 by Sanger and is based on the chain terminator method (Sanger and Coulson, 1975; Sanger et al., 1977). The Sanger sequencing is capable of sequencing up 1 kilobase (kb) of sequence data at a time. The sequencing technology has evolved greatly during the last couple of years. In 2005 the high- throughput sequencing platforms, the so-called next-generation sequencers (NGS), were introduced on the market (Margulies et al., 2005; Shendure et al., 2005). In contrast to Sanger sequencing, the NGS technologies can generate several hundred thousand to tens of millions sequencing reads in parallel (Mardis, 2008a, b; Voelkerding et al., 2009). The drawbacks of the NGS technologies are that they generate shorter reads, i.e., contigs, with lower quality, compared with the Sanger sequencing (Pareek et al., 2011). However, the NGS technologies compensate for this by high sequence coverage at each position (Metzker, 2010; Hui, 2012). Sequencing has several applications for detection and identification, such as targeted sequencing and WGS. The focus in the following will be on the targeted sequencing approach; however, WGS is described.

Target specific DNA sequencing is used for routine identification of bacteria by utilization of conserved genes (Doolittle, 1999; Clarridge, 2004; Petti et al., 2005). The 16S rRNA gene, which is universally presented in all bacteria and minimally affected by horizontal gene transfer (i.e., transfer between distantly related organisms) (Asai et al., 1999). Furthermore, the gene contains hypervariable regions flanked by highly conserved stretches, making these regions suitable for the design of universal primers (Baker et al., 2003). These characteristics features of 16S rRNA gene have made the most used region for bacterial taxonomy and identification (Doolittle, 1999).

The 16S rRNA gene has nine hypervariable regions (V1-V9), where the regions V2, V4, and V6 provides the lowest error rates for assigning taxonomy (Liu et al., 2008). However, the combination of the regions V1-V3 and V7-V9 has been suggested to provide a better characterization (Kumar et al., 2011b). For universal distinguishing of all bacteria at the genus level, the regions of V2 and V3 were identified to be the most suitable (Chakravorty et al., 2007). The advantage of using the 16S rRNA gene for characterization and detection is the variety of established searchable 16S rRNA classification databases, such as the Ribosomal Database Project II (RDP-II) (Cole et al., 2007, 2009), and Greengenes (DeSantis et al., 2006). Other genes such as *recA*, *rpoB*, *gyrA*, and *gyrB* can also be used as a target for sequencing because of their functionally conserved parts with variable regions (Petti, 2007). However, there are some limitations to the use of target-specific sequencing, e.g., difficulties in discriminating between many species within the *Enterobacteriaceae* using the 16S rRNA gene (Chakravorty et al., 2007).

Till date, a total of 2400 *Salmonella* serovars have been identified by 16S rRNA gene sequencing belonging to *S. enteritidis. S. typhimurium, S. heidelberg, S. newport, S. infantis, S. agona, S. Montevideo* (Brenner et al., 2000) with their involvement in food-born outbreaks. In recent years, along with 16S rRNA based PCR methods, 16S–23S rDNA internal transcribed spacer (ITS) region has been the most popular target. It has been seen that the copy number of rRNA gene clusters varies among bacterial species; ten in *Bacillus subtilis,* seven for both *Escherichia coli,* and *S. typhimurium* (Ellwood and Nomura, 1980; Anderson and Roth, 1981; Loughney et al., 1983). Hence the chance of identifying a serovar from a pathogenic species of these bacterial genera becomes error-prone. The ITS regions are located between 16S and 23S rDNA and found to be of different length in different bacteria species. Accordingly, sequences of the bacterial ITS can be used for the designing of genus or species-specific DNA probes and PCR primers. In contrast to targeted sequencing, where only information on the target gene is gained, WGS provides all the available genetic information about the isolate. Using WGS poses one problem, the assembly of all the generated reads into the final genome sequence. Assembly of whole genomes can be applied in two ways; *de novo* assembly or alignment to a reference sequence (also known as comparative assembly). In the comparative assembly approach, the assembly is guided by alignment to a reference sequence of a closely related organism. However, in order to use comparative assembly, prior knowledge on the genome assembled has to be known. The *de novo* assembly approach is used when trying to construct the genome of an organism not similar to any previous sequenced organism (Pop, 2009).

7.3.1.1.1 Historical Resume

During the last decade, rapid new sequencing technologies and their applications have been developed. Interestingly enough, this progress practically started for a new point of care tools. In 1953, Watson and Crick discovered and proposed the DNA double helix structure. After that, it took were one extremely innovative man and the technology of sequencing wreck into life. One of the first DNA sequencing methods was developed in 1977, by Frederick Sanger. Sanger's technique was the base of the modern innovations, and then the technology has advanced rapidly, becoming extremely popular and developing faster. After a huge 13-year effort was decoded the entire human genome in 1990. From the last decade, the developments of new sequencing technology and their applications have been found. Interestingly enough, this progress practically started after the completion of the human genome sequencing. The first human genome sequence was then obtained using first-generation sequencing technology. Despite these advantages, progress remained slow, as the techniques available to researchers (Holley et al., 1961). In 1965, Robert Holley and colleagues were able to produce the first whole nucleic acid sequence, that of alanine tRNA from *Saccharomyces cerevisiae* (Holley et al., 1965).

In the following years, second-generation or next-generation sequencing (NGS) technologies were developed, characterized by massive parallelization, improved automation and speed, and, most importantly, greatly reduced price. The second-generation sequencing differed from existing methods, developed with the large scale dideoxy sequencing. Order to nucleotide identity is through using radio or fluorescently labeled dNTPs or oligonucleotides. Researchers has been discovered luminescent method for measuring pyrophosphate synthesis, this consisted of two-enzymes, sulfurylase is used to convert pyrophosphate into ATP, then it is used as the substrate for luciferase activity, that produce light which is proportional to the amount of pyrophosphate (Nyren, 1985). This method was used to infer sequence on the basis of measuring pyrophosphate production as each nucleotide is washed through the system in turn over the template DNA (Hyman, 1988). Pyrosequencing was the basis of the commercial 'NGS' technology (Ronaghi et al., 1998). While next-generation approaches were being improved, the third or next-next generation technologies appeared, this time characterized by single-molecule sequencing (SMS). Here, third-generation technologies conceive that is capable of sequence single molecules, nullifying the requirement for DNA amplification. Stephen Quake has developed the first SMS technology (Braslavsky et al., 2003; Harris et al., 2008). Helicos BioSciences was commercialized and worked broadly by Illumina in the same manner, but without any bridge amplification of DNA templates. This is relatively slow and expensive, but this was the first technology to allow sequencing of non-amplified DNA, also avoiding all associated biases and errors (Schadt et al., 2010; Pareek et al., 2011). Early 2012 companies took up the third generation, very often expressions like deep sequencing, massively parallel sequencing (MPS) high throughput sequencing (HTS or HT-Seq). Some of these methods were relinquished, and some are still being developed or improved, while others are now routinely used in the laboratories around the world.

7.3.1.1.2 Detection Method (Quantification)

There is a need for accurate and rapid identification of infectious agents in individual illness and global health threat. So many improvements have been done in Sanger sequencing since decade of years. Primarily involved phospho or tritrium-radiolabeling replaced with fluorometric based detection and improved detection through capillary-based electrophoresis. Both of these improvements contributed to the development of automated DNA sequencing machines (Smith et al., 1985; Ansorge et al., 1986) and subsequently the first crop of commercial DNA sequencing machines (Hunkapiller et al., 1991) which were used to sequence the genomes of increasingly complex species. Additional chemical moieties of the increasingly modified dNTPs used for sequence analysis by electrophoresis capillary platform. Capillary sequencers have become very popular for molecular diagnostic laboratories. In sequencing reactions, primers that anneal to a single-stranded DNA template are elongated by DNA Taq polymerase. Fluorescent-labeled deoxynucleotides are introduced one at a time, and the primer is extended in a template-dependent manner. That also helps in terminate the formation of a new DNA strand as they encounter their complementary nucleotides in the target sequence in the same reaction. By this method, multiple sequencing reactions can be loaded onto multiwall plates at the same time.

7.3.1.1.3 DNA Sequencing and Food Safety: Case Study

The developments within sequencing technologies have a major impact on food safety. The increased number of genomes available helps improve detection methods such as the possibility for designing more specific primers, enhancing the specificity of the assays. Furthermore, metagenomics based sequencing can help identify unknown sources of foodborne outbreaks (Kawai et al., 2012; Nakamura et al., 2008).

Two *B. cereus* strains were inoculated in bottled water, from different suppliers, in levels of 10–106 CFU/L. Total DNA extraction directly from water and 16S rRNA gene sequencing was used in combination with principal component analysis (PCA) and multi-curve resolution (MCR) to study the detection level. The results showed detection at levels of 10^5–10^6 CFU/L, depending on the strain and water supplier. The analysis also revealed that background flora in the bottled water varied between the different water suppliers according to their bottling country.

The method evaluated, was further investigated for the use for detection of B. cereus in food and feed samples. Detection of *B. cereus* in food and feed samples has some issue compared with the detection in bottled water. The food and feed samples contain much more inhibitors that might interfere with the different steps of the methods, particularly the PCR amplification of the 16S rDNA. The results showed that the detection of B. cereus was much more difficult as it was possible to obtain sequence results from fewer samples compared with the results obtained in Paper IV. Despite the lower quality of the sequence data, there was some indication of distinguish between the two *B. cereus* strains, however every dependent on the matrix.

WGS is being used more frequently for investigations of food-borne outbreaks as it usually can give higher discrimination of isolates (Wilson et al., 2013). Following an outbreak with *S. montevideo* in peppers for spiced meals in 2009 and 2010 (Lienau et al., 2011). The 34 highly homogeneous S. Montevideo's genome isolates and the outbreak were compared with 24 other S. Montevideo genomes from clinical food associated with the same outbreak (Allard et al., 2012). The isolates associated with the outbreak clustered together in a separate group distinct from other closely related S. Montevideo isolates. The distance between the outbreak associated and unassociated isolates was clearly distinct, which helps to identified isolates linked to the outbreak.

7.3.1.2 ENZYME-LINKED IMMUNOSORBENT ASSAY (ELISA)

ELISA is an immunological technique which is used to detect the presence of an antigens or antibody and proteins in a given sample. ELISA is often used as a screening tool before more in-depth tests are screened. ELISA has wide applications in clinical diagnosis, in pathological studies, and in quality control studies. All microbial species have unique antigens, and such type of antigen used for detection by ELISA. These assays are generally carried out in 96 well plates, allowing multiple samples to be measured in a single experiment. Analysis of these plates required specific absorbance to ensure the antibody or antigen sticks to the surface. So many variations are available, these variant allow us to detect either antigen or antibody, identifying the different strains of microbes at a time.

Five types of major variants have been developed (Figure 7.3) in this assay for detection of antigens and antibodies (Verma et al., 2012), they are:

1. Direct ELISA – use in the detection of antigen;
2. Indirect ELISA – use in the detection of antibody;
3. Sandwich ELISA – use in identification of different epitopes at a time;
4. Competitive ELISA – use in quantifying the antigen/antibody; and
5. Multiplex ELISA – use in identification of multiple antigens/antibodies at a time.

7.3.1.2.1 Historical Resume

Before the 1970's, ELISA known as by radioimmunoassay, in which test radioactively labeled antigens or antibodies were used, provides the reporter signal indicating a specific antigen or antibody is present in the sample. Rosalyn Sussman was first described radioimmunoassay and scientific paper published by Yalow and co-workers in 1960 (Yalow et al., 1960). Latter in 1971, two groups were reported individual paper that synthesized this knowledge into methods to perform EIA/ELISA (Engvall et al., 1971; Weemen et al., 1971). So many studies have been performed by ELISA for rapid detection of food-borne pathogens. One of the examples performed by Kumar et al. (2011a) for detection of pathogenic *Vibrio parahaemolyticus* in seafood by sandwich ELISA, using monoclonal antibodies against the TDH-related hemolysin (TRH) of *V. parahaemolyticus*. The detection limit was slightly high against pathogenic *V. Parahaemolyticus of this assay*. Now ELISA test kits such as BIOLINE *Salmonella* ELISA Test are commercially available which used for the detection of *Salmonella* in food products. The detection limit of this test kit was very high like, 1 CFU/25 g sample (Bolton et al., 2000). ELISA is also commonly used for the detection of *Clostridium perfringens* and *Staphylococcal enteroxins* toxins present in foods (Aschfalk and Mülller, 2002; Zhao et al., 2014).

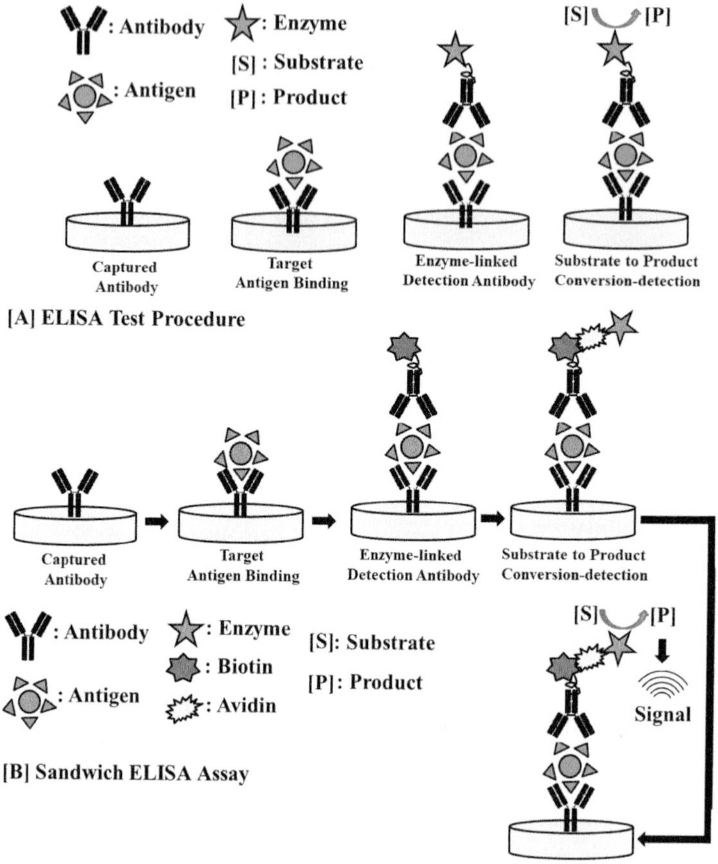

FIGURE 7.3 Enzyme-linked immunosorbent assay (ELISA). A) typical ELISA assay, and B) sandwich ELISA assay.

Source: Adapted from Amani, Mirhosseini, & Imani, 2015.

7.3.1.2.2 Detection Method (Quantification)

The use of different substrates in ELISA has a major advantage as the substrates will bind to the respective conjugates specifically and will develop coloration, which can be read in an ELISA reader in terms of wavelength. The color change is visible to the naked eye. The primary antibody is specific to the target antigen, which labeled onto the wells of the microtiter plate. Apply target samples, present antigen will bind to the primary antibody, and the remaining extra unbound antigens are removed by the washing. After that, an enzyme-conjugated secondary antibody is added which bind to the primary antigen, and remaining unbound antibodies are removed by washing. Secondary antibody (tagged with enzyme) will produce color in the presence of substrate. There are different types of enzymes can be used in ELISA, which includes horseradish peroxidase (HRP), alkaline phosphatase, and β-galactosidase (Yeni et al., 2014). The complex consisting antigen between two antibodies forms a sandwich which can be detected by adding a colorless substrate. Enzyme substrate complexes will be formed, and color will be producing (Zhang, 2013). However, one of the disadvantages is a false positive result, because binding of the chemical and conjugate is very specific, and contamination in the intermediate stages can lead.

7.3.1.2.3 ELISA and Food Safety: Case Study

One of the studies by Zhu et al. (2016) based on rapid detection of *Bacillus Cereus* from food by the double-antibody sandwich ELISA. They developed polyclonal antibody (pAbs) and monoclonal antibodies (mAbs) specific to *B. cereus*. They have generated these antibodies from rabbit antiserum and mouse ascites, respectively. To eliminate cross-reactivity between closely related species such as *B. thuringiensis, B. subtilis, B. licheniformis, and B. perfringens,* IgG-isotype specific monoclonal antibodies were developed. ELISA is a specific, sensitive, and convenient method for measuring polysaccharide, macromolecular protein, and bacteria. ELISA is one of the precise and reproducible methods which not required highly expensive equipment and stable reagents. Therefore, ELISA is relevant to routine uses for detection of microorganism in clinical and research laboratories. Facts of this case study, they used the *B. cereus* whole cells as immunogens to generate a rabbit polyclonal and mouse monoclonal antibodies, which can recognize certain surface components of *B. cereus*. That will be helpful to detect fast, direct, and convenient measurement of *B. cereus* from the food (Zhu et al., 2016).

7.3.1.3 MICROARRAY

Many food-borne and other diseases causing *S. typhimurium* strains share more than 99% genomic content, where the variation is primarily represented by the prophages in the genome. Different phage types lead to variation in phenotype, and some of them show a high degree of variation in host adaptation (Chan et al., 2003; Drahovska et al., 2007). Both the virulence and host specificity to cause disease for *Salmonella* isolates are serotype-specific. Hence a rapid and accurate determination of *Salmonella* serotype is essential to control the epidemics and outbreaks caused by food-borne pathogenic strains (Fierer et al., 2001; Wollin, 2007). Microarray is one of the fastest-growing technologies; it provides assistance for large scale screening systems with simultaneous identification. It is a very powerful tool with greater capacity (100–1000x) compared to other molecular methods (real-time PCR) that can only analyze a small number of targets. Microarrays represent another useful technique in the diagnostic toolbox for food-borne pathogens. In 1995, Schena and co-workers introduced the word "microarray" first time and further development of Southern blotting (Schena et al., 1995). The microarray technology worked as the hybridization of a mixture of target nucleic acids with thousands of individual nucleic acids probes. The probes are immobilized on the array (spot or feature) at specific positions that can be identified in the later analysis process.

Depending on the technique used for printing the spots, the length of the probes varies. The use of a photolithography approach to synthesize the probes directly on the array usually generates probes of 25mers (Singh-Gasson et al., 1999). Another way to produce the arrays is by a spotter. Here, droplets of DNA oligos are deposited on the surface of the array, where after they bind (Lausted et al., 2004). Using a spotter the length of the probes on the array are normally 25–70mers. The use of longer oligos generally gives higher sensitivity, whereas the shorter ones have a higher selectivity (Bates et al., 2005).

Multiplex PCR-based microarray assay provides an accurate and reliable approach for differentiating *Salmonella* and to decipher large-scale epidemiology causes. Microarrays can be used to conduct gene expression profiling and genotyping of the target. Changes in gene expression between two different experimental conditions can be monitored at mRNA levels (based on cDNA) by expression profiling. The total mRNA is extracted from the cells from each condition, and then labeled with different fluorescent dyes, usually red and green dyes. From the fluorescence intensities and the red/green ratio for each spot, correspond to the relative expression of the genes (Trevino et al., 2007; Debnath et al., 2010). For the genotyping microarray the presence of specific DNA sequences of an isolate is analyzed (Pollack et al., 1999). Performing genotyping DNA microarray studies can have two purposes: (i) to investigate the difference in gene content between two organisms or (ii) to investigate the gene content of one organism (Trevino et al., 2007; Lopez-Campos et al., 2012b). See an illustration of the two approaches in Figure 7.4.

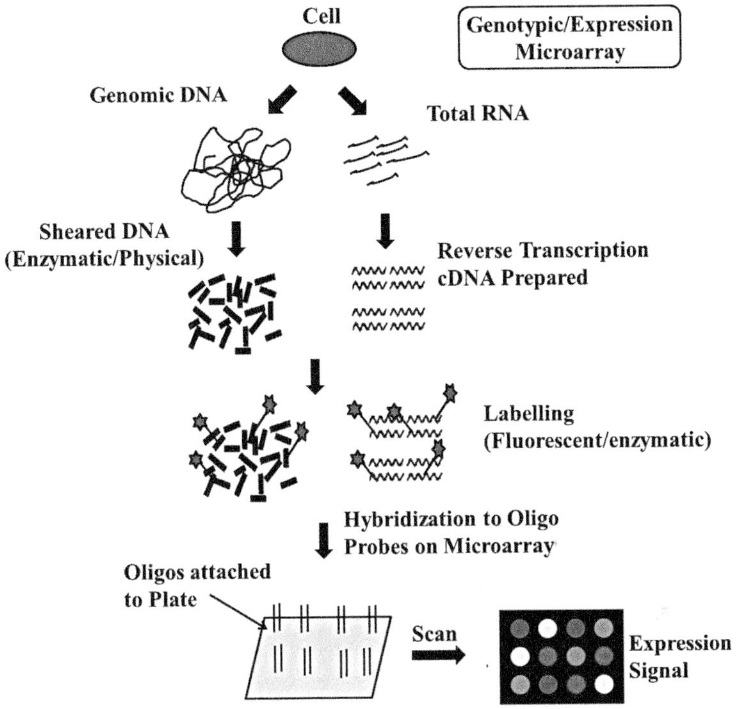

FIGURE 7.4 (See color insert.) Schematic diagrams of genotyping and gene expression microarray experiments.
Source: Adapted from Josefsen, et al., 2012.

7.3.1.3.1 Historical Resume

Microarray technology is a term that refers to the miniaturization of thousands of assays on one small plate. This concept was developed from an earlier concept called ambient analyte immunoassay, which was first introduced by Roger Ekins in 1989 (Sutandy et al., 2013). In the following decade, this concept was successfully transformed into the DNA microarray, a technology that determines mRNA expression levels of thousands of genes in parallel.

The early history of DNA array was created with the colony hybridization method of Grinstein and Hogness (Grinstein and Hogness, 1975). In 1979, this approach was adapted to create ordered arrays by Gergen et al. (1979). In the late 1980's and early 1990's Hans Lehrach's group automated these processes by using robotic systems to rapidly array clones from microtiter plates onto filters (Carig et al., 1990; Lennon and Lehrach, 1991). Microarrays term was introduced in 1995 by Schena et al. (1995), and then used to study gene expression profiles in biologic samples (Lobenhofer et al., 2001). In this technology, 1000's of nucleic acids are bound to a surface and used to measure the relative concentration of nucleic acid sequences in a mixture via hybridization. The two key innovations that lead to an improvement in microarray technology are:

1. Use of solid supports like glass or silicon chips which facilitated miniaturization with fluorescence-based detection which allows an analysis of tens of thousands of genes in a few cm^2; and
2. Development high-density oligonucleotide synthesis method directly on the microarray chip.

Microarrays are also used to study the DNA, RNA, or protein expression. Nucleic acid arrays or more simply DNA arrays are a group of technologies in which specific DNA sequences either deposited or covalently or non-covalently attached on a surface. The term generally refers to cDNA testing for RNA expression used a genomic array and for protein for which the term protein array. The term expression array is usually used to denote testing for RNA expression (Petrik, 2001). Currently available microarrays on nylon membranes allow an analysis of several hundred genes at once.

7.3.1.3.2 Detection Method (Quantification)

Microarray is a simple and high throughput technology that allows detection of thousands of genes simultaneously which consist a solid surface such as a nylon membrane, glass slide, or silicon chips onto that allows attaching small quantities of a DNA (ssDNA) from different known bacterial sources. When ssDNA from as many unknown species is exposed to these DNA chip, complementary strains will bind to their respective sites on the chip because it is the property of complementary nucleic acid sequence to specifically pair with each other by forming hydrogen bonds. The targets are labeled with reporter molecules generally with fluorescent molecules to detect it when binds to the microarray. The reporter fluorescent molecule emits the fluorescence which indicates corresponding probe and the excess of labeled and non-binding targets will be washed away and finally the array is scanned in order to quantify the fluorescence signals for each spot (Lopez-Campos et al., 2012a). The flourish of signals represents the presence of the target sequence of the pathogens.

7.3.1.3.3 Microarray and Food Safety: Case Study

Genotyping with microarray permits the detection of present pathogens in a sample or to examine the difference in gene content of related strains. For instance, in the study, genomic DNA compared from 40 *Salmonella* isolates and detected specific patterns of genes present or absent among the isolates covering different serotypes and source of isolation in a slaughterhouse. Difference in the gene content between the same serovars was observed; however, it was not possible to identify genes that could be linked to the source of isolation.

Reen and co-workers investigated the genomic diversity among *S. enterica* isolates (Reen et al., 2005). They found that the five recognized *Salmonella* pathogenicity islands were present in all the investigated isolates, and they furthermore identified more than 30 genomic regions that varied in their presence among the isolates. Further, they showed that very little genomic diversity exists within particular serovars, even they were widely geographic distributed.

The predominate application of DNA microarrays has been to measure gene expression levels and for comparing levels of gene transcription under various growth conditions, associated with food production (Lopez-Campos et al., 2012b). In a study, Wang, and co-workers compared the global gene expression profiles of *S. typhimurium* over an 8 hour time period for comparing three different growth conditions; swarming, non-swarming and planktonic (Wang et al., 2004). They showed that bacteria grown on the surface of agar had been shown different physiology compared with those grown planktonic. Furthermore, it was possible to identify putative new genes target which are responsible for motility and virulence. The *S. typhimurium* transcriptional response was investigated with the heat treatment and changes of pH. It shows that few of the genes were up-regulated during the heat shock, remained up-regulated after 30 minutes when the temperature got back to normal. In addition, it has been found that the cells after heat shock were more resistant, subsequently heat and acid exposure conditions.

7.3.1.4 POLYMERASE CHAIN REACTION (PCR)

The Polymerase chain reaction (PCR) is the most sensitive technique used for making numerous copies of a particular DNA segment. PCR is the rapid as well as accurate method for detection of microbial pathogens. PCR amplify a single or a few copies of a piece of DNA across several orders generating thousands to millions of copies of a particular DNA sequence. When a specific pathogens are difficult to culture in vitro or require a long cultivation period in that case, PCR have significant diagnostic value. PCR has become the most popular molecular diagnostic approach for detection of food-borne pathogens and is considered to be a valuable alternative to the culture-based detection techniques due to its speed, LOD, sensitivity, and specificity (Maurer, 2011; Rodriguez-Lazaro et al., 2013). The method amplifies specific DNA fragments from small quantities of source DNA material, even when that source DNA is of relatively poor quality. Replication allows amplification of target DNA in the presence of synthetic oligonucleotide primers and a thermostable DNA polymerase (Farber, 1996; Wang et al., 2000). Other components of a PCR reaction such as deoxyribonucleotide triphosphates (dNTPs), $Mgcl_2$, and buffer solutions are used in different concentrations to increase detection limits.

The PCR reaction comprises of three temperature steps. The first step is the denaturation of the double-stranded DNA (dsDNA) to form single-stranded DNA (ssDNA) as a template for the second step. In the second step, primers anneal to the complementary ssDNA target. After annealing with the target DNA, the DNA polymerase extends the primers by adding one complement nucleotide after another. The extension of the primers creates a DNA strand, complementary to the original DNA target. The three steps (denaturation, annealing, and extension) are repeated multiple times in the thermocycler, where each synthesized DNA strand acts as a template for the synthesis of new DNA strands (Mullis et al., 1986; Fairchild et al., 2006). When appropriate conditions are used, each cycle will generate a doubling of the initial number of target (Erlich, 1989). Target gene amplification may occur with a different variance of PCR method according to requirements such as single PCR set or multiple primers (multiplex PCR), real-time PCR, nested PCR, reverse-transcription PCR and many more.

7.3.1.4.1 Real-Time PCR

Real-time PCR allows both detection and quantification during the PCR reaction. The process of real-time amplification is monitored by incorporation of a fluorescently labeled molecular probe or a DNA-binding dye such as SYBR Green I. There exist a large range of different fluorescence probes reviewed by Josefsen et al. (2012), however common for all of them are that they produce a change in the fluorescent signal during the PCR amplification. The DNA-binding dyes bind to dsDNA, thereby releasing a fluorescence signal. The increase in fluorescence, whether it is probe or SYBR Green I based assays, is recorded after each replication cycle, and the amplification of the PCR products can thereby be monitored (Mackay, 2004). Computer software associated with the thermocycler records the amount of fluorescence and the amplification cycle at which the fluorescence predominate a defined threshold level, the background fluorescence, is called as the threshold cycle (Cq, Cp, or CT depending on the software used; in this thesis the designation CT is used), see Figure 7.5. The higher amount of target DNA present in the sample, the earlier will the fluorescence cross the threshold.

The used of real-time PCR has not only decreased the time to perform PCR detection methods, but also reduced the risk of cross-contamination since both the amplification and fluorescent detection occur in the same closed tube (Valasek and Repa, 2005; Barken et al., 2007). RT- PCR one of the fastest and high detection techniques is due to its speed and high specificity and sensitivity.

7.3.1.4.2 Historical Resume

Repeated cycles of DNA synthesis are conducted in order to multiply a target DNA, was first reported in the mid-1980's (Saiki et al., 1985; Mullis et al., 1986). Quantification of a target of interest has been independently developed several times, being described in 1990 and 1991 (Brisco et al., 1991; Sykes et al., 1992) using the term "limiting dilution PCR" and in 1999 (Vogelstein et al., 1999). So profound was the impact of PCR that Kary Mullis was awarded the 1993 Nobel Prize in Chemistry, not even ten years after its introduction. PCR method used to detect down a single target of the pathogen (Batt, 2007). In broad terms, classical PCR can be used for a qualitative or a quantitative purpose, either to study the properties of a target molecule or to determine the number of a target molecule. Digital PCR is used; similarly, the difference being that in digital PCR the sample is partitioned to the level of single molecules, PCR amplification is then performed, an all-or-none, i.e., digital, signal is obtained and either the nature of the target molecule is analyzed or the number of the target molecule is calculated using the Poisson distribution (Morley et al., 2014). Limiting dilution PCR was an improvement over previous techniques, such as competitive PCR, for quantification of PCR targets. It was precise, had a wide dynamic range, and could detect and quantify rare target molecules. No doubt PCR has been revolutionized research and detection not only in the biological sciences and medicine, as well as criminology also. Several major scientific discoveries are, including purification of DNA polymerase and elucidation of the DNA replication mechanism.

7.3.1.4.3 Detection Method (Quantification)

For detection of PCR products, several methods are reported. Gel electrophoresis is one of the traditional electrophoresis methods. Gel electrophoresis is the standard method for analyzing reaction quality and yield. To visualize nucleic acid molecules in agarose gels, ethidium bromide (EtBr) is the commonly used dye. This dye is intercalating with DNA or RNA and glows in the presence of UV light at the 300-nm. The PCR products are loaded on a gel, and electrophoresis is performed. The gel is strained with a fluorescent dye, e.g., ethidium bromide or SYBR Green that binds to the PCR products, which then can be visualized by illumination under ultraviolet (UV) light (Mullis et al., 1986; Aaij and Borst, 1972). Nucleic acid molecules are separated on the basis of size by the disposed electric field where negatively charged molecules move toward anode (positive) pole. The movement and migration of the PCR product is determined exclusively by the molecular weight where small weight molecules migrate faster than the larger ones.

Another method for visualizing is southern hybridization where a specific probe is labeled with a radioisotope or non-radioisotope marker.

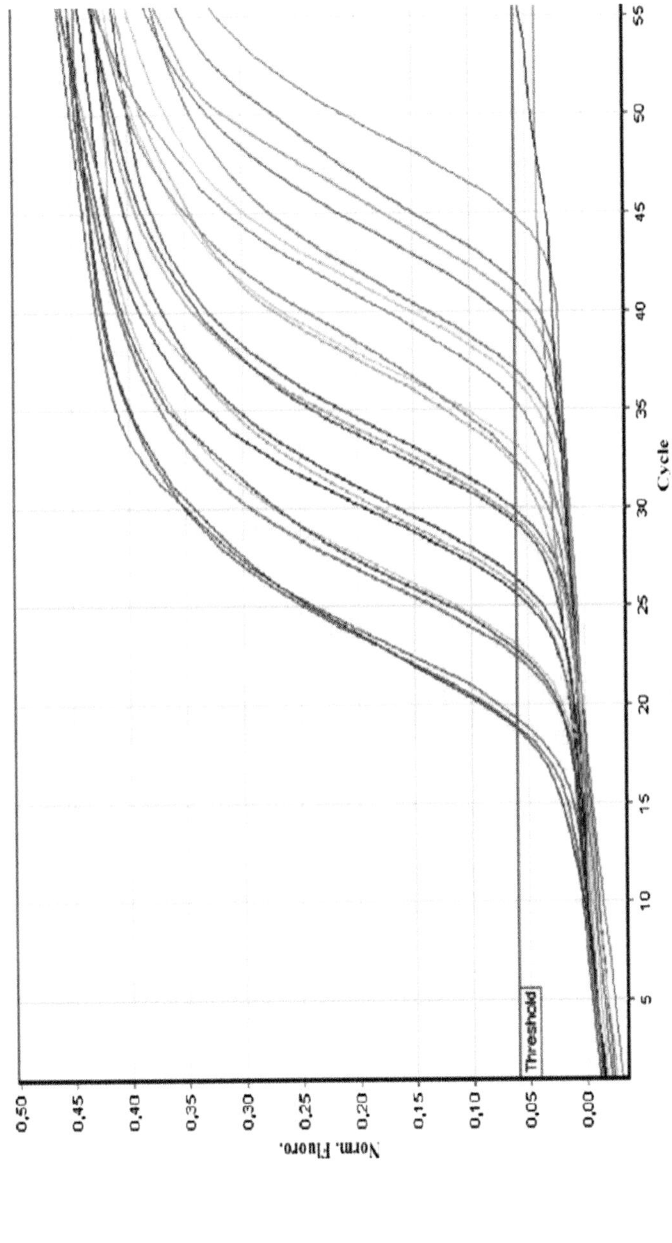

FIGURE 7.5 (See color insert.) An amplification plot obtained by real-time PCR. The y-axis corresponds to the fluorescence emitted from either the labeled probe or DNA binding dye, which rises in each amplification cycle (x-axis). The CT is the amplification cycle where the fluorescence exceeds the background fluorescence. (Reprinted with permission from Hornyák et al., 2012. © 2012 Elsevier.).

This type of detection protocol also gives a higher sensitivity than the ethidium bromide detection method but requires extra hybridization and blotting steps. In this process transfer the DNA from gel to the membrane and hybridize with the specific probe after neutralized the DNA. Probes labeled with the radioisotope or other non-radioisotope enzymes which generate glow after adding substrate.

Another method is for PCR amplicon detection on the basis of digoxigenin (DIG)-PCR. This method involves capturing of amplicons labeled with DIG with the probe immobilized onto the surface of a streptavidin-coated ELISA plate. The bound hybrid is detected with an anti-DIG-peroxidase conjugate and the colorimetric substrate. This ELISA system has given us 10 to 100 times more sensitive results than the traditional electrophoresis method (Newcombe et al., 1996).

The same principle of amplification is employed in real-time PCR. But instead of looking at bands on a gel at the end of the reaction, the process is monitored in "realtime." Literally, the reaction is placed into a real-time PCR machine that watches the occurring reaction with a camera or detector. RT-PCR read-out is given as the Ct value, necessary for achieving a given level of fluorescence. During the initial PCR cycles, the fluorescence signal emitted by SYBR-Green I bounded to PCR product, the fluorescence becomes double at each cycle. Detection of PCR products is enabled by the inclusion of a fluorescent reporter molecule in each of the cycle increased fluorescence with an increasing amount of product DNA (Ponchel et al., 2003). The measured fluorescence is proportional to the total amount of amplicon; the change in fluorescence over time is used to calculate the amount of amplicon produced in each cycle with compare Ct value. The intensity of fluorescent signal is indicating the amplification of target gene as well as the presence of the particular pathogen. Some targeted genes are mentioned in Table 7.4, for higher sensitivity also required a highly conserved specific gene.

7.3.1.4.4 PCR and Food Safety: Case Study

Detection of food-borne pathogens by PCR techniques has the advantage of enabling faster identification, which not only increases the food safety but also the food quality. Furthermore, with qRT-PCR it is possible to study the behavior of pathogens influenced by different environmental conditions by studying the expression of suitable target genes. Several studies have been reported that describes the development of real-time PCR assays for detection of particular food-borne pathogens, including *Salmonella* and *Bacillus cereus* (Table 7.5).

TABLE 7.4 Overview of Some Target Genes Used for Detection of *B. Cereus* and *Salmonella* by PCR

Gene Name	Function	References
B. cereus		
Ces	Cereulide synthetase	Ehling-Schulz et al., 2005
cytK1	Cytotoxin K	Wehrle et al., 2009
Hbl	Haemolysin BL	Wehrle et al., 2009
Nhe	Nonhaemolytic enterotoxin	Wehrle et al., 2009
Pc-plc	Phosphatidylcholine-specific phospholipase C	Martínez-Blanch et al., 2009
Salmonella		
invA	Involved in invasion	Rahn et al., 1992
sefA	Fimbrial subunit	Woodward and Kirwan, 1996
SipB-SipC	Junction of virulence genes	Ellingson et al., 2004
ttrRSBCA	Gene required for the tetrathionate respiration	Malorny et al., 2004

The low detection level of real-time PCR methods gives the possibility to detect pathogens without an enrichment step, leading to shorter identification times. For example, detection of *B. cereus* in gelatine was decreased from two days by the standard culture-based method to 2 hours using real-time PCR (Reekmans et al., 2009). An advantage with real-time PCR based detection is the possibility for multiplexing, enabling the detection of several pathogens simultaneous. Multiplex real-time PCR assays have been developed for detection of *E. coli* 0157:H7, *Salmonella* spp. and *Listeria monocytogenes* in milk and meat samples (Omiccioli et al., 2009; Suo et al., 2010). The short time of the real-time PCR based assays enables the possibility for adding a short enrichment step in order to lower the detection level.

A study by Wang and co-workers on the detection level for *Salmonella* in ground beef was 10^3CFU/g without prior enrichment, but

TABLE 7.5 Examples of Studies with Quantification of *Salmonella* and *B. Cereus* with Real-Time PCR

LOD	Food Matrix	Enrichment	References
B. cereus			
10^3–10^5 CFU/g or10^1–10^3 CFU/g[A]	Rice	No	Fricker et al., 2007
60 CFU/ml	Liquid egg, infant formula	No	Martínez-Blanch et al., 2009
10^5–10^8 CFU/g[B]	Rice pudding, cereal, carrot puree	No	Wehrle et al., 2010
10^1 CFU/g	Rice pudding, cereal, carrot puree	Yes	Wehrle et al., 2010
Salmonella			
4.4×10^2CFU/ml	Pork meat	No	Löfström et al., 2011
20 CFU/g	Seafood	No	Kumar et al., 2010
10 cells/10g	Yogurt	No	D'Urso et al., 2009
1 CFU/125 ml	Milk	Yes	Omiccioli et al., 2009
<18 CFU/10 g	Ground beef	Yes	Suo et al., 2010
10^3CFU/g	Ground beef	No	Wang et al., 2007
10CFU/g	Ground beef	Yes	Wang et al., 2007

[A]Depending on the assay type.

[B]Depending on the matrix and strain.

including a 10-hour enrichment step, 10 CFU/g could be detected (Wang et al., 2007). For *Salmonella* in relation to food safety, qRT-PCR has been used to study, e.g., differentiation of dead and alive *Salmonella* cells, the impact of growth conditions and attachment (Barak et al., 2005; González et al., 2009). An overview of some published studies is shown in Table 7.6.

TABLE 7.6 Overview of Some Published qRT-PCR Studies for *Salmonella* in Relation to Food

Study	Gene(s) Investigated	References
The change in expression of attachment-related genes under different pre-growth conditions	*flhDC, motAB, prgH, fliC, fljB, yhjH*	Hansen et al., 2013
Detection of alive *Salmonella* spp. cells in produce	*invA*	González-Escalona et al., 2009
The role of cellulose and O-antigen capsule in the colonization of plants	*adrA, bcsA*	Barak et al., 2007
Virulence genes that are required for attachment to plant tissue	*agfB*	Barak et al., 2005
Detection of *Salmonella* in contaminated minced beef and whole eggs	*sef*A	Szabo and Mackey, 1999

The qRT-PCR was used to study of different growth conditions that impact on the expression of attachment-related genes in five *S. typhimurium* strains (wildtype, Δ*prg*, Δ*flhDC*, Δ*yhjH*, and Δ*fliC*). The gene expression was measured over a three hours period where the cells were grown on a pork meat surface. The gene expression measurements showed that the growth conditions changed the expression of the investigated attachment-related genes.

Another application of qRT-PCR is to distinguish between alive and dead cells. PCR alone cannot distinguish between DNA originating from live or dead bacteria. Furthermore, DNA can be rather resistant to degradation and be present after the death of the host cell that can lead to false-positive results using PCR (Josephson et al., 1993; Allmann et al., 1995; Wolffs et al., 2005). RNA is degraded more rapidly than DNA, which in principle means that the RNA present in bacterial cells can serve as an indicator for viability. Several reports have been made in the detection of viable cells from *Salmonella* using RT-PCR (Szabo and Mackey, 1999; González-Escalona et al., 2009).

7.3.2 MOLECULAR TECHNIQUES FOR TYPING OF PATHOGENS

Molecular typing is nowadays an integral part of the public health microbiology toolbox. Typing techniques have advantages that they allow the investigation of foodborne outbreaks, give a better understanding to the epidemiology of infections (Arbeit, 1999; Trindade et al., 2003; Adzitey et al., 2013). It indexes subspecies genotypic or phenotypic characters to estimate the genetic relatedness of microbial isolates and infer from it their probability of belonging to the same chain of transmission. It will be helpful to evaluate in term of their performance and pathogenicity that will be the convenience of treatment (Maslow et al., 1993; Struelens, 1996; Trindade et al., 2003). There are numerous molecular techniques that have been developed and used extensively for typing of pathogens. Sero-typing, that is based on the Kauffmann-White scheme, remains the standard for classification of *Salmonella* isolates in outbreak investigations but recently it has been proliferated by range of molecular genotyping methods (Botteldoorn et al., 2004).

Agreements between typing methods are assessed by determining highly similar grouped (Struelens, 1996; Trindade et al., 2003). So many typing methods are available few of them mentioned like, PFGE, random amplified polymorphism DNA (RAPD), multilocus sequence typing (MLST), and plasmid profile analysis.

7.3.2.1 PULSED-FIELD GEL ELECTROPHORESIS (PFGE)

PFGE is a valuable tool for assessing pathogens interrelatedness. PFGE resolves DNA molecules of almost millimeter in length by using pulsed electric fields, which selectively modulate mobilities in a size-dependent fashion. Conventional agarose-gel electrophoresis is almost unable to resolve DNA molecules larger than 40–50 kb because of their size-independent co-migration. Schwartz and Cantor at Columbia University in 1984 (Schwartz et al., 1984), introduced a concept that involves separation of DNA molecules larger than 50KB by using two alternating electric fields (i.e., PFGE). In this method electric current periodically change its direction (three directions) in a gel matrix, unlike the conventional gel electrophoresis where the current flows only in one direction (Schwartz and Cantor, 1984; Arbeit, 1999; Trindade et al., 2003). Chromosomal DNA is digested with restriction enzymes or restriction endonucleases to generate different sizes and patterns of DNA fragments specific to particular species (Shi et al., 2010). This is a standard gold method for pathogen typing and is also known as restriction fragments length polymorphisms; RFLPs. Researchers used this method for food-borne pathogen outbreak investigations and other epidemiological studies (Alonso et al., 2005).

7.3.2.2 MULTILOCUS SEQUENCE TYPING (MLST)

MLSA is based on MLST, a DNA sequence-based method which was first introduced by Maiden et al., in 1998 as a microbial typing method for epidemiological and population genetic studies of pathogenic bacteria (species) (Chan et al., 2001; Maiden et al., 1997). MLST is a straight portable and nucleotide-based method for typing bacteria on the basis of the internal sequences usually house-keeping genes (Maiden et al., 1997; Spratt, 1999; Urwin and Maiden, 2003). MLST was first introduced and validated for *Neisseria meningitides* pathogenic strains for epidemiological studies, where 11 housekeeping genes (each of 470 *bp*) were analyzed and differentiated 107 meningococcal strains (Maiden, 1998). MLST loci (located at diverse chromosomal loci and be widely distributed among the taxa) denotes protein-encoding genes with functions (housekeeping genes) and are more stable with respect to a genetic mutation. Generally, 6–10 loci are recommended for MLST scheme (http://www.mlst.net/ and http://www.pubmlst.org) to sufficiently characterize the strains up to species level, but this number should not be regarded as fixed (Glaesr and Kampfer, 2015). In addition of providing a standardized approach to data collection, by examining the nucleotide sequences of multiple loci encoding housekeeping genes, or fragments of them, MLST data are available freely over the internet to ensure that a uniform nomenclature is instantly available to all those interested in classifying bacteria. Different sequences of distinct alleles and for each house-keeping gene and the alleles at each end of the seven loci, defines the allelic profile or sequence type for an individual isolate (Urwin and Maiden, 2003). Approximately 450–500 *bp* internal fragments of each gene are used and most bacteria have enough variation within the house-keeping genes (Enright and Spratt, 1999; Urwin and Maiden, 2003). The increasing speed and reduced cost of nucleotide sequence determination, together with improved web-based databases and analysis tools, present the prospect of increasingly wide application of MLST. In addition, the data obtained by MLST can be used to address basic questions about the evolutionary and population biology of bacterial species. MLSA can be used explicitly as a method for revealing similarities between strains of different genospecies. For *Salmonella* genus, it was shown that an average difference of 2.5% for concatenated house-keeping genes could be used as a threshold for the distinction of species members, and as a equivalent of 70% DDH value. A common MLST scheme of different combination of house-keeping genes is presented in Tables 7.7 and 7.8. Vandamme and Peeters (2014) referred MLSA to be the replacement of DDH in species delineation. Based on the different recommendations consideration of 4–5 genes (Gene fragments of 400 to less than 600 nucleotides) can be used for single gene MLSA analysis. Even less than four genes can be considered to be sufficient in contrast to the 16S rRNA gene-based phylogenies.3.2.3.

7.3.2.3 RANDOM AMPLIFIED POLYMORPHISM DNA (RAPD)

RAPD is a PCR-based technique. It has been used to verify the existence of population of species that have arisen under different environmental conditions or genetic drift. By the use of random primers of almost 10 bases amplicons throughout the genome are targeted and amplified. Products after amplification are subsequently separated on an agarose gel and stained with ethidium bromide. The genetic variation analysis based on RAPD, permits accurate genetic diversity due to its ability to generate random markers from the entire genome (Lin et al., 2014). This process amplify one or more DNA sequences and produces a set of fingerprinting patterns of different sizes, specific to individual strain (Farber, 1996, Trindade et al., 2003). RAPD is relatively cheap, rapid, readily available, and easy to perform otherwise (Wassenaar and Newell, 2000; Shi et al., 2010; Rezk et al., 2012).

7.3.2.4 PLASMID PROFILE ANALYSIS

Plasmid profile analysis is also a classical molecular technique that is used for epidemiological investigation. The study of plasmids is important to medical microbiology because the plasmids possess the ability to encode genes for virulence factors and antibiotic resistance. Plasmids also serve as markers of vast bacterial strains when a typing system is referred to as plasmid profiling, or plasmid fingerprinting. In this technique, plasmid DNAs are extracted from bacteria's, and the DNA is separated on agarose gel by electrophoresis. Plasmids are mobile extrachromosomal elements that can be spontaneously

TABLE 7.7 MLST Seven House-Keeping Gene Combinations for Species Identification

Function of Gene	Genes	Forward Primer 5'-3' Sequence	Reverse Primer 5'-3' Sequence	Annealing Temperature (°C)
Glycerol uptake facilitator protein	glpF	GCG TTT GTG CTG GTG TAA GT	CTG CAA TCG GAA GGA AGA AG	59
Guanylate kinase	gmk	ATT TAA GTG AGG AAG GGT AGG	GCA ATG TTC ACC AAC CAC AA	56
	gmk_F3	GAG AAG TAG AAG AGG ATT GCT CAT C	GCA ATG TTC ACC AAC CAC AA	55
Dihydroxy-acid dehydratase	ilvD	CGG GGC AAA CAT TAA GAG AA	GGT TCT GGT CGT TTC CAT TC	58
	IlvD_2	AGA TCG TAT TAC TGC TAC GG	GTT ACC ATT TGT GCA TAA CGC	58
Phosphate acetyltransferase	pta	GCA GAG CGT TTA GCA AAA GAA	TGC AAT GCG AGT TGC TTC TA	56
Phosphoribosyl Aminoimidazolecarboxamide	pur	CTG CTG CGA AAA ATC ACA AA	CTC ACG ATT CGC TGC AAT AA	56
Pyruvate carboxylase	pycA	GCG TTA GGT GGA AAC GAA AG	CGC GTC CAA GTT TAT GGA AT	57
Triosephosphate isomerise	tpi	GCC CAG TAG CAC TTA GCG AC	CCG AAA CCG TCA AGA ATG AT	58

TABLE 7.8 Seven House-Keeping Gene Scheme for *Salmonella* MLST Based Species Identification

	Genes Scheme						
Gene Combination	**One Gene**	**Two Genes**	**Three Genes**	**Four Genes**	**Five Genes**	**Six Genes**	**Seven Genes**
1st	*hisD*	*hisD*, *thrA*	*aroC*, *hisD*, *purE*	*aroC*, *hisD*, *purE*, *thrA*	*aroC*, *hemD*, *hisD*, *purE*, *sucA*	*aroC*, *dnaN*, *hisD*, *purE*, *sucA*, *thrA*	*aroC*, *dnaN*, *hemD*, *hisD*, *purE*, *sucA*, *thrA*
2nd	*purE*	*hisD*, *purE*	*hisD*, *purE*, *thrA*	*hisD*, *purE*, *sucA*, *thrA*	*aroC*, *hisD*, *purE*, *sucA*, *thrA*	*aroC*, *dnaN*, *hemD*, *hisD*, *purE*, *sucA*	
3rd	*hemD*	*aroC*, *hisD*	*hemD*, *hisD*, *purE*	*aroC*, *hisD*, *purE*, *sucA*	*dnaN*, *hisD*, *purE*, *sucA*, *thrA*	*aroC*, *hemD*, *hisD*, *purE*, *sucA*, *thrA*	

lost or readily acquired by bacteria. Plasmid DNA of known molecule mass is recognized as a standard. In a second procedure, plasmid DNA that has been cleaved by restriction endonucleases can be easily separated by agarose gel electrophoresis, and the resulting pattern of fragments can be used to verify the identity. So, the unknown isolates that are epidemiologically related can easily exhibit different plasmid profiles (Trindade et al., 2003). Su and co-workers reported that plasmid analysis of Salmonella isolates from duck hatcheries was more diverse than that of goose hatcheries (Su et al., 2011). Same, Tsen and Lin (2001) collected 63 *S. enteritidis* strains isolated from related and unrelated patients suffering from foodborne poisoning during 1991–1997 and were subjected to molecular typing, interestingly they got different strains within a species (Tsen et al., 2001). Furthermore, sixty-three *S. enteritidis* strains isolated from related and unrelated patients suffering from food-borne poisoning were collected and subjected to PFGE in the year 1991–1997.

7.3.2.5 NEXT GENERATION SEQUENCING (NGS)

In the recent past, knowledge on bacterial metabolism, the evolution of resistance genes and their regulation has been the center of attraction for microbiology researchers. Since the microbial potency of tolerating the heavy metals, causing infectious diseases are deep underneath in their genomes, deciphering their amazing genomes is fun-filled and quite exciting. The whole genetic complement of the organisms is termed as genomics. In the context of genomics, WGS is of high value for the understanding of how bacteria function, evolve, and interact with each other with their ecosystems thus providing numerous avenues for newer genomic functions (Mueller et al., 2007). Bacterial genome sequencing work is now nearly two decades old. With reference to microbiology, genetics, molecular biology, the combination of genome sequencing and computational-driven analysis of sequence data (bioinformatics) has transformed our understanding on various mechanisms of drug resistance, antibiosis, infection causes, and genomic plasticity for better niche colonization (Mueller et al., 2006). The process of sequencing has been remarkable these days, so that sequencing projects that used to take years and cost hundreds of thousands of dollars can now be completed in a few days or even in hours. Currently, there are many platforms available in the market launched by various sequencing industries, aimed for different purposes like WGS, meta-genome sequencing, exome, and targeted gene sequencing, amplicon-sequencing as well as total- and meta-RNA sequencing (RNA-Seq). The platforms are commonly known as NGS platforms, and the basic flow diagram of the working function is presented in Figure 7.6. Prior to the use of NGS, genome sequencing was hurdled to be with high-cost, labor-intensive, and time consuming with Sanger's protocol. Roche 454 sequencing system, was the first NGS technology based on pyrosequencing (Margulies et al., 2005), followed by Illumina and Ion torrent. Illumina was originally developed by Solexa, and is based on bridge-amplification (Bentley, 2006), and is provided as bench-top versions. Ion Torrent (Life Technologies) uses a different technology called emulsion PCR for template amplification and semiconductor method that uses a change in pH as detectable signal instead of light. Apart from NGS, third generation technologies (provided by Pacific Biosciences (PacBio), Oxford Nanopore technology) are becoming widely used. PacBio uses single-molecule real-time (SMRT) sequencing with no template amplification, and that occurs in a zero-mode waive-guide after bell template library preparation with long sequencing reads of greater than 10Kb (Eid et al., 2009). With an accuracy of 99%, these NGS platforms provide massive amount of data with a cheaper cost. The output provided by various NGS technologies are called data [mostly in Mb (million base) or Gb (gigabase)] and in the form of reads (nucleotides with quality scores) as FASTQ files. The data obtained are massive and need specific pipelines to be analyzed for gene function and structural based study.

Through various software and computational skills, the data are generally analyzed for interpretation. Along with the development of genome sequencing technologies, massive computational advancement in analysis of genomes has been achieved. There are a numerous number of online (web-based) as well as system-based tools and software for processing the annotated genomes. The detailed pipeline for sequence analysis is presented in Figure 7.7. With the advent of these technologies, WGS has become accessible, affordable, and emerging as a critical tool for gaining a better understanding on microbial genomes and molecular details of metabolic capabilities.

Practically, facts of investigating serovars and strains of *S. enteritidis* and its outbreaks are really challenging because nearly 40% of all isolates show the same PFGE pattern, making epidemiological investigations bit difficult (Allard et al., 2013). To overcome this problem, federal public health and food safety laboratories are applying metagenomic approach (NGS) to define complex outbreak scenarios For example; WGS analysis of *S. Bareilly* isolates cracked its link between outbreak and a specific manufacturing facility in India. Revelation in UK regarding WGS analysis

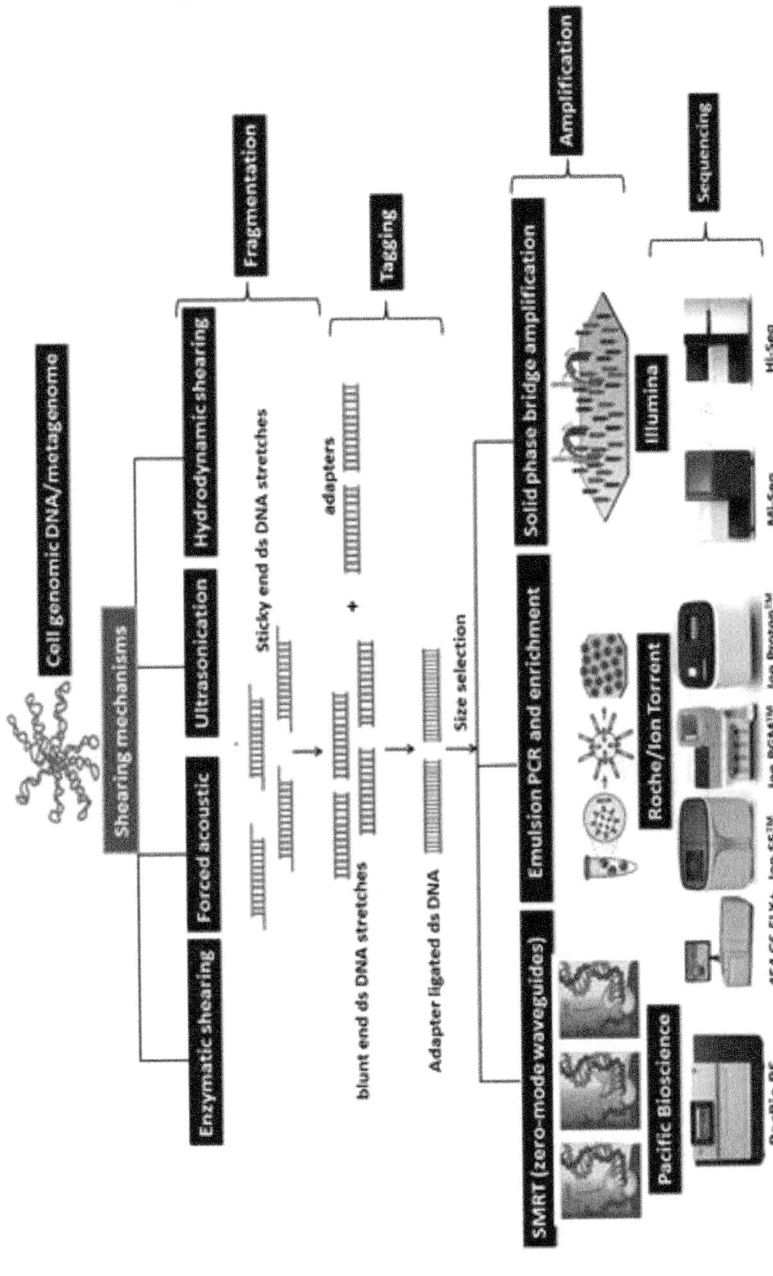

FIGURE 7.6 **(See color insert.)** Workflow showing the NGS based DNA sequencing and sample preparation.

of *S. enteritidis* isolates shows a clear linkage between human, egg, and *S. enteritidis* isolates (Hoffmann et al., 2015; Inns et al., 2015). *Salmonella* outbreaks associated with black pepper (*S. montivideo*), tomato (*S. newport*), cucumber (*S. newport*), watermelon (*S. newport*) and peanut butter (*S. tennessee*) were also identified in this culture-independent NGS based approaches. These recent genome sequencing projects by IMG, JGI, and other organizations are providing vital information in real-time, outbreak causes and prevention strategies. Genomic sequencing is the fast and accurate way to predict the serotypes and drug resistance in *Salmonella* as well as a fine resolution for outbreak analysis and short-term epidemiology (Achtman et al., 2012; Hawkey et al., 2013). Till date, there are approximately 290 complete genome projects for *Salmonella* pathogenic strains that have been indexed in NCBI Genome database (https://www.ncbi.nlm.nih.gov/genome/), indicating GC% range of 51.3–52.4 mol%, average genes and proteins of 4000–5000 and 1–7 rRNA operons possessed by various pathogenic and pathovars of *Salmonella* (Table 7.9).

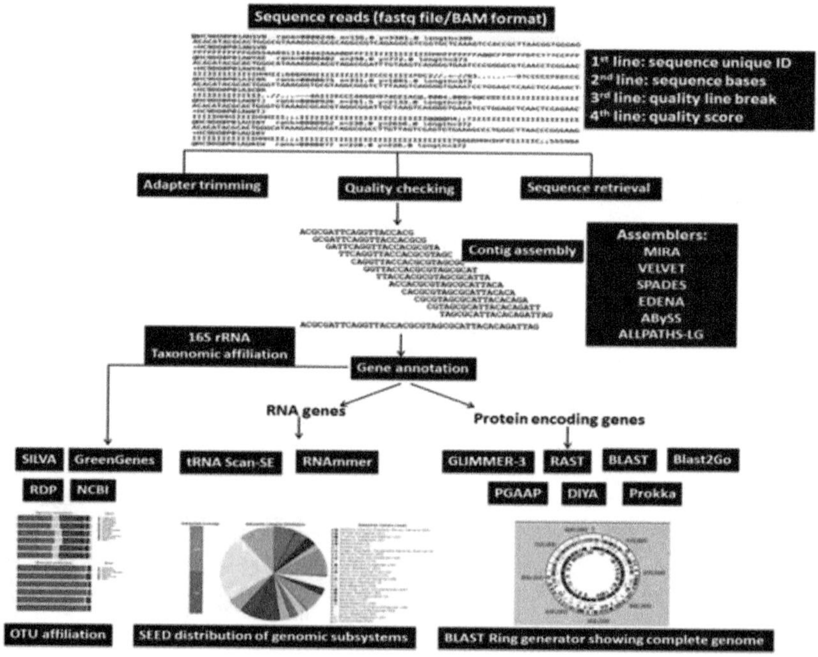

FIGURE 7.7 **(See color insert.)** Detail pipeline for microbial genome and metagenome sequence analysis.

TABLE 7.9 Comparative Genomic Features of Pathogenic *Salmonella* spp. Members

Genome Properties	1	2	3	4	5	6	7	8	9	10	11	12
Genome Size	4E+06	5E+06	5E+06	5E+06	5E+06	5E+06	5E+06	5E+06	5E+06	5E+06	5E+06	5E+06
Gene Count	4418	4563	4509	4996	4652	4511	4398	5712	4502	4598	4801	4816
CRISPR Count	2	2	2	2	2	2	2	4	0	1	2	1
GC Count	2E+06	2E+06	2E+06	3E+06	2E+06	2E+06	2E+06	3E+06	2E+06	2E+06	3E+06	2E+06
Coding Base Count	4E+06	4E+06	4E+06	4E+06	4E+06	4E+06	4E+06	4E+06	4E+06	4E+06	4E+06	4E+06
Coding Base Count%	87.8	86.56	87.9	86.86	87.67	88.19	85.92	89.89	88.32	87.8	88.14	87.63
CDS Count	4243	4489	4340	4782	4488	4301	4319	5646	4412	4406	4608	4641
CDS%	96.04	98.38	96.25	95.72	96.47	95.34	98.2	98.84	98	95.82	95.98	96.37
RNA Count	175	74	169	214	164	210	79	66	90	192	193	175
RNA%	3.96	1.62	3.75	4.28	3.53	4.66	1.8	1.16	2	4.18	4.02	3.63
rRNA Count	9	7	6	22	5	34	7	6	19	13	18	8
16S rRNA Count	7	1	3	7	2	12	1	1	6	1	4	2
tRNA Count	77	67	73	85	66	81	72	60	71	79	79	68
Other RNA Count	89	0	90	107	93	95	0	0	0	100	96	99
Paralogs Count	999	489	452	2665	495	439	454	763	476	1123	1235	1068
COG Count	3247	3359	3268	3353	3371	3433	3318	3455	3327	3432	3457	3344
COG%	73.49	73.61	72.48	67.11	72.46	76.1	75.44	60.49	73.9	74.64	72.01	69.44
KOG Count	804	820	804	821	832	847	817	842	818	846	832	802
KOG%	18.2	17.97	17.83	16.43	17.88	18.78	18.58	14.74	18.17	18.4	17.33	16.65
Pfam Count	3899	3999	3966	4151	4083	3985	3916	4201	3979	4047	4117	4139
Pfam%	88.25	87.64	87.96	83.09	87.77	88.34	89.04	73.55	88.38	88.02	85.75	85.94

TABLE 7.9 *(Continued)*

Genome Properties	1	2	3	4	5	6	7	8	9	10	11	12
TIGRfam Count	1718	1743	1730	1766	1764	1778	1727	1808	1750	1766	1816	1790
TIGRfam%	38.89	38.2	38.37	35.35	37.92	39.41	39.27	31.65	38.87	38.41	37.83	37.17
Enzyme Count	1408	1398	1415	1404	1456	1419	1382	1446	1417	1418	1403	1383
Enzyme%	31.87	30.64	31.38	28.1	31.3	31.46	31.42	25.32	31.47	30.84	29.22	28.72
KEGG Count	1597	1581	1601	1572	1631	1606	1590	1621	1629	1588	1594	1571
KEGG%	36.15	34.65	35.51	31.47	35.06	35.6	36.15	28.38	36.18	34.54	33.2	32.62
KO Count	2972	2973	2992	3001	3060	2972	2958	3058	3017	2983	2981	2936
KO%	67.27	65.15	66.36	60.07	65.78	65.88	67.26	53.54	67.01	64.88	62.09	60.96
Signal Peptide Count	430	454	431	421	422	426	441	441	434	425	443	450
Transmembrane Count	1066	1105	1072	1143	1127	1076	1085	1231	1111	1112	1125	1154
Biosynthetic Cluster	150	200	150	165	149	164	151	162	102	46	50	40

Taxa: 1, S. enterica Paratyphi A str. 00 6712; 2, S. enterica enterica sv. Javiana 10721; 3, S. enterica Paratyphi A str. Banker Type1; 4, S. enterica enterica sv. Choleraesuis SC-B67; 5, S. enterica enterica sv. Enteritidis 08-0047; 6, S. enterica enterica sv. Manhattan 111113; 7, S. enterica enterica sv. Oranienburg 701; 8, S. enterica enterica sv. Newport SL317; 9, S. enterica enterica sv. Eastbourne CFSAN001084; 10, S. typhimurium ASD3-69; 11, S. enterica NGUA-10_S10; 12, S. enterica enterica sv. Typhi ERL052042.

A total of 3090 strains of *Salmonella* whole genomes are available in JGI IMG/M represented from the various geographic location. The genomic size of *Salmonella* isolates ranges between 4.4–5.1 Mb with 4200 to 5400 protein-encoding genes. Genome sequencing and multi-locus typing have enabled to categories 133,200 *Salmonella* strains based on its pathogenesis and MLST schemes has been designed to type based on the genes encoding *aro*C, *dna*N, *hem*D, *his*D, *pur*E, *suc*A, and *thr*A. The average GC content is 35.3%, and genome size of 3.7 Mb. The genomic regions are usually bordered with putative prophages indicating the chance of mobile genetic elements (MGEs) and horizontal transfer events. Genomic information also enables to decipher these *B. cereus* members to be having resistance genes for β-Lactamase, aminoglycoside, tetracycline, macrolide, chloramphenicol, sulfonamide, trimethoprim, quinolone, rifamycin, fosfomycin, and several multi-drug resistance. Additionally, these organisms found to harbor Type-1 to III CRISPR-Cas system, Type I-IV restriction-modification pathways, and Type-I to V toxin-antitoxin system that helps these microorganisms to be more pathogenic and invasive in food-related health problems.

7.4 CONCLUSION

The detection and enumeration of pathogens are in the food and on surfaces that come into contact with foods. It is an important component of any integrated program to ensure the safety of foods. Important is that to identify food pathogens at the initial stage of infection. However, foodborne pathogens are progressively being studied due to their ability to change and surviving at dissimilar environmental conditions. They tend to mutate and survive in the presence of certain antibodies, virulent genes, and other complex defensive mechanisms of the body. In recent time, several food-borne pathogens have becoming important and threat to public health. Surveillance studies data will surely help in better understanding of the existence and spread of food-borne pathogens. Molecular techniques provide better tool for detecting and typing of food-borne pathogens and will help in tracing the source of human infections. A wide range of molecular techniques have been used for detecting, speciating, typing, classifying, and characterizing foodborne pathogens.

7.5 SUMMARY

Food analysis has several key challenges that impede the proper detection of pathogens. These challenges include uneven distribution of bacteria in food, the presence of indigenous microbes that affect the identification of pathogens, and the heterogeneous nature of food matrices. Emphasis on nucleic acid and immunochemical based technologies, provide a very useful alternative against conventional microbiological laboratory methods. Conventional methods involve additive the food sample and performing various media-based metabolic tests which typically require 3–7 days to obtain results. For rapid screening and results required immunological or nucleic acid-based tests such as PCR, ELISA, polymorphism DNA, microarray, DNA sequencing developed for food testing. DNA microarrays are a more sensitive, rapid, and informative detection of pathogens in food and food production sites. These methods utilize the variations in homologous DNA sequences to identify the bacteria. All molecular rapid detection methods have been used to detect and characterize the pathogens, for efficient clinical treatments or epidemiological outbreak investigations. All typing methods are also ultimately useful. The applications and progress of molecular methods of pathogen detection are described above, and the dispute is to find out what methods are useful for specific purposes.

KEYWORDS

- ***B. cereus***
- **DNA sequencing**
- **enzyme-linked immunosorbent assays (ELISA)**
- **food-borne pathogens**
- **metagenomics**
- **microarray**
- **molecular detection**
- **next-generation sequencers (NGS)**
- **PFGE pattern**
- **random amplified polymorphism DNA**
- **real-time PCR**
- ***salmonella***

REFERENCES

Aaij, C., & Borst, P., (1972). The gel electrophoresis of DNA. *Biochemistry and Biophysics Journal – Nucleic Acids and Protein Synthesis, 269*(2), 192–200.

Abdi, R. D., Mengstie, F., Beyi, A. F., Beyene, T., Waktole, H., Mammo, B., Ayana, D., & Abunna, F., (2017). Determination of the sources and antimicrobial resistance patterns of *Salmonella* isolated from the poultry industry in Southern Ethiopia. *BMC Infect Dis., 17*, 352.

Abee, T., Groot, M. N., Tempelaars, M., Zwietering, M., Moezelaar, R., & Van Der Voort, M., (2011). Germination and outgrowth of spores of Bacillus cereus group members: Diversity and role of germinant receptors. *Food Microbiology, 28*(2), 199–208.

Achtman, M., Wain, J., Weill, F. X., Nair, S., Zhou, Z., Sangal, V., et al. (2012). MLST study group. Multilocus sequence typing as a replacement for serotyping in Salmonella enterica. *PLoS Pathog., 8*, e1002776. 10.1371/journal.ppat.1002776.

Adzitey, F., & Nurul, H., (2011). *Campylobacter* in poultry: Incidences and possible control measures. *Res. J. Microbiol., 6*, 182–192.

Adzitey, F., Huda, N., & Ali, G. R., (2013). Molecular techniques for detecting and typing of bacteria, advantages and application to food-borne pathogens isolated from ducks. *3 Biotech., 3*, 97–107.

Adzitey, F., Huda, N., & Gulam, R., (2011). Comparison of media for the isolation of *Salmonella* (XLD and Rambach) and *Listeria* species (ALOA and Palcam) in naturally contaminated duck samples. *Internet J. Food Saf., 13*, 20–25.

Allard, M. W., Luo, Y., Strain, E., Pettengill, J., Timme, R., Wang, C., et al. (2013). On the evolutionary history, population genetics and diversity among isolates of *Salmonella* Snteritidis PFGE pattern JEGX01.0004. *PLoS One, 8*, e55254.

Allmann, M., Höfelein, C., Köppel, E., Lüthy, J., Meyer, R., Niederhauser, C., Wegmüller, B., & Candrian, U., (1995). Polymerase chain reaction (PCR) for detection of pathogenic microorganisms in bacteriological monitoring of dairy products. *Research in Microbiology, 146*(1), 85–97.

Amani, J., Mirhosseini, S. A., & Imani, F. A. A., (2015). A review approaches to identify enteric bacterial pathogens. *Jundishapur Journal of Microbiology, 8*, e17473.

Anderson, P., & Roth, J., (1981). Spontaneous, tandem genetic duplications in *Salmonella Typhimurium* arise by unequal recombination between rRNA (rrn) cistrons *Proc. Natl. Acad. Sci., 78*, 3113–3117.

Andersson, A., & Rönner, U., (1998). Adhesion and removal of dormant, heat-activated, and germinated spores of three strains of bacillus cereus. *Biofouling, 13*(1), 51–67.

Andrews, W. H., Jacobson, A., & Hammack, T. S., (2011). Salmonella. In: *Bacteriological Analytical Manual. U.S. Food and Drug Administration*. http://www.fda.gov/Food/FoodScienceResearch/LaboratoryMethods/ucm070149.htm (accessed on 7 September 2019).

Anonymous, (2002). *Microbiology of Food and Animal Feeding Stuffs - Horizontal Method for the Detection of Salmonella* spp. (pp. 6–127), *ISO 6579:2002*. International Organization of Standardization (ISO).

Anonymous, (2009). The community summary report on food-borne outbreaks in the European Union in 2007. *EFSA J.*

Ansorge, W., Sproat, B. S., Stegemann, J., & Schwager, C., (1986). A non-radioactive automated method for DNA sequence determination. *J. Biochem. Biophys. Methods., 13*, 315–323.

Arbeit, R. D., (1999). Laboratory procedures for the epidemiologic analysis of microorganisms. In: Murray, P. M., Baron, E. J., Pfaller, M. A., Tenover, F. C., & Yolken, R. H., (eds.), *Manual of Clinical Microbiology*. Washington: ASM Press.

Asai, T., Zaporojets, D., Squires, C., & Squires, C. L., (1999). An *Escherichia coli* strain with all chromosomal rRNA operons inactivated: Complete exchange of rRNA genes between bacteria. *Proceedings of the National Academy of Sciences, 96*(5), 1971–1976.

Ashelford, K. E., Chuzhanova, N. A., Fry, J. C., Jones, A. J., & Weightman, A. J., (2005). At least 1 in 20 16S rRNA sequence records currently held in public repositories is estimated to contain substantial anomalies. *Applied and Environmental Microbiology, 71*(12), 7724–7736.

August, C., & Koner, J., (1993). Awarded the honorary chair of anatomy in Halle 100 years ago: Carl Joseph Eberth–discoverer of the typhus pathogen. *Pathology, 14*, 234–236.

Awadallah, M. A., Ahmed, H. A., Merwad, A M., & Selim, M. A., (2016). Occurrence, genotyping, shiga toxin genes and associated risk factors of *E. coli* isolated from dairy farms, handlers and milk consumers. *Vet J., 217*, 83–88.

Bailey, J. S., Fedorka-Cray, P. J., Stern, N. J., Craven, S. E., Cox, N. A., & Cosby, D. E., (2002). Serotyping and ribotyping of *Salmonella* using restriction enzyme *PvuII*. *J. Food Prot., 65*, 1005–1007.

Baker, G. C., Smith, J. J., & Cowan, D., (2003). Review and re-analysis of domain-specific 16S primers. *Journal of Microbiological Methods, 55*(3), 541–555.

Barak, J. D., Gorski, L., Naraghi-Arani, P., & Charkowski, A. O., (2005). Salmonella enterica virulence genes are required for bacterial attachment to plant tissue. *Applied and Environmental Microbiology, 71*(10), 5685–5691.

Barak, J. D., Jahn, C. E., Gibson, D. L., & Charkowski, A. O., (2007). The role of cellulose and Oantigen capsule in the colonization of plants by Salmonella enterica. *Molecular Plant-Microbe Interactions, 20*(9), 1083–1091.

Barken, K. B., Haagensen, J. A., & Tolker-Nielsen, T., (2007). Advances in nucleic acid-based diagnostics of bacterial infections. *Clinica Chimica Acta, 384*(1/2), 1–11.

Bartoszewicz, M., Hansen, B. M., & Swiecicka, I., (2008). The members of *Bacillus cereus* group are commonly present contaminants of fresh and heat-treated milk. *Food Microbiology, 25*, 588–596.

Bates, S. R., Baldwin, D. A., Channing, A., Gifford, L. K., Hsu, A., & Lu, P., (2005). Cooperativity of paired oligonucleotide probes for microarray hybridization assays. *Analytical Biochemistry, 342*(1), 59–68.

Batt, C. A., (2007). Food pathogen detection. *Science, 316*(5831), 1579–1580.

Berry, C., O'Neil, S., Ben-Dov, E., Jones, A. F., Murphy, L., Quail, M. A., Holden, M. T. G., Harris, D., Zaritsky, A., & Parkhill, J., (2002). Complete sequence and organization of pBtoxis, the toxin-coding plasmid of *Bacillus thuringiensis* subsp. israelensis. *Applied and Environmental Microbiology, 68*(10), 5082–5095.

Bolton, F. J., Fritz, E., & Poynton, S., (2000). Rapid enzyme-linked immunoassay for the detection of Salmonella in food and feed products: performance testing program. *J. AOAC Int., 83*, 299–304.

Boonmar, S., Yingsakmongkon, S., Songserm, T., Hanhaboon, P., & Passadurak, W., (2007). Detection of *Campylobacterin* duck using standard culture method and multiplex polymerase chain reaction. *Southeast Asian J. Trop. Med. Pub. Health, 38*, 728–731.

Botteldoorn, N., Herman, L., Rijpens, N., et al. (2004). Phenotypic and molecular typing of salmonella strains reveal different contamination source in two commercial pig slaughterhouses. *Applied and Environmental Microbiology, 70*, 5305–5314.

Bottone, E. J., (2010). *Bacillus cereus*, a volatile human pathogen. *Clin. Microbiol. Rev., 23*(2), 382–398. doi: 10.1128/CMR.00073–09.

Bowers. J., et al. (2009). Virtual terminator nucleotides for next-generation DNA sequencing. *Nat. Methods, 6*, 593–595.

Braslavsky, I., Hebert, B., Kartalov, E. S., & Quake, S. R., (2003). Sequence information can be obtained from single DNA molecules. *Proc. Natl. Acad. Sci., 100*, 3960–3964.

Brenner, F. W., Villar, R. G., Angulo, F. J., Tauxe, R., & Swaminathan, B., (2000). *Salmonella* nomenclature. *J. Clin. Microbiol., 38*, 2465–2467.

Brisco, M. J., Condon, J., Sykes, P. J., Neoh, S. H., & Morley, A. A., (1991). Detection and quantitation of neoplastic cells in acute lymphoblastic leukemia, by use of the polymerase chain reaction. *Br. J. Haematol.*, ***79***, 211–217.

Budd, W., (1873). *Typhoied Fever: Its Nature, Mode of Spreading, and Prevention Longmans* (pp. 610–612). Green, and Co, London.

Carig, A. G., Nizetic, D., Hoheisel, J. D., Zehetner, G., & Lehrach, H., (1990). Ordering of cosmid clones covering the herpes simplex virus type I (HSV-I) genome: A test case for fingerprinting by hybridization. *Nucleic Acids Research, 18*, 2653–2660.

Carretto, E., Barbarini, D., Poletti, F., Capra, F. M., Emmi, V., & Marone, P., (2000). Bacillus cereus fatal bacteremia and apparent association with nosocomial transmission in an intensive care unit. *Scand J. Infect. Dis., 32*, 98–100.

CDC, (2012). *Surveillance for Food-borne Disease Outbreaks*. United States, Annual Report. pp. 1–20, file:///C:/Users/91982/Downloads/foodborne-disease-outbreaks-annual-report-2012–508c.pdf (last retrieved on 16/10/2019)

CDC, (2014). *Multistate Outbreak of Multidrug-Resistant Salmonella Heidelberg Infections Linked to Foster Farms Brand Chicken*. https://www.cdc.gov/salmonella/heidelberg-11–16/index.html (last retrieved on 16/10/2019).

Ceuppens, S., Boon, N., & Uyttendaele, M., (2013). Diversity of bacillus cereus group strains is reflected in their broad range of pathogenicity and diverse ecological lifestyles. *FEMS Microbiology Ecology, Early View, 84*(3), 433–450.

Chakravorty, S., Helb, D., Burday, M., Connell, N., & Alland, D., (2007). A detailed analysis of 16S ribosomal RNA gene segments for the diagnosis of pathogenic bacteria. *Journal of Microbiological Methods, 69*(2), 330–339.

Chan, K., Baker, S., Kim, C. C., Detweiler, C. S., Dougan, G., & Falkow, S., (2003). Genomic comparison of *salmonella enterica serovars* and *salmonella bongori* by use of an *S. enterica serovar typhimurium* DNA microarray. *J. Bacteriol., 185*, 553–563. 10.1128/JB.185.2.553–563.2003.

Chan, M., Maiden, M. C. J., & Spratt, B. G., (2001). Database-driven multi locus sequence typing (MLST) of bacterial pathogens. *Bioinformatics, 17*, 1077–1083.

Cheung, W. H. S., Chang, K. C. K., & Hung, R. P. S., (1990). Health effects of beach water pollution in Hong Kong. *Epidemiol. Infect., 105*, 139–162.

Chon, J. W., Yim, J. H., Kim, H. S., Kim, D. H., Kim, H., Oh, D. H., Kim, S. K., & Seo, K. H., (2015). Quantitative prevalence and toxin gene profile of *bacillus cereus* from ready-to-eat vegetables in South Korea. *Food-Borne Pathog Dis., 12*, 795–799.

Clarridge, J. E., (2004). Impact of 16S rRNA gene sequence analysis for identification of bacteria on clinical microbiology and infectious diseases. *Clinical Microbiology Reviews, 17*(4), 840–862.

Cole, J. R., Chai, B., Farris, R. J., Wang, Q., Kulam-Syed-Mohideen, A. S., McGarrell, D. M., Bandela, A. M., Cardenas, E., Garrity, G. M., & Tiedje, J. M., (2007). The ribosomal database project (RDP-II), introducing myRDP space and quality controlled public data. *Nucleic Acids Research, 35*, 169–172.

Cole, J. R., Wang, Q., Cardenas, E., Fish, J., Chai, B., Farris, R. J., Kulam-Syed, M. A. S., McGarrell, D. M., Marsh, T., Garrity, G. M., & Tiedje, J. M., (2009). The ribosomal database project: Improved alignments and new tools for rRNA analysis. *Nucleic Acids Research, 37*(1), D141–D145.

Communicable Diseases and Public Health, (1999). Vol. 2, pp. 108–113.

Crosa, J. H., Brenner, D. J., Ewing, W. H., & Falkow, S., (1973). Molecular relationships among the *salmonellae*. *J. Bacteriol., 115*, 307–315.

Cunha. B. A., (2004). The death of Alexander the great: Malaria or typhoid fever? *Infect Dis. Clin. North Am., 18*, 53–63.

D'Urso, O. F., Poltronieri, P., Marsigliante, S., Storelli, C., Hernández, M., & Rodríguez-Lázaro, D., (2009). A filtration-based real-time PCR method for the quantitative detection of viable *salmonella enterica* and *listeria monocytogenes* in food samples. *Food Microbiology, 26*(3), 311–316.

Debnath, M., Prasad, G. B. K. S., & Bisen, P. S., (2010). Microarray. In: *Molecular Diagnostics: Promises and Possibilities* (pp. 193–208). Springer Netherlands.

DeSantis, T. Z., Hugenholtz, P., Larsen, N., Rojas, M., Brodie, E. L., Keller, K., Huber, T., Dalevi, D., Hu, P., & Andersen, G. L., (2006). Greengenes, a chimera-checked 16S rRNA gene database and workbench compatible with ARB. *Applied and Environmental Microbiology, 72*(7), 5069–5072.

Diane, G. N., Marion, K., Linda, V., Erwin, D., & Awa, A. K., (2011). Food-borne diseases: The challenges of 20 years ago still persist while new ones continue to emerge. *International Journal of Food Microbiology, 145*(2/3), 493.

Dierick, K., Coillie, E. V., Swiecicka, I., Meyfroidt, G., Devlieger, H., Meulemans, A., Hoedemaekers, G., Fourie, L., Heyndrickx, M., & Mahillon, J., (2005). Fatal family outbreak of bacillus cereus-associated food poisoning. *J. Clin. Microbiol., 43*(8), 4277–4279. doi: 10.1128/JCM.43.8.4277–4279.

Doolittle, W. F., (1999). Phylogenetic classification and the universal tree. *Science, 284*(5423), 2124–2128.

Drahovska, H., Mikasova, E., Szemes, T., Ficek, A., Sasik, M., Majtan, V., & Turna J, (2007). Variability in occurrence of multiple prophage genes in *salmonella typhimurium* strains isolated in Slovak Republic. *FEMS Microbiol. Lett., 270*, 237–244. 10.1111/j.1574–6968.2007.00674.x.

Drobniewski, F. A., (1993). *Bacillus cereus* and related species. *Clin. Microbiol. Rev., 6*, 324–338.

Dwivedi, H. P., & Jaykus, L. A., (2011). Detection of pathogens in foods: the current state-of-the-art and future directions. *Critical Reviews in Microbiology, 37*(1), 40–63.

EFSA, (2005). Opinion of the scientific panel on biological hazards on *Bacillus cereus* and other *Bacillus spp.* in foodstuffs. *EFSA Journal, 175*, 1–48.

EFSA, (2013). The European Union summary report on trends and sources of zoonoses, zoonotic agents and food-borne outbreaks in 2011. *EFSA Journal, 11*(4), 3129.

Ehling-Schulz, M., Fricker, M., & Scherer, S., (2004). Bacillus cereus, the causative agent of an emetic type of food-borne illness. *Molecular Nutrition & Food Research, 48*(7), 479–487.

Ehling-Schulz, M., Fricker, M., Grallert, H., Rieck, P., Wagner, M., & Scherer, S., (2006). *Cereulide synthetase* gene cluster from emetic *bacillus cereus*: Structure and location on a mega virulence plasmid related to *Bacillus anthracis* toxin plasmid pXO1. *BMC Microbiology, 6*(1), 20.

Ehling-Schulz, M., Vukov, N., Schulz, A., Shaheen, R., Andersson, M., Märtlbauer, E., & Scherer, S., (2005). Identification and partial characterization of the nonribosomal peptide synthetase gene responsible for cereulide production in emetic Bacillus cereus. *Applied and Environmental Microbiology, 71*, 105–113.

Ellingson, J. L. E., Anderson, J. L., Carlson, S. A., & Sharma, V. K., (2004). Twelve hour real-time PCR technique for the sensitive and specific detection of Salmonella in raw and ready-to-eat meat products. *Molecular and Cellular Probes, 18*(1), 51–57.

Ellwood, M., & Nomura, M., (1980). Deletion of a ribosomal ribonucleic acid and operon in *Escherichia coli*. *J. Bacteriol., 143*, pp. 1077–1080.

Engvall, E., & Perlmann, P., (1971). Enzyme-linked immunosorbent assay (ELISA) quantitative assay of immunoglobulin G. *Immunochemistry, 8*, 871–874.

Eriksson, E., & Aspan, A., (2007). Comparison of culture, ELISA and PCR techniques for Salmonella detection in faecal samples for cattle, pig and poultry. *BMC Veterinary Research, 3*(1), 21.

Erlich, H., (1989). Polymerase chain reaction. *Journal of Clinical Immunology, 9*(6), 437–447.

Fairchild, A., Lee, M., & Maurer, J., (2006). PCR basics. In: Maurer, J., (ed.), *PCR Methods in Foods* (pp. 1–25). Springer US.

Farber, J. M., (1996). An introduction to the hows and whys of molecular typing. *J. Food Prot., 59*, 1091–1110.

Ferreira, M. A. S., & Lund, B. M., (1987). The influence of pH and temperature on initiation of growth of *Salmonella* spp. *Letters in Applied Microbiology, 5*(4), 67–70.

Fierer, J., & Guiney, D. G., (2001). Diverse virulence traits underlying different clinical outcomes of salmonella infection. *The Journal of Clinical Investigation, 107*(7), 775–780.

Fricker, M., Messelhäusser, U., Busch, U., Scherer, S., & Ehling-Schulz, M., (2007). Diagnostic real-time PCR assays for the detection of emetic *Bacillus cereus* strains in foods and recent food-borne outbreaks. *Applied and Environmental Microbiology, 73*, 1892–1898.

Gahring, L. C., Heffron, F., Finlay, B. B., & Falkow, S., (1990). Invasion and replication of Salmonella typhimurium in animal cells. *Infection and Immunity, 58*(2), 443–448.

Galán, J. E., & Curtiss, R., (1989). Cloning and molecular characterization of genes whose products allow *Salmonella typhimurium* to penetrate tissue culture cells. *Proceedings of the National Academy of Sciences, 86*(16), 6383–6387.

Gergen, J. P., Stern, R. H., & Wensink, P. C., (1979). Filter replicas and permanent collections of recombinant DNA plasmids. *Nucleic Acids Research, 7*, 2115–2136.

Giffel, M. C., & Beumer, R. R., (1999). *Bacillus cereus*: A review. *The Journal of Food Technology in Africa, 4*, 7–13.
Ginocchio, C., Pace, J., & Galán, J. E., (1992). Identification and molecular characterization of a *Salmonella typhimurium* gene involved in triggering the internalization of salmonellae into cultured epithelial cells. *Proceedings of the National Academy of Sciences, 89*(13), 5976–5980.
González-Escalona, N., Hammack, T. S., Russell, M., Jacobson, A. P., De Jesús, A. J., Brown, E. W., & Lampel, K. A., (2009). Detection of live Salmonella sp. cells in produce by a TaqMan-based quantitative reverse transcriptase real-time PCR targeting invA mRNA. *Applied and Environmental Microbiology, 75*(11), 3714–3720.
Granum, P. E., & Lund, T., (1997). Bacillus cereus and its food poisoning toxins. *FEMS Microbiology Letters, 157*(2), 223–228.
Griffiths, M. W., (2010). *Pathogens and Toxins in Foods: Challenges and Interventions* (pp. 1–19). ASM Press, Washington, DC.
Grimont, A. D., & Weill, F. X., (2007). *Antigenic Formulae of the Salmonella Serovars* (9th edn., pp. 1–166). WHO Collaborating Centre for Reference and Research on Salmonella.
Grinstein, M., & Hogness, D. S., (1975). Colony hybridization: A method for the isolation of cloned DNAs that contain a specific gene. *Proceedings of the National Academy of Sciences of the United States of America, 72*, 3961–3965.
Guinebretière, M. H., Auger, S., Galleron, N., Contzen, M., De Sarrau, B., De Buyser, M. L., et al. (2013). *Bacillus cytotoxicus* sp. nov. is a novel thermotolerant species of the *bacillus cereus* group occasionally associated with food poisoning. *International Journal of Systematic and Evolutionary Microbiology, 63*(1), 31–40.
Hald, T., Wingstrand, A., Swanenburg, M., Von Altrock, A., & Thorberg, B. M., (2003). The occurrence and epidemiology of Salmonella in European pig slaughterhouses. *Epidemiology & Infection, 131*(03), 1187–1203.
Hansen, T., Riber, L., Nielsen, M. B., Vigre, H., Thomsen, L. E., Hoorfar, J., & Löfström, C., (2013). Influence of contact time and pre-growth conditions on the attachment of *salmonella enterica serovar typhimurium* to pork meat. *Submitted to Food Microbiology.*
Harris, T. D., et al. (2008). Single-molecule DNA sequencing of a viral genome. *Science (New York, N.Y.), 320*, 106–109.
Hawkey, J., Edwards, D. J., Dimovski, K., Hiley, L., Billman-Jacobe, H., Hogg, G., & Holt, K. E., (2013). Evidence of microevolution of *Salmonella Typhimurium* during a series of egg-associated outbreaks linked to a single chicken farm. *BMC Genomics, 14*, 800. 10.1186/1471–2164–14–800.
Hernan, R., Chang, L. H., Jeyaseelan, L. K., & Earnest, E. S., (1997). Phylogenetic relationships of salmonella typhi and salmonella typhimunum based on 16s rRNA sequence analysis. *International Journal of Systematic Bacteriology,* 1253–1254.
Hoffmann, M., Luo, Y., Monday, S. R., Gonzalez-Escalona, N., Ottesen, A. R., Muruvanda, T., et al. (2015). Tracing origins of the *salmonella* bareilly strain causing a food-borne outbreak in the United States. *J. Infect Dis., 213*, 502–508.
Holley, R. W., Apgar, J., Merrill, S. H., & Zubkoff, P. L., (1961). Nucleotide and oligonucleotide compositions of the alanine, valine, and tyrosine-acceptor soluble ribonucleic acids of yeast. *J. Am. Chem. Soc., 83*, 4861–4862.
Holley, R. W., et al. (1965). Structure of a ribonucleic acid. *Science, 147*, 1462–1465.

Hood, S. K., & Zottola, E. A., (1997). Adherence to stainless steel by food-borne microorganisms during growth in model food systems. *International Journal of Food Microbiology, 37*(2/3), 145–153.

Hornyák, A., Bálint, A., Farsang, A., Balka, G., Hakhverdyan, M., Rasmussen, T. B., Blomberg, J., & Belák, S., (2012). Detection of subgenomic mRNA of feline coronavirus by real-time polymerase chain reaction based on primer-probe energy transfer (P-sg-QPCR). *J. Virol. Methods., 181*, 155–163.

Hunkapiller, T., Kaiser, R., Koop, B., & Hood, L., (1991). Large-scale and automated DNA sequence determination. *Science, 254*, 59–67.

Hyman. E. D., (1988). A new method for sequencing DNA. *Anal. Biochem., 174*, 423–436.

Inns, T., Lane, C., Peters, T., Dallman, T., Chatt, C., McFarland, N., et al. (2015). A multi-country *Salmonella* Enteritidis phage type 14b outbreak associated with eggs from a German producer: 'Near real-time' application of whole genome sequencing and food chain investigations, United Kingdom. *Eurosurveillance, 20*, 15–22.

Irvine, W. N., Gillespie, I. A., Smyth, F. B., Rooney, P. J., McClenaghan, A., Devine, M. J., & Tohani, V. K., (2009). Outbreak control team. Investigation of an outbreak of *salmonella enterica* serovar newport infection. *Epidemiol. Infect., 137*, 1449–1456.

Jantsch, J., Chikkaballi, D., & Hensel, M., (2011). Cellular aspects of immunity to intracellular *Salmonella enterica. Immunological Reviews, 240*(1), 185–195.

Jasson, V., Jacxsens, L., Luning, P., Rajkovic, A., & Uyttendaele, M., (2010). Alternative microbial methods: An overview and selection criteria. *Food Microbiology, 27*(6), 710–730.

Jensen, A. N., Dalsgaard, A., Stockmarr, A., Nielsen, E. M., & Baggesen, D. L., (2006). Survival and transmission of *Salmonella enterica* serovar Typhimurium in an outdoor organic pig farming environment. *Applied and Environmental Microbiology, 72*(3), 1833–1842.

Jeßberger, N., Dietrich, R., Bock, S., Didier, A., & Märtlbauer, E., (2014). *Bacillus cereus* enterotoxins act as major virulence factors and exhibit distinct cytotoxicity to different human cell lines. *Toxicon., 77*, 49–57. doi: 10.1016/j.toxicon.2013.10.028.

Josefsen, M. H., Löfström, C., Hansen, T., Reynisson, E., & Hoorfar, J., (2012). Instrumentation and fluorescent chemistries used in qPCR. In: Filion, M., (ed.), *Quantitative Real-Time PCR in Applied Microbiology* (pp. 27–52) Caister Academic Press, UK.

Joseph, B., Otta, S. K., Karunasagar, I., & Karunasagar, I., (2001). Biofilm formation by *Salmonella spp.* on food contact surfaces and their sensitivity to sanitizers. *International Journal of Food Microbiology, 64*(3), 367–372.

Josephson, K. L., Gerba, C. P., & Pepper, I. L., (1993). Polymerase chain reaction detection of nonviable bacterial pathogens. *Applied and Environmental Microbiology, 59*(10), 3513–3515.

Juozaitis, A., Willeke, K., Grinshpun, S. A., & Donnelly, J., (1994). Impaction onto a glass slide or agar versus impingement into a liquid for the collection and recovery of airborne microorganisms. *Appl. Environ. Microbiol., 60*, 861–870.

Kawai, T., Sekizuka, T., Yahata, Y., Kuroda, M., Kumeda, Y., Iijima, Y., Kamata, Y., Sugita-Konishi, Y., & Ohnishi, T., (2012). Identification of Kudoa septempunctata as the causative agent of novel food poisoning outbreaks in japan by consumption of paralichthys olivaceus in raw fish. *Clinical Infectious Diseases, 54*(8), 1046–1052.

Keim, P., Gruendike, J. M., Klevytska, A. M., Schupp, J. M., Challacombe, J., & Okinaka, R., (2009). The genome and variation of Bacillus anthracis. *Molecular Aspects of Medicine, 30*(6), 397–405.

Kotiranta, A., Lounatmaa, K., & Haapasalo, M., (2000). Epidemiology and pathogenesis of *Bacillus cereus* infections. *Microbes Infect., 2,* 189–198.

Kumar, B. K., Raghunath, P., Devegowda, D., Deekshit, V. K., Venugopal, M. N., & Karunasagar, I., (2011a). Development of monoclonal antibody based sandwich ELISA for the rapid detection of pathogenic *Vibrio parahaemolyticus* in seafood. *Int. J. Food Microbiol., 145,* 244–249.

Kumar, P. S., Brooker, M. R., Dowd, S. E., & Camerlengo, T., (2011b). Target region selection is a critical determinant of community fingerprints generated by 16S pyrosequencing. *PLoS One, 6*(6), e20956.

Kumar, R., Surendran, P. K., & Thampuran, N., (2010). Rapid quantification of salmonella in seafood using real-time PCR assay. *Journal of Microbiology and Biotechnology, 20*(3), 569–573.

Kusunoki, J., Kai, A., Yanagawa, Y., Monma, C., Shingaki, M., Obata, H., Itoh, T., Ohta, K., Kudoh, Y., & Nakamura, A., (1997). Biochemical and molecular characterization of *Salmonella* serovar enteritidis phage type 4 isolated from food poisoning outbreaks in Tokyo. *Kansenshogaku Zasshi., 71,* 730–737.

Lahiri, A., Lahiri, A., Iyer, N., Das, P., & Chakravortty, D., (2010). Visiting the cell biology of *Salmonella* infection. *Microbes and Infection, 12,* 809–818.

Lapidus, A., Goltsman, E., Auger, S., Galleron, N., Ségurens, B., Dossat, C., et al. (2008). Extending the *bacillus cereus* group genomics to putative food-borne pathogens of different toxicity. *Chemico-Biological Interactions, 171*(2), 236–249.

Lausted, C., Dahl, T., Warren, C., King, K., Smith, K., Johnson, M., Saleem, R., Aitchison, J., Hood, L., & Lasky, S., (2004). POSaM: A fast, flexible, open-source, inkjet oligonucleotide synthesizer and microarrayer. *Genome Biology, 5*(8), R58.

Lennon, G., Auffray, C., Polymeropoulos, M., & Soares, M. B., (1996). The IMAGE Consortium: An integrated molecular analysis of genomes and their expression. *Genomics, 33,* 151–152.

Lin, T. H., Lin, L. Y., & Zhang, F., (2014). Review on molecular typing methods of pathogens. *Open Journal of Medical Microbiology, 4,* 147–152.

Lindsay, D., Brözel, V. S., Mostert, J. F., & Von Holy, A., (2000). Physiology of dairy-associated Bacillus spp. over a wide pH range. *International Journal of Food Microbiology, 54*(1/2), 49–62.

Liu, Z., DeSantis, T. Z., Andersen, G. L., & Knight, R., (2008). Accurate taxonomy assignments from 16S rRNA sequences produced by highly parallel pyrosequencers. *Nucleic Acids Research, 36*(18), e120.

Lo, F., Wong, D. M. A., Hald, T., Van Der Wolf, P. J., & Swanenburg, M., (2002). Epidemiology and control measures for Salmonella in pigs and pork. *Livestock Production Science, 76*(3), 215–222.

Lobenhofer, E. K., et al. (2001). Progress in the application of DNA microarrays. *Environ. Health Perspect., 109,* 881–891.

Löfström, C., Schelin, J., Norling, B., Vigre, H., Hoorfar, J., & Rådström, P., (2011). Culture independent quantification of *salmonella enterica* in carcass gauze swabs by flotation prior to real-time. *International Journal of Food Microbiology, 145,* 103–109.

Logan, N. A., & De Vos, P., (2009). Bacillus. In: De Vos, P., Garrity, G. M., Jones, D., Krieg, N. R., Ludwig, W., Rainey, F. A., & Whitman, W. B., (eds.), *Bergey's Manual of Systematic Bacteriology* (pp. 21–128).

Logan, N. A., (2012). Bacillus and relatives in food-borne illness. *Journal of Applied Microbiology, 112*(3), 417–429.

López-Campos, G., Martínez-Suárez, J., Aguado-Urda, M., & López-Alonso, V., (2012a). DNA microarrays: Principles and technologies. In: *Microarray Detection and Characterization of Bacterial Food-Borne Pathogens* (pp. 33–47). Springer US.

López-Campos, G., Martínez-Suárez, J., Aguado-Urda, M., & López-Alonso, V., (2012b). Applications of DNA Microarrays to study bacterial food-borne pathogens. In: *Microarray Detection and Characterization of Bacterial Food-Borne Pathogens* (pp. 93–114). Springer US.

Loughney, K., Lund, E., & Dahlberg, J. E., (1983). Deletion of an rRNA gene set in *bacillus subtilis. J. Bacteriol., 154*(), pp. 529–532.

Lowther, S. A., Medus, C., Scheftel, J., Leano, F., Jawahir, S., & Smith, K., (2011). Food-borne outbreak of *Salmonella* subspecies IV infections associated with contamination from bearded dragons. *Zoonoses Public Health, 58*, 560–566.

Magistrado, P., Carcia, M., & Raymundo, A., (2001). Isolation and polymerase chain reaction base detection of *Campylobacter jejuni* and *Campylobacter colifrom* poultry in the Philippines. *Int. J. Food Microbiol., 70*, 194–206.

Maiden, M. C. J., Bygraves, J. A., Feil, E., Morelli, G., et al. (1997). Multilocus sequence typing: A portable approach to the identification of clones within populations of pathogenic microorganisms. *Proceedings of the National Academy of Sciences of the United States of America, 95*, 3140–3145.

Maiden, M. C., Bygraves, J. A., Feil, E., Morelli, G., Russel, J. E., Urwin, R., et al. (1998). Spratt multilocus sequence typing: A portable approach to the identification of clones within populations of pathogenic microorganisms *Proc. Natl. Acad. Sci. U.S.A., 95*, 3140–3145.

Majowicz, S. E., Musto, J., Scallan, E., Angulo, F. J., Kirk, M., O'Brien, S. J., et al. (2010). The global burden of nontyphoidal Salmonella Gastroenteritis. *Clinical Infectious Diseases, 50*(6), 882–889.

Malorny, B., Bunge, C., Guerra, B., Prietz, S., & Helmuth, R., (2007). Molecular characterization of Salmonella strains by an oligonucleotide multiprobe microarray. *Molecular and Cellular Probes, 21*(1), 56–65.

Malorny, B., Paccassoni, E., Fach, P., Bunge, C., Martin, A., & Helmuth, R., (2004). Diagnostic real-time PCR for detection of salmonellain food. *Applied and Environmental Microbiology, 70*, 7046–7052.

Mardis, E. R., (2008a). The impact of next-generation sequencing technology on genetics. *Trends in Genetics, 24*(3), 133–141.

Mardis, E. R., (2008b). Next-generation DNA sequencing methods. *Annual Review of Genomics and Human Genetics, 9*, 387–402.

Margulies, M., Egholm, M., Altman, W. E., Attiya, S., Bader, J. S., Bemben, L. A., et al. (2005). Genome sequencing in microfabricated high-density picolitre reactors. *Nature, 437*(7057), 376–380.

Martínez-Blanch, J. F., Sánchez, G., Garay, E., & Aznar, R., (2009). Development of a real-time PCR assay for detection and quantification of enterotoxigenic members of

Bacillus cereus group in food samples. *International Journal of Food Microbiology, 135*, 15–21.

Mead, P. S., Slutsker, L., Dietz, V., McCaig, L. F., Bresee, J. S., Shapiro, C., Griffin, P. M., & Tauxe, R. V., (1999). Food-related illness and death in the United States. *Emerging Infectious Diseases, 5*, 607–625.

Messelhäusser, U., Kampf, P., Fricker, M., Ehling-Schulz, M., Zucker, R., Wagner, B., Busch, U., & Holler, C., (2010). Prevalence of emetic *bacillus cereus* in different ice creams in Bavaria. *Journal of Food Protection, 73*(2), 395–399.

Metzker, M. L., (2010). Sequencing technologies-the next generation. *Nature Reviews Genetics, 11*(1), 31–46.

Miodrag, G. J., (2013). The history of DNA sequencing. *Med. Biochem., 32*, 301–312.

Moorhead, S. M., & Bell, R. G., (1999). Psychrotrophic *clostridia* mediated gas and botulinal toxin production in vacuum-packed chilled meat. *Lett. Appl. Microbiol., 28*, 108–112.

Morild, R. K., Olsen, J. E., & Aabo, S., (2011). Change in attachment of Salmonella *Typhimurium, Yersinia enterocolitica*, and *Listeria monocytogenes* to pork skin and muscle after hot water and lactic acid decontamination. *International Journal of Food Microbiology, 145*(1), 353–358.

Morley, A. A., (2014). Digital PCR: A brief history. *Biomol. Detect Quantif., 1*, 1–2.

Mullis, K., Faloona, F., Scharf, S., Saiki, R., Horn, G. T., & Erlich, H. A., (1986). Specific enzymatic amplification of DNA *in vitro*: The polymerase chain reaction. *Cold Spring Harbor Symposia on Quantitative Biology, 51*, 263–273.

Nadon, C., Van Walle, I., Gerner-Smidt, P., et al. (2017). PulseNet international: Vision for the implementation of whole genome sequencing (WGS) for global food-borne disease surveillance. *Eurosurveillance, 22*(23), 30544. doi: 10.2807/1560–7917.ES.2017.22.23.30544.

Nakamura, S., Maeda, N., Miron, I. M., Yoh, M., Izutsu, K., Kataoka, C., et al. (2008). Metagenomic diagnosis of bacterial infections. *Emerging Infectious Diseases, 14*(11), 1784–1786.

Newcombe, J, Cartwright, K., Palmer, W. H., & McFadden, J., (1996). PCR of peripheral blood for diagnosis of meningococcal disease. *J. Clin. Microbiol., 34*, 1637–1640.

Nielsen, M. B., Knudsen, G. M., Danino-Appleton, V., Olsen, J. E., & Thomsen, L. E., (2013). Comparison of heat stress responses of immobilized and planktonic *salmonella enterica serovar typhimurium. Food Microbiology, 33*(2), 221–227.

Nyren, P. I., (1985). Lundin Enzymatic method for continuous monitoring of inorganic pyrophosphate synthesis. *Anal. Biochem., 509*, 504–509.

Oliveira, K., Oliveira, T., Teixeira, P., Azeredo, J., Henriques, M., & Oliveira, R., (2006). Comparison of the adhesion ability of different Salmonella Enteritidis serotypes to materials used in kitchens. *Journal of Food Protection, 69*(10), 2352–2356.

Omiccioli, E., Amagliani, G., Brandi, G., & Magnani, M., (2009). A new platform for real-time PCR detection of *salmonella spp.*, listeria monocytogenes and *Escherichia coli* O157 in milk. *Food Microbiology, 26*(6), 615–622.

Østensvik, Ø., From, C., Heidenreich, B., O'Sullivan, K., & Granum, P. E., (2004). Cytotoxic Bacillus spp. belonging to the B. cereus and B. subtilis groups in Norwegian surface waters. *Journal of Applied Microbiology, 96*(5), 987–993.

Palomino-Camargo, C., & González-Muñoz, Y., (2014). Molecular techniques for detection and identification of pathogens in food: advantages and limitations. *Rev. Peru Med. Exp. Salud. Publica., 31*, 535–546.

Papagrigorakis, M. J., Yapijakis, C., Synodinos, P. N., & Baziotopoulou-Valavani, E., (2006). DNA examination of ancient dental pulp incriminates typhoid fever as a probable cause of the Plague of Athens. *Int. J. Infect Dis., 10*, 206–214.

Pareek, C. S., Smoczynski, R., & Tretyn, A., (2011). Sequencing technologies and genome sequencing. *J. Appl. Genet., 52*, 413–435.

Peng, J. S., Tsai, W. C., & Chou, C. C., (2002). Inactivation and removal of *bacillus cereus* by sanitizer and detergent. *International Journal of Food Microbiology, 77*(1/2), 11–18.

Pergola, S., Franciosini, M. P., Comitini, F., Ciani, M., De Luca, S., Bellucci, S., Menchetti, L., & Casagrande, P. P., (2017). Genetic diversity and antimicrobial resistance profiles of *Campylobacter coli* and *Campylobacter jejuni* isolated from broiler chicken in farms and at time of slaughter in central Italy. *J. Appl. Microbiol., 122*, 1348–1356.

Petrik, J., (2001). Microarray technology: The future of blood testing? *Vox. Sang., 80*, 1–11.

Petti, C. A., (2007). Detection and identification of microorganisms by gene amplification and sequencing. *Clinical Infectious Diseases, 44*(8), 1108–1114.

Petti, C. A., Polage, C. R., & Schreckenberger, P., (2005). The role of 16S rRNA gene sequencing in identification of microorganisms misidentified by conventional methods. *Journal of Clinical Microbiology, 43*(12), 6123–6125.

Pezzoli, L., Elson, R., Little, C. L., Yip, H., Fisher, I., Yishai, R., et al. (2008). Packed with *Salmonella* investigation of an international outbreak of *Salmonella Senftenberg* infection linked to contamination of prepacked basil in 2007. *Food-Borne Pathog. Dis., 5*, 661–668.

Pollack, J. R., Perou, C. M., Alizadeh, A. A., Eisen, M. B., Pergamenschikov, A., Williams, C. F., Jeffrey, S. S., Botstein, D., & Brown, P. O., (1999). Genome-wide analysis of DNA copy-number changes using cDNA microarrays. *Nat. Genet., 23*(1), 41–46.

Ponchel, F., Toomes, C., Bransfield, K., Leong, F. T., Douglas, S. H., Field, S. L., et al. (2003). Real-time PCR based on SYBR-green I fluorescence: An alternative to the TaqMan assay for a relative quantification of gene rearrangements, gene amplifications and micro gene deletions. *BMC Biotechnol., 13*, 3–18.

Pop, M., (2009). Genome assembly reborn: Recent computational challenges. *Briefings in Bioinformatics, 10*(4), 354–366.

Popoff, M. Y., & Le Minor, L. E., (2005). *Salmonella*. In: Garrity, G. M., Brenner, D. J., Krieg, N. R., & Staley J. R., (eds.), *Bergey's Manual of Systematic Bacteriology* (Volume 2, Part B, pp. 578.). Springer-Verlag.

Prouty, A. M., & Gunn, J. S., (2003). Comparative analysis of *Salmonella enterica serovar Typhimurium biofilm* formation on gallstones and on glass. *Infection and Immunity, 71*(12), 7154–7158.

Rahn, K., Grandis, S. A. D., Clarke, R. C., McEwen, S. A., Galán, J. E., Ginocchio, C., Curtiss, III, R., & Gyles, C. L., (1992). Amplification of an invA gene sequence of *Salmonella typhimurium* by polymerase chain reaction as a specific method of detection of salmonella. *Molecular and Cellular Probes, 6*(4), 271–279.

Rasko, D. A., Altherr, M. R., Han, C. S., & Ravel, J., (2005). Genomics of the *Bacillus cereus* group of organisms. *FEMS Microbiology Reviews, 29*(2), 303–329.

Rasooly, R., & Do, P. M., (2010). *Clostridium botulinum* neurotoxin type B is heat-stable in milk and not inactivated by pasteurization. *J. Agric. Food Chem., 58*, 12557–12561.

Reekmans, R., Stevens, P., Vervust, T., & Vos, P. D., (2009). An alternative real-time PCR method to detect the Bacillus cereus group in naturally contaminated food gelatine: A comparison study. *Letters in Applied Microbiology, 48*(1), 97–104.

Reen, F. J., Boyd, E. F., Porwollik, S., Murphy, B. P., Gilroy, D., Fanning, S., & McClelland, M., (2005). Genomic comparisons of *Salmonella enterica serovar* Dublin, Agona, and Typhimurium strains recently isolated from milk filters and bovine samples from Ireland, using a Salmonella microarray. *Applied and Environmental Microbiology, 71*(3), 1616–1625.

Ronaghi, M., Uhlen, M., & Nyrén, P. I., (1998). A sequencing method based on real-time pyrophosphate. *Science, 281*, 363–365.

Roth, C. M., & Yarmush, M. L., (1999). Nucleic acid biotechnology. *Annu. Rev. Biomed. Eng.*, 265–297.

Ryu, J. H., & Beuchat, L. R., (2005). Biofilm formation and sporulation by Bacillus cereus on a stainless steel surface and subsequent resistance of vegetative cells and spores to chlorine, chlorine dioxide, and a peroxyacetic acid-based sanitizer. *Journal of Food Protection, 68*(12), 2614–2622.

Saiki, R. K., Scharf, S., Faloona, F., Mullis, K. B., Horn, G. T., Erlich, H. A., & Arnheim, N., (1985). Enzymatic amplification of beta-globin genomic sequences and restriction site analysis for diagnosis of sickle cell anemia. *Science, 230*(4732), 1350–1354.

Sanger, F., & Coulson, A. R., (1975). A rapid method for determining sequences in DNA by primed synthesis with DNA polymerase. *Journal of Molecular Biology, 94*(3), 441–448.

Sanger, F., Nicklen, S., & Coulson, A. R., (1977). DNA sequencing with chain-terminating inhibitors. *Proceedings of the National Academy of Sciences, 74*(12), 5463–5467.

Schadt, E. E., Turner, S., & Kasarskis, A., (2010). A window into third-generation sequencing. *Hum. Mol. Genet., 19*, 227–240.

Schena, M., Shalon, D., Davis, R. W., & Brown, P. O., (1995). Quantitative monitoring of gene expression patterns with a complementary DNA microarray. *Science, 270*(5235), 467–470.

Scher, K., Romling, U., & Yaron, S., (2005). Effect of heat, acidification, and chlorination on *salmonella enterica serovar Typhimurium* cells in a biofilm formed at the air-liquid interface. *Applied and Environmental Microbiology, 71*(3), 1163–1168.

Schoeni, J. L., & Wong, A. C. L., (2005). Bacillus cereus food poisoning and its toxins. *Journal of Food Protection, 68*(3), 636–648.

Schwartz, D. C., & Cantor, C. R., (1984). Separation of yeast chromosome-sized DNAs by pulsed field gradient gel electrophoresis. *Cell, 37*, 67–75.

Sedgwick, W. T., & Hazen, A., (1892). Typhoid fever in Chicago. *Eng. News Am. Rail. J.*, 1–21.

Shendure, J., Porreca, G. J., Reppas, N. B., Lin, X., McCutcheon, J. P., Rosenbaum, A. M., Wang, M. D., Zhang, K., Mitra, R. D., & Church, G. M., (2005). Accurate multiplex polony sequencing of an evolved bacterial genome. *Science, 309*(5741), 1728–1732.

Simmonds, P., Mossel, B. L., Intaraphan, T., & Deeth, H. C., (2003). Heat resistance of Bacillus spores when adhered to stainless steel and its relationship to spore hydrophobicity. *Journal of Food Protection, 66*(11), 2070–2075.

Singh-Gasson, S., Green, R. D., Yue, Y., Nelson, C., Blattner, F., Sussman, M. R., & Cerrina, F., (1999). Maskless fabrication of light-directed oligonucleotide microarrays using a digital micromirror array. *Nature Biotechnology, 17*(10), 974–978.

Smith, L. M., Fung, S., Hunkapiller, M. W., Hunkapiller, T. J., & Hood, L. E., (1985). The synthesis of oligonucleotides containing an aliphatic amino group at the 5′ terminus synthesis of fluorescent DNA primers for use in DNA sequence analysis. *Nucleic Acids Res., 13*, 2399–2412.

Stefanie, P. G., & Peter, K., (2015). Multilocus sequence analysis (MLSA) in prokaryotic taxonomy. *Systematic and Applied Microbiology, 38*(4), 237–245. ISSN 0723–2020, http://dx.doi.org/10.1016/j.syapm.2015.03.007.

Stenfors, A. L. P., Fagerlund, A., & Granum, P. E., (2008). From soil to gut: *Bacillus cereus* and its food poisoning toxins. *FEMS Microbiol. Rev., 32*, 579–606.

Stromberg, Z. R., Johnson, J. R., Fairbrother, J. M., Kilbourne, J., Van Goor, A., Curtiss, R. R. D., & Mellata, M., (2017). Evaluation of *Escherichia coli* isolates from healthy chickens to determine their potential risk to poultry and human health. *PLoS One, 12*, e0180599.

Su, Y. C., Yu, C. Y., Lin, J. L., Lai, J. M., Chen, S. W., Tu, P. C., & Chu, C., (2011). Emergence of *Salmonella enterica* Serovar Potsdam as a major serovar in waterfowl hatcheries and chicken eggs. *Avian Dis., 55*, 217–222.

Suo, B., He, Y., Tu, S. I., & Shi, X., (2010). A multiplex real-time polymerase chain reaction for simultaneous detection of *Salmonella spp., Escherichia coli* O157, and *Listeria monocytogenes* in meat products. *Food-borne Pathogens and Disease, 7*(6), 619–628.

Sutandy, F. X., Qian, J., Chen, C. S., & Zhu, H., (2013). Overview of protein microarrays. *Current Protocols in Protein Science, 72*(1), 27.1.1–27.1.16.

Swanenburg, M., Urlings, H. A. P., Snijders, J. M. A., Keuzenkamp, D. A., Van Knapen, F., (2001). Salmonella in slaughter pigs: Prevalence, serotypes and critical control points during slaughter in two slaughterhouses. *International Journal of Food Microbiology, 70*(3), 243–254.

Sykes, P. J., Neoh, S. H., Brisco, M. J., Hughes, E., Condon, J., & Morley, A. A., (1992). Quantitation of targets for PCR by use of limiting dilution. *Biotechniques, 13*, 444–449.

Szabo, E. A., & Mackey, B. M., (1999). Detection of *salmonella enteritidis* by reverse transcription-polymerase chain reaction (PCR). *International Journal of Food Microbiology, 51*(2/3), 113–122.

Tewari, A., Singh, S. P., & Singh, R., (2015). Incidence and enterotoxigenic profile of *Bacillus cereus* in meat and meat products of Uttarakhand, India. *J. Food Sci. Technol., 52*, 1796–1801.

Tietjen, M., & Fung, D. Y., (1995). Salmonellae and food safety. *Critical Reviews in Microbiology, 21*(1), 53–83.

Tindall, B. J., Grimont, P. A. D., Garrity, G. M., & Euzéby, J. P., (2005). Nomenclature and taxonomy of the genus Salmonella. *International Journal of Systematic and Evolutionary Microbiology, 55*(1), 521–524.

Tompkins, D. S., Hudson, M. J., Smith, H. R., Eglin, R. P., Wheeler, J. G., Brett, M. M., et al. (1999). A Study of Infectious Intestinal Disease in England: Microbiological Findings in Cases and Controls. *Communicable Disease and Public Health, 2*(2), 108–113.

Trevino, V., Falciani, F., & Barrera-Saldaña, H. A., (2007). DNA microarrays: A powerful genomic tool for biomedical and clinical research. *Molecular Medicine, 13*(9/10), 527–541.

Tsen, H. Y., & Lin, J. S., (2001). Analysis of *Salmonella Enteritidis* strains isolated from food-poisoning cases in Taiwan by pulsed field gel electrophoresis, plasmid profile and phage typing. *Journal of Applied Microbiology, 91*, 72–79.

Uyttendaele, M., Vanwildemeersch, K., & Debevere, J., (2003). Evaluation of real-time PCR vs. automated ELISA and a conventional culture method using a semi-solid medium for detection of Salmonella. *Letters in Applied Microbiology, 37*(5), 386–391.

Valasek, M. A., & Repa, J. J., (2005). The power of real-time PCR. *Advances in Physiology Education, 29*(3), 151–159.

Van der Zee, H., & Huisint, V. J. H., (2000). Methods for the rapid detection of Salmonella. In: Wray, C., & Wray, A., (eds.), *Salmonella in Domestic Animals* (pp. 373–392). Cabi Publishing.

Van Hoek, A. H. M., De Jonge, R., Van Overbeek, W. M., Bouw, E., Pielaat, A., Smid, J. H., Malorny, B., Junker, E., Löfström, C., Pedersen, K., Aarts, H. J. M., & Heres, L., (2012). A quantitative approach towards a better understanding of the dynamics of Salmonella spp. in a pork slaughter-line. *International Journal of Food Microbiology, 153*(1/2), 45–52.

Vandamme, P., & Peeters, C., (2014). *Time to Revisit Polyphasic Taxonomy Antonie Van Leeuwenhoek, 106*, 57–65.

Verma, J., Saxena, S., & Babu, S. G., (2012). ELISA-based identification and detection of microbes. *Analyzing Microbes, Springer Protocols Handbooks*, 169–186.

Vestby, L. K., Møretrø, T., Langsrud, S., Heir, E., & Nesse, L. L., (2009). Biofilm forming abilities of salmonella are correlated with persistence in fish meal- and feed factories. *BMC Veterinary Research, 5*(20), 20.

Vignaud, M. L., Cherchame, E., Marault, M., Chaing, E., Le Hello, S., Michel, V., Jourdan-Da Silva, N., Lailler, R., Brisabois, A., & Cadel-Six, S., (2017). MLVA for *salmonella enterica* subsp. *enterica Serovar* Dublin: Development of a method suitable for inter laboratory surveillance and application in the context of a raw milk cheese outbreak in France in 2012. *Front Microbiol., 8*, 295.

Vilas-Boas, G. T., Peruca, A. P. S., & Arantes, O. M. N., (2007). Biology and taxonomy of *Bacillus cereus*, *Bacillus anthracis*, and *Bacillus thuringiensis*. *Can. J. Microbiol., 53*, 673–687.

Visscher, C. F., Klein, G., Verspohl, J., Beyerbach, M., Stratmann-Selke, J., & Kamphues, J., (2011). Serodiversity and serological as well as cultural distribution of salmonella on farms and in abattoirs in Lower Saxony, Germany. *International Journal of Food Microbiology, 146*(1), 44–51.

Voelkerding, K. V., Dames, S. A., & Durtschi, J. D., (2009). Next-generation sequencing: From basic research to diagnostics. *Clinical Chemistry, 55*(4), 641–658.

Vogelstein, B., & Kinzler, K. W., (1999). P.C.R. Digital. *Proc. Natl. Acad. Sci., U.S.A., 96*, 9236–9241.

Von Stetten, F., Mayr, R., & Scherer, S., (1999). Climatic influence on mesophilic Bacillus cereus and psychrotolerant Bacillus weihenstephanensis populations in tropical, temperate and alpine soil. *Environmental Microbiology, 1*(6), 503–515.

Wang, H., Ng, L. K., & Farber, J. M., (2000). Detection of campylobacter *Jejuni* and *thermophilic Campylobacter spp.* from foods by polymerase chain reaction. In: Spencer,

J. F. T., & Ragout, S. A. L., (eds.), *Methods in Biotechnology, Food Microbiology Proto-Cols* (Vol. 14, pp. 95–106).

Wang, L., Li, Y., & Mustapha, A., (2007). Rapid and simultaneous quantitation of *Escherichia coli* O157:H7, *Salmonella*, and *Shigella* in ground beef by multiplex real-time PCR and immunomagnetic separation. *Journal of Food Protection, 70*(6), 1366–1372.

Wang, Q., Frye, J., McClelland, M., & Harshey, R. M., (2004). Gene expression patterns during swarming in Salmonella typhimurium: Genes specific to surface growth and putative new motility and pathogenicity genes. *Molecular Microbiology, 52*(1), 169–187.

Weemen Van, B. K., & Schuurs, A. H., (1971). Immunoassay using antigen enzyme conjugates. *FEBS Letters, 15*, 232–236.

Wegener, H. C., Hald, T., Lo Fo Wong, D., Madsen, M., Korsgaard, H., Bager, F., Gerner-Smidt, P., & Mølbak, K., (2003). Salmonella control programs in Denmark. *Emerging Infectious Diseases, 9*(7), 774–780.

Wehrle, E., Didier, A., Moravek, M., Dietrich, R., & Märtlbauer, E., (2010). Detection of Bacillus cereus with enteropathogenic potential by multiplex real-time PCR based on SYBR green I. *Molecular and Cellular Probes, 24*, 124–130.

Wehrle, E., Moravek, M., Dietrich, R., Bürk, C., Didier, A., & Märtlbauer, E., (2009). Comparison of multiplex PCR, enzyme immunoassay and cell culture methods for the detection of enterotoxinogenic Bacillus cereus. *Journal of Microbiological Methods, 78*, 265–270.

Weile, J., & Knabbe, C., (2009). Current applications and future trends of molecular diagnostics in clinical bacteriology. *Analytical and Bioanalytical Chemistry, 394*(3), 731–742.

WHO (World Health Organization), (2016). *WHO Estimates of the Global Burden of Food-Borne Diseases*. Technical report. Geneva: WHO. Available from: http://www.who.int/foodsafety/publications/food-borne_disease/fergreport/en/ (accessed on 7 September 2019).

Wijman, J. G. E., De Leeuw, P. P. L. A., Moezelaar, R., Zwietering, M. H., & Abee, T., (2007). Air-liquid interface biofilms of Bacillus cereus: Formation, sporulation, and dispersion. *Applied and Environmental Microbiology, 73*(5), 1481–1488.

Wolffs, P., Norling, B., & Rådström, P., (2005). Risk assessment of false-positive quantitative real-time PCR results in food, due to detection of DNA originating from dead cells. *Journal of Microbiological Methods, 60*(3), 315–323.

Wollin, R., (2007). A study of invasiveness of different Salmonella serovars based on analysis of the Enter-net database. *Euro Surveill., 12*, E070927.3.

Wong, H. S., Townsend, K. M., Fenwick, S. G., Trengove, R. D., & O'Handley, R. M., (2010). Comparative susceptibility of planktonic and 3-day-old Salmonella Typhimurium biofilms to disinfectants. *Journal of Applied Microbiology, 108*(6), 2222–2228.

Woodward, M. J., & Kirwan, S. E., (1996). Interlaboratory diagnostic accuracy of a salmonella specific PCR-based method. *The Veterinary Record, 138*(17), 411–413.

Wysok, B., Wiszniewska-Łaszczych, A., Uradziński, J., & Szteyn, J., (2011). Prevalence and antimicrobial resistance of *Campylobacter* in raw milk in the selected areas of Poland. *Pol. J. Vet. Sci., 14*, 473–477.

Yalow, R., Berson, S., & Solomon, A., (1960). Immunoassay of endogenous plasma insulin in man. *The Journal of Clinical Investigation, 39*, 1157–1175.

Yeni, F., Acar, S., Polat, O. G., Soyer, Y., & Alpas, H., (2014). Rapid and standardized methods for detection of food-borne pathogens and mycotoxins on fresh produce. *Food Cont., 40*, 359–367.

Zeigler, D. R., (2011). The genome sequence of Bacillus subtilis subsp. spizizenii W23: Insights into speciation within the *B. subtilis* complex and into the history of *B. subtilis* genetics. *Microbiology, 157*, 2033–2041.

Zhang, G., (2013). Food-borne pathogenic bacteria detection: An evaluation of current and developing methods. *Meducator., 1*, 15.

Zhao, X., Lin, C. W., Wang, J., & Oh, D. H., (2014). Advances in rapid detection methods for food-borne pathogens. *J. Microbiol. Biotechn., 24*, 297–312.

Zhu, L., He, J., Cao, X., Huang, K., Luo, Y., & Xu, W., (2016). Development of a double-antibody sandwich ELISA for rapid detection of *bacillus cereus* in food. *Sci. Rep., 6*, 16092.

Zottola, E. A., & Sasahara, K. C., (1994). Microbial biofilms in the food processing industry—should they be a concern? *International Journal of Food Microbiology, 23*(2), 125–148.

Index

C

D

J

K

L

T

U

V

W

Y

Z